The Herbaceous Layer in Forests of Eastern North America

The Herbaceous Layer in Forests of Eastern North America

Edited by

Frank S. Gilliam

Mark R. Roberts

2003

OXFORD
UNIVERSITY PRESS

Oxford New York
Auckland Bangkok Buenos Aires Cape Town Chennai
Dar es Salaam Delhi Hong Kong Istanbul Karachi Kolkata Kuala Lumpur
Madrid Melbourne Mexico City Mumbai Nairobi São Paulo
Shanghai Taipei Tokyo Toronto

Published by Oxford University Press, Inc.
198 Madison Avenue, New York, New York 10016

www.oup.com

Library of Congress Cataloging-in-Publication Data
Gilliam, Frank S., 1954–
The herbaceous layer in forests of eastern North America / by Frank S. Gilliam and Mark R. Roberts.
p. cm.
Includes bibliographical references.
ISBN 0-19-514088-5
1. Forest plants—East (U.S.) 2. Forest plants—Ecology—East (U.S.) 3. Forest plants—Canada, Eastern. 4. Forest plants—Ecology—Canada, Eastern. I. Roberts, Mark R., 1950– II. Title.
~~QK115 .G56 2003~~
581.7'3'0974—dc21 2002070438

9 8 7 6 5 4 3 2 1

Printed in the United States of America
on acid-free paper

This book is dedicated to the memory of Elizabeth Gilliam Himes—daughter, mother, aunt, wife—and sister. And to the memory of her daughter, Camille Siragy Fournier—granddaughter, sister, cousin, wife—and niece.

F.S.G.

This book is also dedicated to the memory of Bryce Gerard McInnis—graduate student, friend, Best Man.

M.R.R.

Preface

One of the first characteristics one may notice when entering a forest of eastern North America is the diverse array of low-growing plants comprising numerous species and life forms, including spring ephemeral flowering plants, herbaceous perennials, small shrubs, and seedlings of trees. In the past several years, there has been a growing awareness among ecologists, conservationists, and forest managers that this herbaceous layer serves a special role in maintaining the functional integrity of forest ecosystems. Because of this importance, the herb layer has been the focus of many studies in eastern forests, and we have learned much concerning the ecological dynamics of the herbaceous layer in these forest ecosystems.

Examples of our emerging knowledge of herb layer dynamics include the following: (1) Of all forest strata, species diversity is highest in the herb layer. (2) Competitive interactions within the herb layer can determine the initial success of plants occupying higher strata, including regenerating stems of overstory dominant species. (3) The herb layer can form a linkage with the overstory through parallel responses to similar environmental gradients, a relationship that may vary spatially (i.e., among forest types) and temporally (i.e., through secondary succession). (4) The herbaceous layer responds sensitively to disturbance across broad spatial and temporal scales, from microsite alterations (e.g., tip-up mounds) to forest management practices (e.g., harvesting and herbicide treatments). (5) Herb layer diversity can respond to macrosite differences, including parent materials and environmental gradients related to topography. Although much is known about the ecology of the herb layer of eastern forests, this has been based largely on unconnected studies, underlining the need for a synthesis of complementary data and ideas.

To this end, the overall purpose of this book is to synthesize information concerning herb layer structure, composition, and dynamics of a variety of forest ecosystem types in eastern North America. Not unlike other books organized around a relatively circumscribed subject, another purpose of this book is to answer some specific questions about ecological interactions and dynamics within the herbaceous stratum in forests of eastern United States and Canada. However, we also want to raise general questions as a guide to further research, based on what we see as needs and gaps in our knowledge of the ecology of the herb layer of eastern forests. The potential strength of this approach is in bringing together in a single volume plant ecologists with expertise that varies with respect to levels of ecological organization (e.g., from ecophysiologists to ecosystem ecologists) and represents many (although certainly not all) of the various forest ecosystem types of eastern North America.

The advantage of producing an edited volume on this (or any) subject is obviously in the greater amount of expertise and research experience that can be brought to each topic by inviting such qualified individuals. The disadvantage is just as obviously the potential lack of cohesiveness that can result when numerous ecologists with different perspectives and writing styles are asked to contribute one part to a whole that they cannot see. The result, all too often, is a book that reads more like a collection of individual papers merely bound together into a single volume than like a unified work focused on a clear theme.

Our intent with this book was to maximize this advantage while minimizing the disadvantage. To achieve the latter, we have endeavored to provide chapter authors with as much of the perspective of the whole that they cannot see as possible. We did not require the authors to conform to a single writing style or mode of expression, largely to avoid the hubris of assuming that only we know the correct way to write.

Thus, we have tried to give the reader a useful volume that provides in-depth presentation and analysis of topics we felt germane to improving our understanding of the ecology of the herbaceous layer. Readers can expect to find numerous instances of cross-referencing between chapters, along with a single Reference section at the end of the book, rather than Reference sections at the end of each chapter.

This book is an outgrowth of a symposium entitled, "Ecology and Dynamics of the Herbaceous Layer of Forest Communities of Eastern North America," that we organized for the 83rd annual meeting of the Ecological Society of America (ESA), Baltimore, Maryland, 6 August 1998. Based on general attendance, the number of questions asked, and the amount of discussion it generated, the topic of the symposium was timely and relevant. We attribute its success largely to the expertise of the individuals who agreed to participate. Many of the contributors to this volume were presenters in that symposium, whereas others have been invited to address additional issues as necessary. We hope that this book will be as well received as the symposium was.

In addition to being the direct extension of the 1998 ESA symposium, this book is also the culmination of two additional formal gatherings of vegetation scientists. The first of these occurred in 1993 when we organized a symposium for the 78th annual meeting of the ESA in Madison, Wisconsin, entitled, "Effects of Forest Management on Plant Biodiversity: Patterns and Mechanisms in Forest Ecosystems," which resulted in the 1995 publication of a special feature ("Plant Diversity in Managed Forests") in *Ecological Applications*. We organized a session of the 2nd North American Forest Ecology Workshop entitled, "Response of Understory Plant Diversity to Forest Practices," which was held in Orono, Maine, in June 1999, a time when we were initially developing ideas for this book. Many of the papers presented in our session were published in January 2002 as part of a special issue ("Forest Ecology in the Next Millennium: Putting the Long View into Practice") of *Forest Ecology and Management*; others have been published independently in other journals.

The exchange of ideas that occurred at each of these three sessions offered invaluable insights into elements that we ultimately believed were important in developing in this book. Accordingly, we thank the ESA for the opportunity to present the 1993 and 1998 symposia. We also thank the organizers of the 2nd North American Forest Ecology Workshop, especially Alison Dibble, who served as chair.

We are indebted to a great many individuals, without whom completion of this volume would have been much more difficult or impossible. We appreciate the participation of individuals in the symposia and workshop, and particularly their thoughts and ideas on herbaceous layer dynamics. We extend deep-felt appreciation to and respect for the chapter authors for their invaluable contributions to this volume. We are in awe of the artistic skills of Dr. Susan Power of Marshall University, who drew the pen-and-ink depiction of the forest scene for the book cover. Finally, we thank Kirk Jensen, executive editor, Oxford University Press, for his initial confidence in this project and for his exceptional patience and insight in guiding us through the publication process.

Individually, F.S.G. thanks Leonard Deutsch, dean of the Marshall University Graduate College, for providing needed summer support and release time from an otherwise heavy teaching load to allow me to work on the book. I also thank Nicole Turrill Welch and Staci Smith Aulick, whose graduate research was part of the initial work at Fernow Experimental Forest, West Virginia. Thanks go as well to Mary Beth Adams, U.S.D.A. Forest Service, for faithfully providing logistical support of my Fernow research. Without the research, this book might not have been possible; without these three women, the research would not have been possible. Finally, I must acknowledge the critical role of my family in this process. I owe virtually everything I have accomplished professionally (and personally, for that matter) to my parents, Randy and Sara Rainey Gilliam, for their sustaining love that led to, among other things, a willingness to indulge my interest in nature at an early age. Thanks go to my wife, Laura, for her unswerving emotional support of love and patience, and to our children, Rachel and Ian, for their constant source of comic relief for a father badly in need of it, and for being a reminder of the future and why we should protect these precious plants.

M.R.R. thanks Mark Dijkstra, Lixiang Zhu, and Barbara Ramovs, whose graduate research provided much of the background for chapter 13. I extend special thanks to my parents, Richard and Janet Roberts, for passing on to me their love of nature, and to my wife, Monique Roy, for her love and support, particularly during those times when deadlines were pressing.

Huntington, West Virginia F.S.G.
Fredericton, New Brunswick, Canada M.R.R.

Contents

II Population Dynamics of the Herbaceous Layer

III Community Dynamics of the Herbaceous Layer across Spatial and Temporal Scales

IV Community Dynamics of the Herbaceous Layer and the Role of Disturbance

Contributors

Wendy B. Anderson
Department of Biology
Drury College
900 N. Benton
Springfield, MO 65802

Fakhri Bazzaz
Department of Organismal and Evolutionary Biology
16 Divinity Avenue
Harvard University
Cambridge, MA 02138

Susan Beatty
Department of Geography, CB 260
University of Colorado
Boulder, CO 80309-0260

Yves Bergeron
Groupe de Recherche en Écologie Forestiére
Université du Québec
Montréal, Québec H3C 3P8
Canada

Catherine Boudreault
Ressources naturelles Canada
Service canadien des forêts
Centre de foresterie des Laurentides
1055, rue du P.E.P.S., C.P. 3800
Sainte-Foy, Québec G1V 4C7
Canada

Norman L. Christensen
Nicholas School of the Environment
Duke University
Durham, NC 27708-0328

Lisa George
7938 Clarion Way
Houston, TX 77040

Frank S. Gilliam
Department of Biological Sciences
Marshall University
Huntington, WV 25755-2510

Louis De Grandpré
Ressources naturelles Canada
Service canadien des forêts
Centre de foresterie des Laurentides
1055, rue du P.E.P.S., C.P. 3800
Sainte-Foy, Québec G1V 4C7
Canada

Pierre Grondin
Ministère des Ressources naturelles du Québec
Direction de la Recherche forestière
2 700 rue Einstein
Sainte-Foy, Québec G1P 3W8
Canada

Claudia L. Jolls
S-107A Howell Science Complex
Department of Biology
East Carolina University
Greenville, NC 27858

James O. Luken
Department of Biology
Coastal Carolina University
755 Highway 544
Conway, SC 29526

Brian C. McCarthy
Department of Environmental and Plant Biology
317 Porter Hall
Ohio University
Athens, OH 45701-2979

Robert N. Muller
Santa Barbara Botanic Garden
1212 Mission Canyon Road
Santa Barbara, CA 93105

Howard Neufeld
Department of Biology
Appalachian State University
Boone, NC 28608

Mark R. Roberts
Faculty of Forestry and Environmental Management
University of New Brunswick
Fredericton, NB E3B 6C2
Canada

Thuy Nguyen
Groupe de Recherche en Écologie Forestiére
Universté du Québec,
Montréal, Québec H3C 3P8
Canada

Donald Young
Department of Biological Sciences
Virginia Commonwealth University
Richmond, VA 23284-2012

The Herbaceous Layer in Forests of Eastern North America

1

Introduction

Conceptual Framework for Studies of the Herbaceous Layer

Frank S. Gilliam
Mark R. Roberts

Forest ecosystems have always been an integral part of human existence, whether as a source of food, fiber, and habitat, as an essential component in maintaining the atmospheric balance of O_2 and CO_2, or as a source of musical, artistic, or poetic inspiration. Yet, our image of forests often comes from the broad brush of a landscape perspective, whereby we see only the grandeur of the predominant vegetation—the trees. Such a distortion figuratively and literally masks the vegetation that, though of lesser stature, contains the most diverse and spatially and temporally dynamic assemblage of forest plants. Often called the *herbaceous layer* (other synonyms are discussed later in this chapter), this stratum of forest vegetation carries with it an ecological significance to the structure and function of the forest ecosystem that belies its physical stature.

In a synthesis of species richness among several general plant and animal taxa of North America, Ricketts et al. (1999) found that species richness of non-tree vascular plants (a taxon representing a high percentage of herbaceous layer species) correlated much better with richness of several animal taxa (including birds, butterflies, and mammals) than did richness of tree species. They also found that richness of non-tree vascular plant species in eastern North America was higher than that in any other region of North America and was more than 13 times that of tree species richness in the region (Ricketts et al. 1999). However, of the plant species currently listed by The Nature Conservancy/Association for Biodiversity Information as either "extinct," "missing and possibly extinct," or "extinct/missing in the wild, but still extant in cultivation," virtually none are tree species (Stein et al. 2000). Indeed, herbaceous plant species have extinction rates that are more than twice those of woody species (Levin and Wilson 1976; Levin and Levin 2001). Thus, this diverse assemblage of forest vegetation also contains some of the more sensitive, threatened, and endangered plant species.

We introduce this edited volume by summarizing our general knowledge and understanding of the ecology and dynamics of forest herbaceous layers. Because the literature contains numerous terms used synonymously with the term *herbaceous layer*, we begin with a discussion of the terminology and definitions that have been commonly applied to the herbaceous layer. Next, we develop a simple conceptual framework for understanding the spatial and temporal dynamics of the herbaceous layer. Finally, we describe the organization of the book.

Terminology

Our survey of the ecological literature revealed numerous synonyms for the term *herbaceous layer* used by ecologists, presenting a challenge to experienced and beginning researchers alike. Whereas we have adopted the term *herbaceous layer* for the title of this book (and will use it interchangeably with the more abbreviated *herb layer*), other authors use such terms as *herbaceous* (or *herb*) *stratum*, *herbaceous understory*, *ground layer*, *ground vegetation*, and *ground flora*. In addition, foresters and others interested in forest management sometimes refer to it as the *regeneration layer* (e.g., Waterman et al. 1995; Baker and van Lear 1998). This latter term arises from both an interest in patterns of regeneration of overstory dominant species and an awareness that successful regeneration of such species can be determined largely by interactions among plant species in this stratum (chapter 11, this volume). When we were graduate students at Duke University; the professor of the summer dendrology course referred to plants of the herb layer as *step-overs* while walking through the Duke Forest. Such a hyperbolic term emphasizes the lack of importance given to the herb layer by some foresters, at least in the late 1970s, as comprising plants unworthy of study and thus were to be stepped over while learning about trees. There are likely still other synonyms we have not encountered, so this is not intended to be an exhaustive list. Rather, our goal is to provide some idea of the diversity of terms one should expect to find in the literature.

We have summarized the results of a search of *Ecological Abstracts* for citations from the past 20 years that have the herb layer synonyms mentioned either in the title, as key words, or in the abstract (table 1.1). The search represents articles from some 3000 journals and 2000 other publications, including books and monographs, and thus provides an indication of the frequency with which one might expect to encounter the various terms in the literature. It should be noted that the number of occurrences are not necessarily mutually exclusive among terms. That is, it is possible that one article may have used, for example, *herbaceous layer* in the title and *ground*

Table 1.1. Synonyms for herbaceous layer and number of occurrences in the ecological literature in the last 20 years.

Synonym	Occurrences		
	1980–1989	1990–1999	Total
Herbaceous/herb layer	73	164	237
Herbaceous/herb stratum	10	11	21
Herbaceous understory	4	20	24
Ground layer	40	64	104
Ground vegetation	56	161	217
Ground flora	27	68	95
Step-overs	0	0	0
Total	210	488	698

Information taken from *Ecological Abstracts* (Elsevier Science Ltd.), representing approximately 3000 primary journals and 2000 other publications, including books, monographs, reports, and theses.

layer as a key word; this would result in one occurrence in each of the two synonym categories.

Clearly, *herbaceous/herb layer* and *ground vegetation* are the more commonly used terms, together representing about 65% of the nearly 700 occurrences since 1980. For reasons that are not immediately apparent, North American studies tend to use *herbaceous/herb layer*, whereas non-North American (particularly European) studies tend to use *ground vegetation*.

We are not suggesting that a single, consensus term be used. In fact, as editors of this volume, we have not required that all authors use identical terminology. Rather, we would like to point out, particularly to researchers just beginning in this field, that there are several terms that one must expect to encounter in the ecological literature. Accordingly, from a practical standpoint, one performing searches for herb layer studies (e.g., using web search engines) would be strongly advised to use either several terms (but especially *herbaceous/herb layer* and *ground vegetation*) or focus the appropriate term toward the geographical area of interest.

It is also notable that the number of occurrences more than doubled during 1990–1999 compared to 1980–1989. In other words, nearly 70% of the occurrences from the past 20 years came in the past decade. This substantial increase is indicative of greatly increased interest in the herb layer of forest ecosystems among plant ecologists, forest ecologists, conservation biologists, and resource managers.

Definitions

It is not surprising that there are almost as many definitions of the herb layer of forests as there are investigators who study it. The more commonly used definitions of the herbaceous layer emphasize its physical aspects as an assemblage of forest vegetation, with the focus on height, rather than on growth form. We have defined the herb layer, as have numerous studies, as the forest stratum composed of all vascular species that are ≤1 m in height. The maximum height limit, however, varies greatly among studies, as does exclusion/inclusion of nonvascular plant species. In one of the earlier quantitative studies of the herb layer, Siccama et al. (1970) used 0.5 m as an upper limit for Hubbard Brook Experimental Forest. More recently, Yorks and Dabydeen (1999) used 1.37 m to delimit the herb layer in clearcut hardwood stands of western Maryland. Using the terms *understory* and *inferior layer* interchangeably, Rogers (1981) defined this stratum as comprising vascular plants <2 m in height for mature, mixed mesophytic stands from Minnesota, Wisconsin, and Michigan.

Although it is rarely immediately evident why different studies use different height limits, the variation likely results from a combination of research inertia (i.e., "well, that's the way we've always done it in this lab"), along with true variation among forest types in the structure of vegetation. For example, mature, second-growth hardwood forests, such as that found in Watershed 6 of Hubbard Brook, often lack a prominent shrub component, so use of 0.5 m as the upper height limit by Siccama et al. (1970) was certainly justified. Yorks and Dabydeen (1999) used the term *vascular understory* along with their height limit of 1.37 m. Although they provide no reason for such a distinct height limit, it corresponds to the breast height often used in conjunction with dbh (diameter at breast height). Other studies include nonvascular plants in their definition (e.g., Bisbee et al. 2001). Although such studies are relatively uncommon, they generally occur in forests where bryophyte cover can be prevalent (e.g., boreal forests; chapter 10, this volume). Still other studies fail to specify a maximum height to distinguish the herb layer from other forest vegetation strata.

Just as we sought no consensus on a single term to be used for studies of the herb layer of forest ecosystems, it is similarly not our intention in this book to establish a uniform definition of the herb layer. For the very reasons brought out here (particularly the great intersite differences in the physical structure of forest vegetation), vegetation scientists should feel the freedom of adapting their definitions appropriately. However, we do suggest that researchers base their definition on a careful consideration of the biological and physical structure of the forest system and articulate specifically their working definition of the herb layer, along with a justification of their definition, especially if it departs greatly from the typical height range of 0.5–1.0 m.

Conceptual Framework for Studies of the Herbaceous Layer

Because the plant kingdom comprises species of an impressive array of physical growth forms, life-history characteristics, and patterns of resource use, botanists and plant ecologists have long endeavored to group plant species into categories based on shared characteristics. This serves the dual purpose of decreasing the complexity and increasing the understanding of the ecological significance of those characteristics. One of the earliest such attempts was made by the Danish botanist Christen Raunkiaer, whose pioneering work was published in the early twentieth century and later translated into English in the classic book, *The Life Forms of Plants and Statistical Plant Geography* (Raunkiaer 1934). As the title implies, he classified plants into life forms (also called *growth forms*), a classification he based on the location of the structure that allows a plant to exist from one growing season to the next (i.e., the perennating structure—buds, rhizomes, seeds). Still in use today, Raunkiaer's life forms represent one of the more successful endeavors to place plant species into ecologically meaningful categories.

Categories such as these are essentially groups of plant species based on common ecological functions. Appropriately, then, in more recent literature they are often referred to as *plant functional groups*, and their ecological relevance has been expanded to include such phenomena as maintenance of biodiversity and stability of ecosystems and effects on nutrient cycling (Huston 1994; Hooper and Vitousek 1998; Díaz and Cabido 1997). Other terms in the literature synonymous with functional groups include *guilds* and *functional types* (Wilson 1999b). Körner (1994) discussed criteria for determining levels of organization within functional groups, and suggested that such levels represent a gradient of integration from sub-cellular structures up to ecosystems, and that ecological relevance increases along this spatially-expanding gradient at the expense of precision (Körner 1994).

Resident versus Transient Species

We propose a simple conceptual framework for the forest herbaceous layer, composed of two functional groups: resident species and transient species. *Resident species* are those with life-history characteristics that confine them to above-ground heights of 1–1.5 m (or perhaps others, depending the height distinction used in one's definition of the herb layer). These species would include, for example, annuals, herbaceous perennials, and low-growing shrubs. *Transient species* comprise plants whose existence in the herb layer is temporary because they have the potential to develop and emerge into higher strata (e.g., shrub, understory, and overstory layers). This group would include larger shrubs and trees. Juveniles (i.e., seedlings and sprouts) of regenerating overstory species must pass through this layer and compete as transient species with resident species (Morris et al. 1993; Wilson and Shure 1993). Because resident species play an important role in competition among themselves (Muller 1990) and with seedling and sprouting individuals of potential forest canopy dominants (Maguire and Forman 1983; Davis et al. 1998, 1999), we view the herb layer as a dynamic assemblage of these two groups.

We should emphasize that our use of the term *transient* has a specific temporal and physical connotation that should not be confused with Grime's (1998) classification of plant species into dominant, subordinate, and transient species. His classification is based on the different roles species have in linking plant diversity to ecosystem function. Thus, his transient species are so called because they are transient in abundance and persistence, not in the strata of forest vegetation. In this sense, Grime's (1998) transient species are closely analogous to the *satellite* species of Hanski's (1982) core and satellite species hypothesis (see Gibson et al. 1999 for an excellent synthesis of both Grime's and Hanski's concepts).

As transient species emerge from the resident species, they become members of the other, overlying forest strata. These higher strata compete with the herbaceous layer through shading and utilization of moisture and nutrients (Maguire and Forman 1983). In addition, higher strata affect substrates for the herbaceous layer through inputs of litter and creation of tip-up mounds (chapter 7, this volume). Thus, it is important to understand the interactions between the herbaceous layer and other forest strata (chapter 8, this volume). Although Parker and Brown (2000) called into question the usefulness of applying the term *stratification* to forest canopies, we find considerable ecological justification for it, considering the widely contrasting height-growth strategies seen among plant species of forest communities. Indeed, there may be a large number of forest strata, including several canopy layers, epiphytes and lianas within the tree canopy, shrubs, the herbaceous layer, and the thallophyte (nonvascular plant) layer (Harcombe and Marks 1977; Kimmins 1996; Oliver and Larson 1996).

The dynamic balance of resident and transient species in forest herbaceous layers, in terms of both numbers of species (i.e., richness) and cover, is mediated by (1) competitive interactions, (2) responses to disturbances, such as windthrow of canopy trees, herbivory, and harvesting, and (3) responses to environmental gradients, such as soil moisture and fertility, and other factors that vary spatially and temporally. Working in mature mesophytic stands from Minnesota to Michigan, Rogers (1981) found that the ratio of transient species cover to resident species cover in stands with high *Fagus grandifolia* Ehrh. codominance in the overstory was nearly twice that in stands with little or no *F. grandifolia* (0.78 vs 0.40, respectively). Gilliam et al. (1995) found that relative cover of resident species was significantly higher in early successional stands than in mature stands of central Appalachian hardwood forests (71% vs 54%, respectively).

Resident versus Transient Species: Reproduction and Dispersal

Among the unique aspects of the herbaceous layer, then, is the intimate spatial and temporal coincidence of resident and transient species, which are two otherwise disparate plant groups. The distinction between them is manifested not only in the more obvious differences in growth form, but also in the factors that determine their distribution and patterns of reproduction. Transient (in particular, tree) species are generally limited in their distribution by various combinations of disturbance patterns (Loehle 2000), and indeed have the potential for rapid migration (Clark 1998). In contrast, the distribution of resident species (predominantly woodland, or forest, herbs) is determined more by availability of suitable habitats, the likelihood of seeds to be dispersed to those habitats, and the successful germination (and subsequent growth) of seeds that reach them (Ehrlén and Eriksson 2000, Verheyen and Hermy 2001). Seed size can be an important variable in these latter two factors. Ehrlén and Eriksson (2000) found that seed size was negatively correlated with likelihood of reaching suitable habitat, but positively correlated with probability of successful germination. Furthermore, a disproportionate number of resident species are cryptophytes and hemicryptophytes (chapter 5, this volume) with the capability of asexual (clonal) reproduction (especially in the absence of disturbance), whereas far fewer transient species use this reproductive mode in the absence of disturbance. Singleton et al. (2001) found that only 7 of 50 forest herb taxa from central New York lacked clonal expansion. McLachlan and Bazely (2001) suggested that knowledge of dispersal mechanisms of understory herbs could be applied to their use as indicators of recovery of deciduous forests after disturbance.

There are also sharp contrasts between transient versus resident species in their respective mechanisms of seed dispersal. For transient species (again, tree species in particular), the predominant mechanisms are wind and vertebrate herbivores (e.g., birds and rodents) (Cain et al. 1998; Clark et al. 2001). In contrast, the predominant dispersal vectors for resident species are invertebrates, particularly the phenomenon of myrmecochory, or seed dispersal by ants (Handel et al. 1981; Kalisz et al. 1999). Pakeman (2001) examined an additional dispersal vector for woodland herbs, large mammalian herbivores, and distinguished between endozoochory (seeds consumed and passed through the gut) and ectozoochory (seeds carried externally) as mechanisms for dispersal. He concluded that endozoochory could be an important mechanism for long-distance dispersal of herb species. Two mammalian herbivore species Pakeman considered, white-tailed deer (*Odocoileus virginianus* Zimmermann) and moose (*Alces alces* L.), have particular relevance for the herb layer of eastern North American forests (chapter 13, this volume).

Based on a recent survey of literature, Cain et al. (1998) concluded that most woodland herbaceous species are substantially limited in their seed-dispersal capabilities (chapter 5, this volume). Whitney and Foster (1988) cited poor colonizing ability (based largely on limited dispersal) as one of several factors that leads to the uniqueness of regional herb layer floras. Matlack (1994b) also demonstrated both slow clonal growth (asexual reproduction, e.g., via rhizomes) and low rates of plant migration via seed dispersal for forest herbs in hardwood forests of the Delaware/Pennsylvania Piedmont.

When seed dispersal mechanisms are compared between resident and transient species, seeds are dispersed much greater distances for transient species.

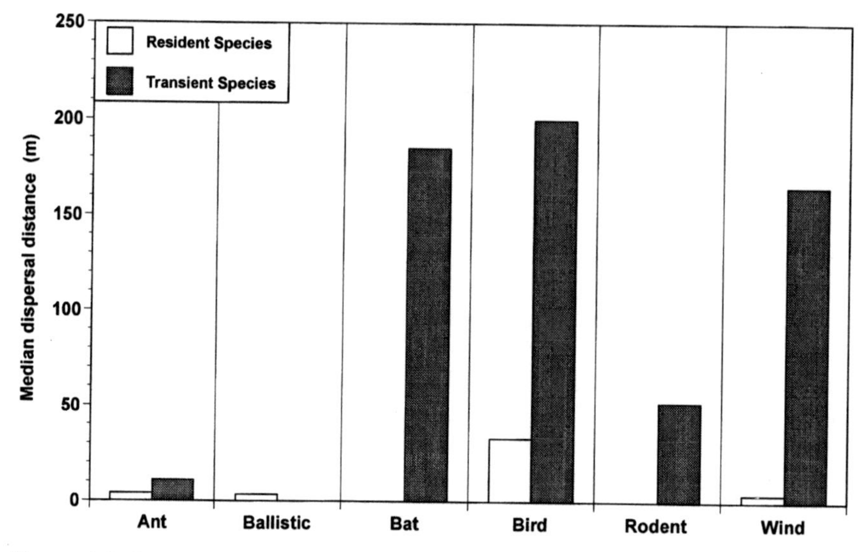

Figure 1.1. Distances of dispersal by biotic and abiotic vectors for resident versus transient species of the herbaceous layer. Based on data from Cain et al. (1998).

This is not a surprising result, considering the respective sizes of seed-bearing individuals of the two groups. Figure 1.1 summarizes a subset of data from Cain et al. (1998) to make direct comparisons between transient species (trees and shrubs in the original article) and resident species (woodland herbs in the article) for both the mechanism and the distance of seed dispersal. Ballistic dispersal was the only mechanism unique to resident species, whereas bat and rodent dispersal were unique to transient species. For mechanisms common to both, median seed dispersal distances were consistently far greater for transient species, by factors of 2.5, 6.0, and nearly 40 for ant, bird, and wind dispersal, respectively (fig. 1.1).

In conclusion, these numerous differences in resident versus transient species in the herbaceous layer of forest ecosystems create a forest stratum with impressive spatial and temporal variability, the very dynamic nature of vegetation originally articulated by Cowles (1899). Some of the substantial increase in herb layer research documented in table 1.1 likely has arisen from an increasing awareness among plant ecologists of the excitement and challenge of understanding the complex ecology of this important vegetation stratum and of the urgency of applying this knowledge toward the wise, sustainable use of forest resources that will conserve herb layer species. Such complexity can be seen at all levels of ecological organization, from species-specific differences in light and nutrient use, to response of herb communities, to disturbances to the forest canopy. We have taken a cue from this hierarchy of organization as a general approach to organizing this book.

Organization of the Book

Our own research on the herbaceous layer of eastern North American forests has generally been on the level of the ecological community, as has that of many of our colleagues and collaborators. The 1998 symposium that led to this book (see the Preface to this volume) even had the term *forest communities* in its title. It is not surprising, then, that this book has a decidedly community-level orientation in its approach to examining the ecology of the herb layer within this broad region. As already discussed, however, the herb layer comprises plant species with widely varying responses to environmental factors and with widely varying population dynamics. Although seemingly inconsequential in biomass relative to trees, the herb layer has several important roles in maintaining structure and function of forest ecosystems. Accordingly, we wanted to address the herb layer on all levels of ecological organization, from ecophysiological and population levels to community and ecosystem levels, much as one would find in a college ecology course. We have sought ecologists with noted expertise in each of these fields to be contributors to this book.

The book is divided into five major sections. Part I addresses aspects of the environment in which plants of the herbaceous layer grow, including nutrient relations and light in chapters 2 and 4 (Robert Muller and Wendy Anderson), and ecophysiological adaptations of herbaceous species to environment in chapter 3 (Howard Neufeld and Donald Young). In part II, Claudia Jolls discusses population dynamics, with a particular focus on conservation ecology and rare species. Community dynamics of the herbaceous layer is the subject of part III. Chapters 6–8 of part III deal with mechanisms of herbaceous layer dynamics, with emphasis on old-growth forests (Brian McCarthy), habitat heterogeneity (Susan Beatty), and linkages between the herbaceous layer and overstory (Frank Gilliam and Mark Roberts). Chapters 9 and 10 are syntheses of studies of community dynamics in two widely contrasting forest types, oak-hickory forests of the North Carolina Piedmont (Norm Christensen and Frank Gilliam) and the boreal forest of Québec (De Grandpré and others). The focus of part IV is community dynamics of the herbaceous layer and the role of disturbance, including competitive interactions between the herbaceous layer and tree seedlings (chapter 11, Lisa George and Fakhri Bazzaz), impacts of invasive species (chapter 12, James Luken), and an overview of the interactions of the herbaceous layer with disturbance (chapter 13, Mark Roberts and Frank Gilliam). Finally, in part V we attempt to assess our state of knowledge with respect to the herbaceous layer in eastern forests, summarize and synthesize some of the key ideas presented in previous chapters, and suggest areas for further research (chapter 14, Frank Gilliam and Mark Roberts).

I

The Environment of the Herbaceous Layer

2

Nutrient Relations of the Herbaceous Layer in Deciduous Forest Ecosystems

Robert N. Muller

The contributions of deciduous forest herbs to ecosystem-level processes are frequently ignored due to low contributions of this stratum to overall biomass. Thus, we are familiar with studies that treat the herbaceous layer as a dependent variable, responding to light and throughfall (R.C. Anderson et al. 1969), soil and biotic influences on nutrient availability (Snaydon 1962, Crozier and Boerner 1984), and local patterns of topographic variation (e.g., windthrow pits and mounds; Beatty 1984; chapter 7, this volume). Studies treating deciduous forest herbs as independent agents influencing community and ecosystem processes are less common and frequently focus on isolated circumstances in which processes such as control of pollinators (Thompson 1986), allelopathy (Gliessman 1976; Horsley 1977b), or exploitative competition (Meekins and McCarthy 1999) can be demonstrated. Yet, in considering ecosystem-level influences of the herbaceous layer, we need only look at the effectiveness of understory removal on increasing wood production in plantation forestry to appreciate the importance of the herbaceous layer. For instance, a 12-year study of loblolly pine seedling response to herbaceous weed control demonstrated a 30–99% increase in growth (volume) over the period (Glover et al. 1989). Clearly, in these early successional stands, there are important competitive relationships among strata of mesic forests. In considering the potential for deciduous forest herbs to influence ecosystem-level processes, and the limitations of that influence, it is important to focus on the unique characteristics of the herbaceous layer that distinguish it from the overstory. Aside from size and the reduced proportion of woody tissues, these characters include differing tissue chemistry, phenology, and proportional allocation of fixed carbon to ephemeral tissues.

The importance of the herbaceous layer in nutrient cycles of temperate forests has its roots in the life histories of individual species and interactions of those species with site and environment. In this chapter I discuss the attributes of deciduous forest herbs that significantly contribute to ecosystem-level nutrient dynamics. I attempt to distinguish among characteristics that are unique to some or all herbaceous species. Nutrient content and seasonal patterns of nutrient accumulation are compared among groups of deciduous forest herbs and with overstory species. I consider site quality (nutrient availability) as a determinant of herbaceous nutrient accumulation and discuss patterns of internal cycling (retranslocation) and decomposition of ephemeral materials. Finally, a fresh outlook on the influence of herbaceous populations on ecosystem-level nutrient cycling is discussed. In particular, I discuss the idea that spring ephemerals function to retard nutrient loss during spring runoff (a vernal dam; Muller 1978) with regard to its potential use and limitations in understanding the complex nature of deciduous forest ecosystems.

Herbaceous Nutrient Concentrations

Most forest ecosystems are nutrient limited (i.e., exhibit growth response to increased nutrient availability). Thus, in the context of Chapin (1980), plants of forested ecosystems might be expected to exhibit a suite of attributes consistent with survival on infertile soils, including slow growth rate, low rate of nutrient absorption by roots, long root life, and low tissue concentration of mineral nutrients. However, nutrient limitation within forests is not static, and both temporal and spatial gradients of nutrient availability exist. Thus, plant adaptations to the nutrient environment of forests must be considered in the context of varying intensities of nutrient limitation.

Nutrient concentrations of forest herbs exhibit distinct differences from woody components of the vegetation, suggesting that, within the fertility level of a given site, life form or position in the canopy may play an important role in mineral nutrition and, hence, in ecosystem relations. It has been commonly observed that concentrations of some foliar nutrients of forest herbs are higher than in woody vegetation from the same site (Bard 1945, 1949; Scott 1955; Gerloff et al. 1964; Likens and Bormann 1970; Henry 1973; Garten 1978). Whereas variation certainly exists among studies, data from Hubbard Brook, New Hampshire, are reasonably representative (fig. 2.1; Likens and Bormann 1970). Potassium is consistently 2–3 times more concentrated in foliage of herbs than of trees (fig. 2.1). Similarly, magnesium concentrations of herbaceous foliage can be up to two times as high as in woody foliage (Likens and Bormann 1970). These patterns are less consistent for nitrogen, phosphorus, and calcium. However, spring ephemeral herbs appear to have significantly greater concentrations of foliar nitrogen than trees and other herbaceous groups. Among herbaceous life forms, there appears to be a greater concentration of calcium in summer-green herbs than

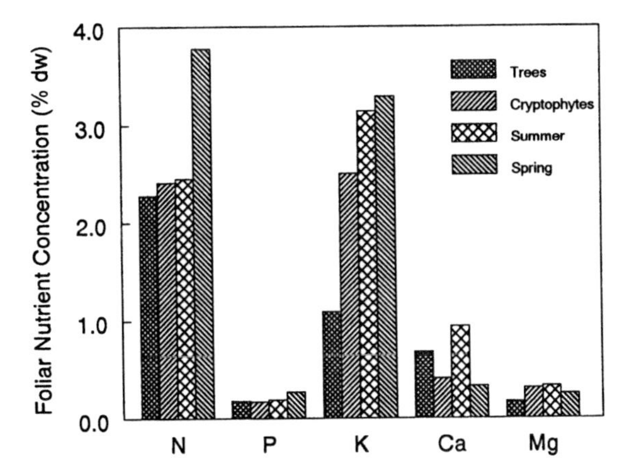

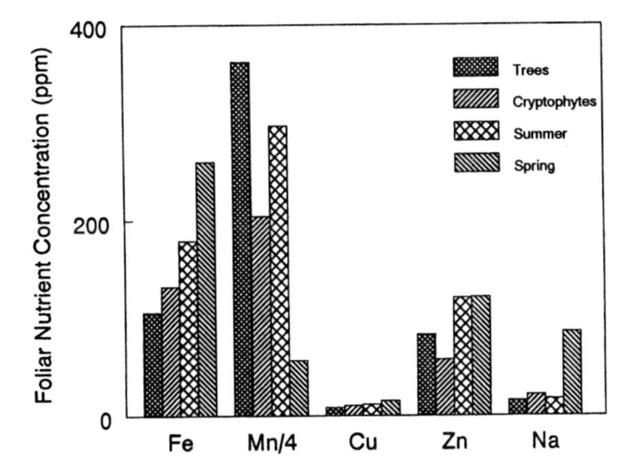

Figure 2.1. Average nutrient concentration in foliage of trees, cryptophytes (ferns), summer-green herbs (completing photosynthesis during the summer growing season), and spring herbs (completing photosynthesis before canopy development) in the northern hardwood forest at Hubbard Brook, New Hampshire. Foliage was collected in mid-growing season for the respective phenological groups. Data from Likens and Bormann (1970). Nitrogen data for spring ephemerals is from a single species, *Erythronium americanum* (Muller 1978).

in either spring herbs or cryptophytes. Among micronutrients, foliar concentrations of iron are consistently higher in herbs than in woody species (fig. 2.1; Likens and Bormann 1970; Henry 1973), whereas manganese is lower.

Siccama et al. (1970) noted that among herbaceous layer species, cryptophytes appear to have lower concentrations of phosphorus, calcium, potassium, zinc, and iron than summer-green herbs, but also have higher concentrations of sodium. However, inclusion of spring ephemerals in the analysis (data of Likens and Bormann 1970) suggests a more complex pattern. Relative to summer-green herbs, spring ephemerals have marginally higher phosphorous concentrations, considerably lower calcium and manganese concentrations, and much higher sodium and iron concentrations.

The biochemical and physiological role of the mineral constituents of plants is the subject of an extensive literature (Marschner 1995). However, some comments concerning the ecological ramifications of these patterns are important in understanding the role of herbaceous species in ecosystem relations. The strikingly high potassium concentrations suggest an important role in the fitness of herbaceous species. Potassium plays a number of roles in the biochemistry and physiology of plants; one of its more significant roles is in the osmoregulatory maintenance of cell turgor and especially the function of stomatal guard cells (Marschner 1995). High foliar concentrations of potassium in experimental settings have been found to result in better water relations and greater growth of herbaceous species (see Grewal and Singh 1980). The rooting volume of forest herbs is generally limited to surficial soils (Bauhus and Messier 1999), and, though many forest herbs are mycorrhizal (Brundrett and Kendrick 1990; Widden 1996), their access to water supplies is limited to those surface soil volumes. In contrast, trees have much more extensive rooting volumes (Stone and Kalisz 1991) and can tap deeper water sources. Herbaceous foliage, especially that of nongrass forbs, is frequently poor in structural polysaccharides (lignin; Melin 1930; Taylor et al. 1989; Wise and Schaefer 1994), which provide structural stability during periods of water stress. Additionally, limited studies of leaf ultrastructure suggest that vascular bundle extensions, which provide additional structural stability, are limited or nonexistent in herbaceous species (McLendon 1992).

In the forest floor environment, three primary resources for herbaceous growth—light, nutrients, and moisture—exist in limited supply. In the allocation of limited supplies of fixed carbon, the energetic costs of lignin production and construction of vascular bundle extensions to support leaf structure may be excessive. Consequently, alternative mechanisms of maintaining leaf structure, such as utilizing potassium to promote leaf turgor, may be favored. The known capacity of sodium to substitute for potassium in a variety of functions, but especially in maintaining water status, further suggests that leaf structure is maintained primarily by osmotic mechanisms. The rather succulent tissue of several spring ephemerals is suggestive of *Commelina benghalensis* L., in which the replacement of sodium for potassium in water relations has been demonstrated (Raghavendra et al. 1976). To the extent that structural compounds (lignin and cellulose) are reduced by the use of an alternative mechanism to maintain leaf structure (improved cell turgor as conferred by increased potassium and sodium concentration), the foliage of forest herbs would be expected to decompose at more rapid rates, thereby turning over nutrients more rapidly than other organic materials.

The relatively high concentrations of iron in foliage of all herbaceous species may well reflect its function in photosynthesis in low-light environments (Marschner 1995). However, it is notable that the highest iron (as well as nitrogen) concentrations are observed in spring ephemeral species, which are active during the spring period when light is greatly increased relative to summer. The apparent discrimination against manganese by forest herbs, especially spring ephemerals, is also notable. Newell and Peet (1998) found a positive correlation between soil manganese and diversity and abundance of the herbaceous layer. The implications of that correlation remain a mystery; however, manganese may be an indicator of general soil fertility, which may support greater biomass and species diversity.

Site-to-Site Variation

Foliar nutrient concentrations of vascular plants frequently vary in response to environmental nutrient availability. Numerous studies in controlled laboratory or cropland situations have demonstrated a close relationship between foliar nutrient concentration and soil nutrient availability (Barber 1995). However, this relationship is less obvious in forest herbs growing in uncontrolled wildland settings. Some work (Tyler 1975; Gilliam 1988) has found correlations between foliar nutrient concentrations of forest herbs and soil nutrient availability. Fertilization studies have also found some, but limited, response of foliar chemistry (Eickmeier and Schussler 1993; W. B. Anderson and Eickmeier 1998). To the extent that foliar chemistry responds to site quality, the influence of the herbaceous layer on ecosystem properties will be a function of both foliar concentration and herbaceous abundance, both of which would be enhanced on nutrient-rich sites. However, Chapin (1980) noted that enhanced growth response of wildland plants on fertile sites will effectively dilute foliar nutrient concentration, and a number of other studies have suggested that foliar nutrients of forest herbs vary little in relation to site quality. Gagnon et al. (1958) concluded that the lack of variation occurs because individual herbaceous species are site specific and consequently are not exposed to widely varying soil characteristics. In a precursor to contemporary views of positive feedbacks between vegetation and site quality, Gagnon et al. also suggested that forest plants, including herbaceous species, influence soil fertility such that little relationship would be observed between foliar chemistry and site quality. However, Bard (1949) found that even in herb species common to three distinct soils in central New York, little variation in average foliar nutrient concentrations occurred (fig. 2.2). Foliar potassium showed the greatest variation among sites, but was unrelated to soil-available potassium. Foliar concentrations of other elements showed little variation among sites. Whether the distribution of individual species is restricted to particular sites or not, the relative constancy of foliar quality suggests that the influence of the herbaceous layer on ecosystem properties is primarily a function of abundance (and, consequently, total nutrient content) on the site.

The mechanisms accounting for the apparent lack of relationship between foliar chemistry and site quality remain unclear. Gilliam and Adams (1996b) suggested that foliar chemistry was strongly related to soil resources in early successional stands, but not later in succession. They suggested that in late successional stands, light becomes the limiting resource, thus eliminating the foliar chemistry–soil nutrient availability relationship. However, apparently consistent relationships among mineral cations in the foliage of a wide range of vascular and nonvascular plants (Garten 1976, 1978) suggest that wild-

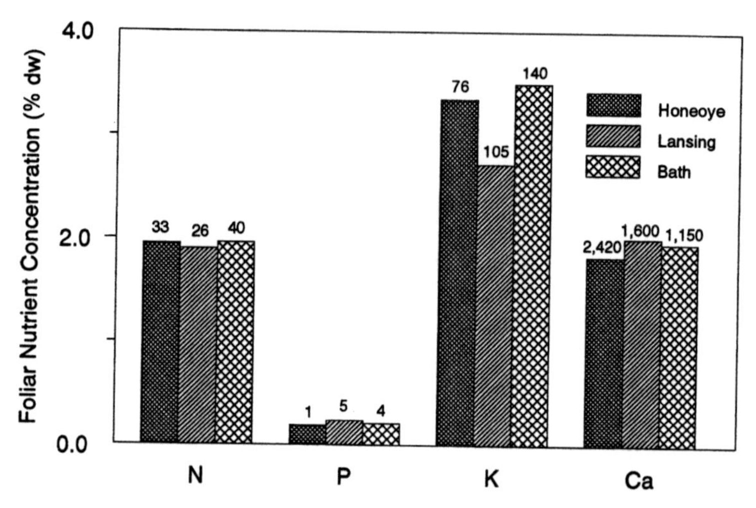

Figure 2.2. Average foliar nutrient concentrations of three herbs (*Galium triflorum* Michx., *Smilacina racemosa* (L.) Desf., and *Arisaema triphyllum* (L.) Torr.) common to three differing soil types in central New York. Values above bars are concentrations of soil extractable nutrients (µg/g). Soil pH varied from 5.00 (Bath), to 5.92 (Lansing), to 6.01 (Honeoye). Data from Bard (1949).

land species have fixed nutrient requirements, both as individual ion concentrations and in proportion to other chemical constituents.

Luxury consumption has not been intensively studied in spring herbs but appears to be limited in wildland settings (W. B. Anderson and Eickmeier 1998). Boerner (1986) has suggested that varying mycorrhizal infection with soil fertility may account, in part, for the seemingly fixed chemical composition of many forest herbs. Mycorrhizal infection of *Geranium maculatum* L. and *Polygonatum pubescens* Pursh. declined over a gradient of increasing soil fertility. Boerner (1986) suggested that forest herbs face a trade-off in allocation of fixed carbon to mycorrhizal support or plant growth. Under conditions of nutrient limitation, fixed carbon is allocated to mycorrhizal support, thereby enhancing nutrient uptake and meeting plant nutrient demands. Under conditions of nutrient abundance, fixed carbon is allocated to plant growth, thereby diluting accumulated nutrients and limiting apparent luxury consumption. Similarly, Crick and Grime (1987) noted plasticity in allocation of resources to root systems of forest herbs and concluded that increased allocation on nutrient-poor sites enabled increased "foraging," which supported uniform foliar nutrient concentrations.

Seasonal Nutrient Dynamics

It is well-recognized that nutrients in tissues of forest herbs are dynamic and change seasonally in response to phenological development and physiological demand. Most forest herbs are perennial, and these seasonal changes are frequently interpreted in the context of changing physiological demand created by fluctuating sinks of biomass (Chapin 1980). Within this framework, nutrient concentration should remain reasonably constant, and patterns of uptake should reflect immediate physiological need. An alternative interpretation considers physiological need in relation to environmental availability (Chapin 1980; Hommels et al. 1989; Lipson et al. 1996). Long-term success of plants that are beginning a period of active growth in a resource-limited environment may depend on uptake of an abundance of nutrients during periods of less limited availability. Subsequent internal reallocation meets the demands of changing tissue requirements, and tissue nutrient concentration may vary widely, whereas total nutrient content remains reasonably constant.*

Limited data seem to fit the latter interpretation. During early growth, foliar concentrations of most essential nutrients are high, followed by a gradual decline throughout the remainder of the leaf's life (Ferguson and Armitage 1944; Muller 1978; Grigal and Ohmann 1980; Whigham 1984). This pattern is most recognizable for nitrogen and is seemingly independent of phenological adaptations of species, as shown for a spring ephemeral (*Erythronium americanum* Ker.) and a winter-green herb (*Tipularia discolor* Nutt.) in figure 2.3. The almost immediate reduction of foliar nutrient concentration after leaf emergence is simply a dilution due to rapidly increasing biomass. Thus, most of the nitrogen used for photosynthesis is translocated into the young leaf early in development, rather than later when an increasing biomass sink might be viewed as creating additional demand. Phosphorus, potassium, and, to a lesser extent, magnesium show similar concentration patterns. However, calcium concentration (and content) increases throughout the foliar period, apparently in response to increasing biomass, reflecting the well-known structural function of this element in cell wall binding (Marschner 1995).

Subsequent patterns of allocation among plant parts and physiological function vary depending on nutrient and species. In *Allium tricoccum* Ait., total plant content of nitrogen, phosphorus, potassium, and magnesium remains reasonably constant after an initial spring increase (fig. 2.4; Nault and Gagnon 1988). Internal re-translocation, however, shifts these elements among plant parts, such that much of the nitrogen, which had been contained in leaves, is translocated to bulb and reproductive biomass upon foliar senescence. In contrast, calcium is concentrated in foliar biomass. As a less

*Such an interpretation would seem to argue for a reevaluation of earlier comments on luxury consumption (previous section). However, seasonal patterns of nutrient accumulation are probably well-synchronized from one site to another. Most analyses of luxury consumption involve collections of material from contrasting sites at the same time. Thus, it is not surprising that strong seasonal patterns of nutrient uptake (indeed, a different form of luxury consumption) may exist, while luxury consumption in the traditional sense is not apparent in populations from sites of different nutrient regimes.

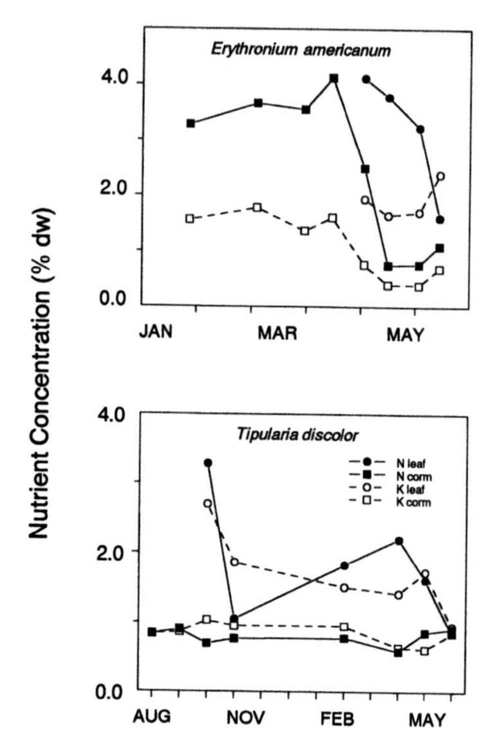

Figure 2.3. Seasonal patterns of nutrient concentrations in foliage and below-ground organs of *Erythronium americanum* (top) and *Tipularia discolor* (bottom). Legend in the bottom panel applies to top panel as well. Data from Muller (1978; top) and Whigham (1984; bottom).

mobile element, its total plant content reflects seasonal biomass patterns and peaks as foliar biomass peaks.

Two important patterns of mineral nutrition of forest herbs emerge from these studies. First, re-translocation of stored nutrients is an important aspect of survival in the nutrient-poor and highly competitive environment of forest soils (Chapin 1980). Re-translocation, of course, reduces dependence on root absorption of nutrients in an environment in which supply is unpredictable. Second, absorption of nutrients is responsive to availability. Thus, rapid uptake occurs in spring and fall when nutrients are most available (reduced uptake by trees or increased decomposition of recently released organic material) and leads to an apparent luxury consumption at the beginning of the growing season (but see W. B. Anderson and Eickmeier 1998). The balance between translocation and absorption (Whigham 1984) varies by species and environment. However, no species is completely efficient in its utilization and re-translocation of nutrients among plant parts. Uptake at the beginning of the growing season and release with decomposing plant parts at the end of the season creates the potential for impacts on ecosystem-level nutrient cycles.

Chapin (1980) has suggested a strong dichotomy between crop plants and plants of resource-rich, ruderal environments on the one hand and wildland plants of limited resource environments on the other hand. In this dichotomy, wildland plants are presumed to be slow growing, with slow rates of nutrient

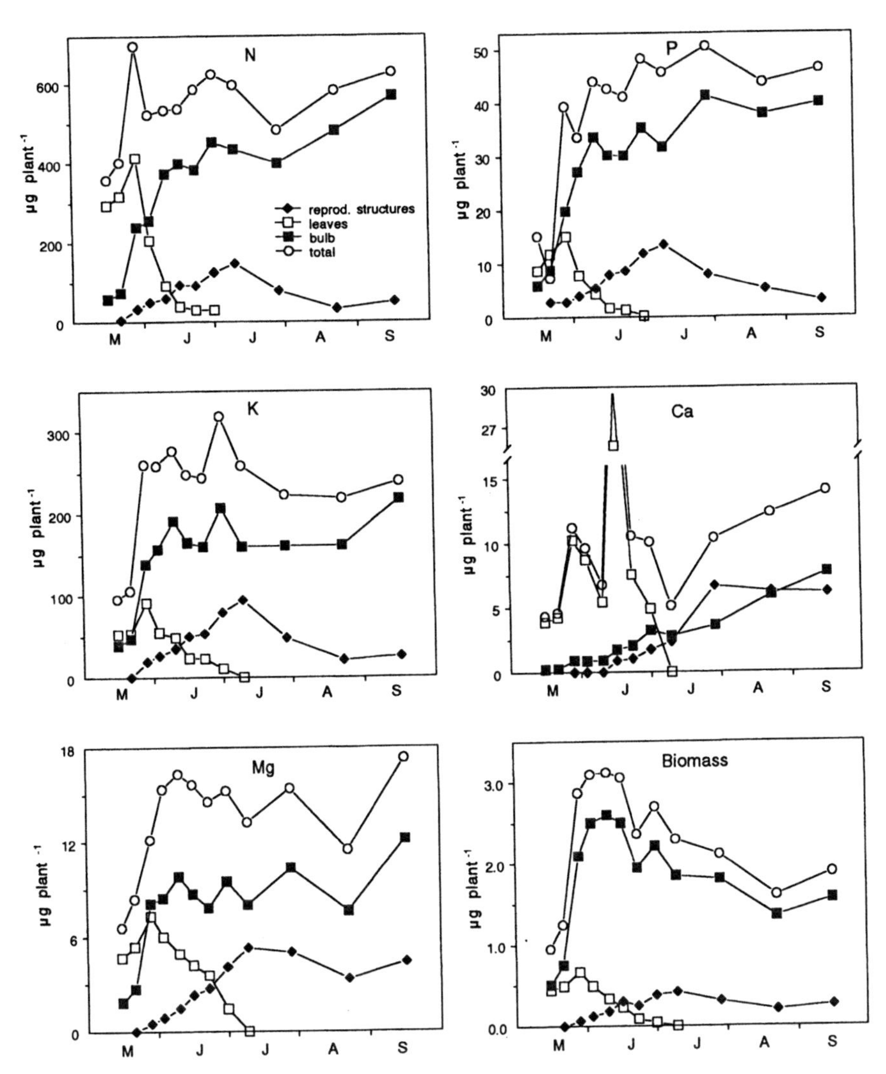

Figure 2.4. Seasonal nutrient and biomass accumulation in *Allium tricoccum*. Redrawn from Nault and Gagnon (1988), with permission. Letters along X-axis indicates the months May, June, July, August, and September.

absorption and little evidence of luxury consumption. However, these infertile sites may also be characterized by reasonably predictable nutrient flushes in the spring and to some extent in the fall. The rate of nutrient absorption in wildland plants is slow and does not change in response to changing availability of nutrients during flushes. Thus, plants of resource-poor environments should be expected to exhibit greater total uptake of nutrients during a flush. The evidence presented in figures 2.3 and 2.4 suggests that woodland herbs

Table 2.1. Above-ground living biomass and foliar litter fall of northern temperate zone forests.

Location	Forest type	Above-ground living biomass: Overstory g/m²	Above-ground living biomass: Understory g/m²	Above-ground living biomass: Herbaceous g/m²	Overstory foliage g/m²	Annual foliar litter fall g/m²/year	Page
Ontario	Spruce	46,223	1	3	2,122	229	578
Ontario	Spruce	16,908	1	16	1,002	158	579
Denmark	Beech	21,570	540	16	210	285	581
Japan	Larch	16,439	403[a]	96	359	455	600
Poland	Oak	22,599	3,549	100	201	392	609
Romania	Beech-fir	29,470	0	20	1,410	375	613
Spain	Pine-holly	19,269	1,122	22	1,148	267	615
Sweden	Oak-birch	7,892	743	67	161	240	616
Sweden	Beech	32,390	0	10	390	367	617
Sweden	Beech	22,530	0	1	330	241	619
Sweden	Oak	15,500	4,510	20	300	346	620
Sweden	Beech	31,380	0	90	480	365	621
Russia	Oak	29,953	617	70	360	446	627
Great Britain	Oak	11,243	1,480	72	269	396	647
New Hampshire	Maple, birch	10,110	21	7	315	349	651
Oregon	Douglas fir	76,230	870	7	1,329	224	656
Tennessee	Y-Poplar	13,117	853	28	323	401	657
Wisconsin	Oak	26,470	86	18	386	429	663
Germany	Beech	27,423	0	1	308	299	666

Annual foliar litter fall has been adjusted to include herbaceous litter, which is usually not accounted for in litter collectors. Data from Woodlands Data Set of De Angelis et al. 1981. Page number references specific data source.

[a]Includes standing dead.

do respond in this manner to spring increases in nutrient availability. However, W. B. Anderson and Eickmeier's (1998) finding of limited response to fertilizer application suggests that there are limits to the amount of nutrients that can be taken up during this period.

Biomass/Productivity

In considerations of ecosystem dynamics (energy flux and nutrient transfers), the herbaceous layer of deciduous forests is frequently ignored because of its small contribution to the overall biomass of the system. Above-ground living biomass of a number of mature forests of the temperate zone of the Northern Hemisphere has been summarized by DeAngelis et al. (1981; table 2.1). Although widely varying in absolute amount, herbaceous-layer standing biomass is never >1% of the total above-ground living biomass, and, as a component of energy flow, the herbaceous layer is usually <5% of above-ground net primary productivity. There are no strong trends in these patterns with latitude within the temperate zone, and biomass and productivity distribution among forests at latitudes of 40–60° N are similar to those of forests at 30–40° N latitude (table 2.2).

Biomass is not a singular material. In terms of the dynamics of energy flow and nutrient release, the biotic material produced in forested ecosystems is variably resistant to decomposition. Herbaceous layer biomass is important because it is a component of the much smaller pool of transient materials that decomposes and returns nutrients to available nutrient pools quickly. One measure of this, albeit a coarse one, is to consider herbaceous biomass as a component of annual foliar litter fall (table 2.1). Most herbaceous species of temperate forests are deciduous, and, with some exceptions, above-ground stem and foliar biomass of all herbaceous species is replaced at least annually. As a component of annual foliar litter fall, herbaceous species may account for as much as 20–25% of the total, although on average this is 10–15% and can be considerably less (table 2.2).

When the rapid decomposition of herbaceous litter is taken into account (see below), it is appropriate to consider herbaceous biomass as part of a labile pool of biological materials that decompose readily. The proportion of biotic material that can be considered to be labile varies with the origin of that material. As a generalization, the labile component of tree foliage is in the range of 20–45% ($\bar{x}$ = 32.5%; Currie and Aber 1997), whereas bark and wood have only trace amounts of labile extractives. Using the Hubbard Brook ecosystem as a model, above-ground net organic return (Gosz et al. 1973) has been partitioned by stratum into labile and nonlabile pools (table 2.3). Also shown in table 2.3 is the amount of material that could be expected to decompose in the first year after entry into the dead organic pools of the forest. The comparison assumes that the labile component of wood is of little significance and that herbaceous material is entirely labile. Whereas the herbaceous layer accounts for 3.2% of the total organic matter return of the forest, it is 15% of the labile fraction of that material. Calculation of the amount of material decomposed during the first year after entry into the ecosystem suggests a similar significance. The important point in this analysis is that, whereas the herbaceous layer represents a small fraction of total biomass and annual energy fixation in a forest, it can account for an important component of turnover of senesced materials returned to the forest floor each year. This is increasingly important in forests with large herbaceous populations relative to overstory litter fall (table 2.1).

Ecosystem Nutrient Dynamics

Whatever influences the herbaceous layer may have on nutrient dynamics of forests, they are the product of the biology of the species involved, their interactions with other species and their seasonal and spatial place in the ecosystem. As already noted, herbaceous foliage frequently contains higher

Table 2.2. Distribution of biomass and productivity in cold and warm temperate zone forests.

	Forest biomass distribution				Forest productivity			Herbaceous litter as a component of total litter Fall	
	Biomass (g/m^2)	Overstory (%)	Understory (%)	Herbs (%)	Total NPP ($g/m^2/year$)	Herbaceous NPP ($g/m^2/year$)	Herb %	(Percent)	(Range)
Northern temperate zone forests (40–60° N lat)	24,432	96.3	3.5	0.2	1050	41	3.9	15.9	0.4–28.8
Southern temperate zone forests (30–40° N lat)	21,145	96	3.8	0.2	863	36	4.2	9.1	1.8–20.2

Data are averages of values from DeAngelis et al. (1981). NPP, net primary productivity.

Table 2.3. Vertical distribution of the origin of above-ground litter fall within the northern hardwood forest at Hubbard Brook, New Hampshire (from Gosz et al. 1973)

	Litter fall		Labile fraction		1st Year decomposition	
	g/m²/year	%	g/m²/year	%	g/m²/year	%
Overstory						
Foliage	341.9	57.5	111.1	83.9	79.9	65.8
Wood	216.9	36.5	trace	trace	20.6	17.0
Understory						
Foliage	5.1	0.8	1.7	1.3	1.2	1.0
Wood	1.5	0.2	trace	trace	0.1	0.1
Herbs	19.6	3.2	19.6	14.8	19.6	16.1
Total	595.0		132.4		121.4	

The labile fraction was calculated using average labile fractions from Currie and Aber (1997). Wood was assumed to have only a trace of labile material in its biomass, and herbaceous materials were assumed to be 100% labile. First-year decomposition was calculated using a weighted average decay constant (k) for hardwood foliage from Gosz et al. (1973) and an average 10-month decay constant for woody material. The latter was chosen because dramatic weight losses during the following 2 months were reported to have come from the physical loss of buds and bark. Herbaceous material was assumed to completely decompose within the first year (see table 2.4).

concentrations of nutrients than foliage of woody species. Additional influences on ecosystem nutrient dynamics include decomposition, influences on throughfall, and phenological patterns of growth and senescence.

Decomposition

Whereas the majority of decomposition studies have focused on foliage and wood of trees, limited work on herbaceous materials has addressed rapidity of decomposition, interactions with soil organisms, and possible stimulatory effects of mixed litters on decomposition. However, cursory observation of herbaceous materials (foliage and stems) during the fall and winter in a deciduous woodland suggests rapid disappearance from the forest floor after senescence. In one of the first analyses of decomposition, Melin (1930) used CO_2 evolution in laboratory incubations to evaluate patterns of decomposition in fresh foliar litter of herbs and trees and noted variable rates of decomposition over the first 27 days of incubation. He attributed this variation to initial litter quality (nitrogen concentration), although this relationship was imperfect. In this study, decomposition of *Aralia nudicaulis* L. was more rapid than *Acer saccharum* Marsh. and *Fagus grandifolia* Ehrb., but not faster than *Fraxinus americana* L. Likely the rapid decomposition exhibited by all species reflected CO_2 evolution from decomposition of labile materials. These details of the very first stages of the decomposition process do not emerge from more conventional litter-bag studies.

More recent work, however, suggests that decomposition of foliar materials can be readily classified into materials whose decomposition is completed in less than a year and those that require longer than a year for completion. Most tree foliage has a decomposition constant (k) <1. Whereas foliage of some tree species approaches $k = 1$, most are in the range of 0.30–0.8 (table 2.4). A clear exception to this trend has been observed in foliage of *Cornus florida* L. (Cromack and Monk 1975), whose calcium-rich foliage decomposes rapidly ($k \cong 1$). In contrast, decomposition constants for herbaceous material (including stems) are generally >1, and in some cases considerably greater (table 2.4). An apparent exception occurs in the bulked herbaceous litter of a 29-year-old mixed hardwood stand in New Brunswick, Canada (k = 0.43–0.41; MacLean and Wein 1978). However, this stand was dominated by *Pteridium aquilinum* L. and *Vaccinium* L. sp., whose lower decomposition rates may have modified those of other herbaceous species in the bulked sample.

Thirty years of intense analysis of decomposition has amply demonstrated that, holding site variables and microclimate constant, decomposition is controlled most directly by litter quality, including principally concentrations of inorganic nutrients, lignin, and cellulose (Melillo et al. 1989). The higher concentrations of inorganic nutrients observed in herbaceous litters (fig. 2.1) and the apparently low concentrations of lignin and cellulose in those tissues (Dwyer and Merriam 1984) would certainly explain their rapid turnover. Most studies of decomposition of herbaceous species note that herbaceous material is fully decomposed within 6 months (or less) after senescence. The combination of initial leaching of nutrients after senescence and rapid decomposition of any remaining organic material leads to rapid disappearance and incorporation of inorganic nutrients and soluble organic materials into the forest floor. An extreme case is observed in *Erythronium americanum*, where leaf cells lyse during senescence, beginning at the leaf apex and moving basipetally (Muller 1978). This lysing pattern not only appears to enhance resorption of nutrients into the perennating organ but also encourages subsequent leaf decomposition. With the first rains after to lysis, all vestiges of the foliar material are lost.

Throughfall

The influence of the herbaceous layer on throughfall chemistry, as on other nutrient processes, is heavily dependent on its density. In dense, well-developed understory layers, the herbaceous stratum may have a leaf area index of 0.5–2 (one side: Werger and van Laar 1985; Kriebitzsch 1992c; Gratani 1997), providing an important surface for chemical exchange with incident rainfall as it passes through the canopy to the forest floor. Analyses are limited, but existing studies suggest that the herbaceous layer can significantly modify the characteristics of throughfall. The nature of this modification appears variable, depending on species composition of the understory and time of year.

Carlisle et al. (1967; data also presented in Brown 1974) present annual

Table 2.4. Decomposition constants (*k*) and half-life of foliar litter from herbs and trees from a number sites in temperate zone forests

Species	Location	Forest Type	Duration (months)	k/year	Half-life (years)	Reference
Cornus canadensis L.	Washington, USA	23-year *Abies amabilis*	12	1.02	0.68	Vogt et al. (1983)
			24	0.72	0.96	
Clintonia uniflora (Schult.) Kunth.		180-year *Abies amabilis*	12	3.31	0.21	
			24	1.96	0.35	
Bulked herbs	New Brunswick,	7-year mixed hardwoods	11	1.25	0.55	MacLean and Wein (1978)
Bulked herbs	Canada	29-year mixed hardwoods	12	0.61	1.1	
			24	0.43	1.6	
Aralia nudicaulis			4	1.39	0.50	Melin (1930)
Anemone nemorosa	Germany	*Fagus sylvatica*	12	1.55	0.45	Wise and Shaefer (1994)
Mercurialis perennis			12	2.03	0.34	
Ash-maple			17	0.97	0.71	
Beech			17	0.31	2.23	
Allium tricocum	Ottawa,	Beech-maple	6	1.90	0.36	Dwyer and Merriam (1984)
Caulophyllum thalictroides (L.) Michx.	Canada		6	1.08	.64	
Beech-maple			6	1.04	.67	
Acer saccharum	New Hampshire,	Northern hardwoods	12	0.51	1.36	Gosz et al. (1973)
Fagus grandifolia	USA		12	0.37	1.87	
Betula lutea Michx. f.			12	0.85	0.82	
Quercus montana Wild.	North Carolina,	Oak forest		0.61	1.14	Cromack and Monk (1975)
Quercus alba L.	USA			0.72	.96	
Acer rubrum L.				0.77	.90	
Cornus florida				1.26	0.55	

throughfall estimates for an oak woodland (*Quercus petraea* (Matt.) Liebl.) with a dense understory of *Pteridium aquilinum* (L.) Kuhn (table 2.5). The herbaceous layer, which accounted for 26.7% of total foliar litter fall, added as much as 37.0% of the potassium and as little as 0.6% of the calcium contributed by both layers to throughfall. The high potassium contribution would seem to be a consequence of its higher concentrations in herbaceous foliage as well as its greater mobility in organic tissues than other elements. In contrast, the reduced contributions of the herbaceous layer to calcium cycling reflect its role in cell wall structure and binding and in maintaining leaf form in the absence of the structural materials of lignin and cellulose. For all elements other than potassium, the proportionate impact of the herbaceous layer was reduced over what might be expected from its contribution to litter fall alone. The herbaceous layer accounted for 14.1% and 14.5%, respectively, of the total vegetation contribution to throughfall phosphorus and magnesium. Similarly, the herbaceous layer accounted for only 9.1% of the organic nitrogen contributed by vegetation to throughfall and also accounted for 21.3% of the removal of inorganic nitrogen from incident precipitation. The absorption of inorganic nitrogen would seem to be responsive simply to the amount of foliar biomass in the herbaceous layer. However, the proportionately reduced contribution of herbs to throughfall modification for all other elements, except potassium, would seem to reflect more conservative cycling of those elements in this stratum.

The idea of conservation of nutrients within the herbaceous layer is supported by other studies suggesting a seasonal aspect to the relative influence of canopy and understory on throughfall. Andersson (1992) suggested a strong tendency for foliar absorption by an understory of *Mercurialis perennis* L. in a *Quercus robur* L. woodland. During late summer/early fall, the herbaceous layer reduced throughput of calcium by 27%, magnesium by 33%, potassium by 27%, and nitrogen by 39%. The cations in this analysis also showed a seasonal influence, with greatest absorption during the growing season, and no apparent net influence during the period of foliar senescence. Yarie (1980) suggested similar trends in coniferous forests (*Tsuga mertensiana* Carr. and *Abies amabilis* Dougl. ex J. Forbes) of coastal British Columbia. Whereas significant only for phosphorus and nitrogen, the general trend was for absorption of all elements during the growing season and release of cations during senescence.

Conservation of nutrients with respect to throughfall and, indeed, absorption of nutrients from throughfall, has several implications for the herbaceous layer. Regardless of origin of nutrients in canopy throughfall (leaching vs. washoff), the potential for absorption of throughfall nutrients suggests a mechanism of direct cycling, which reduces the need for extensive root systems to locate and exploit nutrients. Further, to the extent that potassium contributes to positive water balances of forest herbaceous species, foliar absorption of this element also reduces dependence on extensive root systems to tap limited water supplies. The case can be made that the forest understory is a strongly resource-limited environment in which light, moisture, and

Table 2.5. Canopy and herbaceous layer modifications of throughfall in an oak woodland, Lancashire, England (after Carlisle et al. 1967)

		Precipitation and throughfall quality							
	Litter fall	Inorganic N	Organic N	Total N	P	K	Ca	Mg	Na
Incident rainfall	—	6.65	2.07	8.71	0.28	2.84	6.72	6.10	50.76
Canopy throughfall	4032.1 (73.3%)	4.32 (−78.7%)	5.49 (90.7%)	9.60 (145.9%)	0.55 (85.9%)	18.86 (63.0%)	19.01 (99.4%)	11.70 (85.5%)	82.78 (94.9%)
Herbaceous throughfall	1470.0 (26.7%)	3.69 (−21.3%)	5.84 (9.3%)	9.32 (−46.0%)	0.92 (14.1%)	28.26 (37.0%)	19.08 (0.6%)	12.65 (14.5%)	84.49 (5.1%)

The canopy was dominated by *Quercus petraea*; the herbaceous layer was a monospecific stand of *Pteridium aquilinum*. Data present annual foliar litter fall from the canopy and herbaceous layer (kg/ha/year), and nutrient content of water in incident rainfall, throughfall from the forest canopy and throughfall from the herbaceous layer canopy (kg/ha/year). Values in parentheses are contribution of each vegetation layer to the net change in incident precipitation reaching the soil surface.

nutrients all are in short supply. In considering the allocation of fixed carbon, any process that reduces the apparent limitations of one or another resource enables increased allocation to obtain another resource that continues to be limited. Hence, conservation and potential capture of nutrients in canopy throughfall by the herbaceous layer reduces the demand for extensive root systems and supports greater shoot growth, thereby enhancing energy capture and, as well, increased foliar absorption of nutrients (Andersson 1992). Plants of shaded conditions (e.g., forest herbs) generally have low root:shoot ratios (Simonovich 1973; Zavitkovsky 1976; Bloom et al. 1985; Chapin et al. 1993; Andersson 1997).

From an ecosystem standpoint, the role of the herbaceous layer in modification of throughfall may strongly influence the amount of nutrients returned to the forest floor through this vector. Subsequent influences on microbial populations and uptake by roots of other species have not been examined.

Vernal Dam Revisited

In the northern hardwood forest, a suite of circumstances occurs that has suggested that one component of the herbaceous layer may play an important role in reducing nutrient loss during a period of high mobility. Snowmelt, which may account for as much as 30% of annual streamflow (Likens et al. 1977), occurs during the spring months of March–May. Nutrient removal from these forests is most directly related to volume of water in streamflow. Hence, elevated streamflow during the spring snowmelt creates a time of significant nutrient loss. Overall biotic activity during this same period is much reduced compared to midsummer. Yet, some elements of the biota are active during the spring between snowmelt and canopy leafout. Among the herbs, the vernal component completes its entire above-ground life cycle during this 6-week period (Mahall and Bormann 1978; Muller 1978), and rising soil temperatures enhance microbial activity at this same time (Zogg et al. 1997).

The concept of the vernal dam (Muller and Bormann 1976; Muller 1978) builds upon this juxtaposition of high streamflow, rapid biomass accumulation by spring herbs, and subsequent senescence and decomposition of the nonperennating biomass of those species. During the spring period of high runoff, biomass accumulation by spring herbs, which represents the total net annual productivity of those species, involves the uptake and fixation of nutrients that might otherwise be lost in streamflow. Above-ground senescence of these species coincides with canopy closure of the overstory, and release of nutrients from the vernal species by mineralization occurs when streamflow levels are considerably reduced and activity of summer-green species (including the overstory) is high (fig. 2.5). Thus, vernal herbs may serve as a temporary biotic dam that retards the loss of nutrients and enhances internal biogeochemical cycling. At the Hubbard Brook Experimental Forest

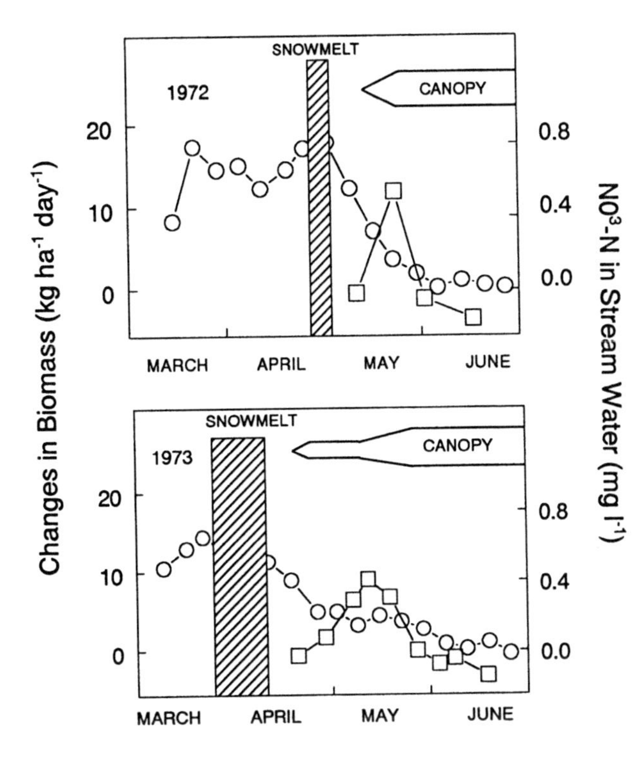

Figure 2.5. Stream water nitrate concentrations (circles) and biomass increase (squares) in *Erythronium americanum*. Biomass increase was calculated as the rate of biomass change between any two consecutive harvest dates. Redrawn from Muller and Bormann (1976), with permission.

in New Hampshire, the spring flora (most of which is *Erythronium americanum*) appears to serve in this capacity, taking up and releasing potassium and nitrogen at the same order of magnitude as that lost in spring streamflow (table 2.6).

Since first proposed, the concept of the vernal dam has received some scrutiny, exploring the validity of the concept, its limitations, and potential ramifications. Blank et al. (1980) and Peterson and Rolfe (1982) verified the concept in two midwestern forests and found that, because of significantly increased production of the spring flora of those ecosystems, the potential impact was as much as an order of magnitude greater than in New Hampshire (table 2.6). In both of these studies the greater diversity and productivity of the spring herbs was attributed to significantly richer soils of central Indiana and Illinois. Eickmeier and others (Eickmeier and Schussler 1993; W. B. Anderson and Eickmeier 1998, 2000) have evaluated the contribution of plasticity to the magnitude of the vernal dam. Increased productivity and tissue nutrient concentrations in *Claytonia virginica* L. that had been artificially fertilized suggested that a degree of plasticity would enable this and perhaps other species to respond to annual variation in spring nutrient availability (Eickmeier and Schussler 1993). However, W. B. Anderson and Eickmeier (1998) found that this plastic response had limitations such that above certain levels of nutrient availability, no further increase in total spring ephem-

Table 2.6. Net primary productivity (g/m²/year) and nutrient uptake (kg/ha/year) by the spring herbaceous community in three regions of the deciduous forest of the eastern United States

	New Hampshire[a]	Indiana[b]	Illinois[c]	Tennessee[d]	
				Unfertilized	Fertilized
NPP	5.1	66.8	68.5	21.1	74.5
N	0.9 (1.1)	5.5	10.6	0.8	3.8
P	nd	0.1	1.6	0.1	0.5
K	0.6 (0.7)	4.5	18.9	nd	nd
Ca	0.1 (3.6)	1.8	3.9	nd	nd
Mg	0.1 (0.8)	0.6	1.7	nd	nd

For comparison, gross spring stream water losses of nutrients (April and May; kg/ha) in New Hampshire are provided. Similar data from the other sites are not available. In the Tennessee study, weekly additions of aqueous fertilizer solutions totaled annual inputs of 30 g N/m²/year and 29 g P/m²/year. NPP, net primary productivity; nd, not determined.

[a]Data from Muller and Bormann (1976) and Muller (1978).
[b]Data from Blank et al. (1980).
[c]Data from Peterson and Rolfe (1982).
[d]Data calculated from W.B. Anderson and Eickmeier (1998).

eral nutrient pool size was observed. Nonetheless, they found a potential maximum vernal nutrient pool size significantly greater than had been previously reported (table 2.6).

Further analyses of the vernal dam, particularly the interactions of spring ephemeral herbs and soil microbial populations, have raised important questions concerning the significance of the contributions of the spring vascular flora to the conservation of nutrients in deciduous forests. Zak et al. (1986) noted a close correlation in the distribution of herbaceous species in deciduous forests to rates of nitrogen cycling. Diversity and abundance of spring herbs were consistently related to nitrification potential of soils on a microscale. However, potential nitrate utilization by two prominent herbaceous species of the spring flora in those forests was strikingly low (Zak and Pregitzer 1988). *Allium tricoccum* and *Asarum canadense* L. demonstrated particularly low nitrate reductase activity, implying limited ability to utilize the ionic form of nitrogen most susceptible to leaching during the spring period. In addition, analysis of the soil microflora during the spring has suggested a dynamic system with considerable potential for microbial immobilization of nitrogen in addition to that utilized by plants (Groffman et al. 1993). Indeed, Groffman et al. (1993) suggested a competitive environment in which microbial populations are significantly more important than plants as nitrogen immobilizers. Tracer studies using ^{15}N suggest that on sites with abundant spring herbaceous growth, uptake of nitrogen by microbial biomass is significantly greater than uptake by active plants (table 2.7). Zak et al. (1990) suggested that spring ephemeral herbs and spring microbial populations are competing for soil nitrogen resources and that both may play a role in nitrogen retention

Table 2.7. Percent recovery of ^{15}N in soil pools in two forests of southern Michigan

	Acer saccharum, Tilia americana, Allium tricoccum[a]		*Fagus americana, Acer saccharum, Quercus rubra, Claytonia virginica*[b]	
	$[^{15}N]NH_4^+$	$[^{15}N]NO_3^-$	$[^{15}N]NH_4^+$	$[^{15}N]NO_3^-$
Herbaceous biomass	1	3	<1	<1
Microbial biomass	22	32	21	20
Soil organic N	6	13	40	28
NH_4-N	3	2	10	0
NO_3-N	5	5	7	15

Applications and recovery of ^{15}N were made shortly after the spring ephemeral species had reached peak leaf area in mid-April.

[a]Data from Zak et al. (1986).
[b]Data from Groffman et al. (1993).

during that period. These results suggest that spring microbial populations may serve an even more important function than plants in reducing nutrient losses during the spring when water movement is high and before the principal period of summer growth. As competitors, the net influences of herbs and microbes are not necessarily additive, and the absence of herbs may not necessarily imply a net reduction in retention of nutrients.

In a recent experimental test of the vernal dam hypothesis, Rothstein (2000) concluded that the influence of spring ephemerals in limiting nutrient loss from deciduous forests is less important than originally envisioned. Removal of *Allium tricoccum* Ait. in a sugar maple-basswood forest resulted in no observable increase in nitrate loss during the vernal period. Rothstein (2000) suggested that microbial populations dominate spring nitrogen dynamics such that any nitrogen that might be utilized by spring ephemerals would otherwise be absorbed by the soil microflora. Thus, if increased hydrologic movement during the spring does represent a time of potential loss, the vernal dam may in fact be important but may be moderated by immobilization in increasing microbial biomass rather than in the biomass of spring herbs. In this study, the reciprocal experiment (removal of microbial influence) was not conducted.

In an interesting twist, Jandl et al. (1997) suggested that rich spring herb assemblages may actually enhance nutrient loss from forest ecosystems. In a beech forest dominated by a spring herbaceous layer of *Allium ursinum* L. (net primary productivity [NPP] = 200–300 g/m/year), the presence of senescent *Allium* foliar tissue in early summer is a nutrient-rich (particularly nitrogen) substrate for mesofauna grazing on the forest floor. Significantly enhanced populations of mesofauna (primarily collembolans) were observed on microsites dominated by *Allium*. Fragmentation of leaf litter by large and diverse mesofauna populations stimulated mineralization by soil microorganisms, creating even more favorable environments for soil mesofauna. Peak concentrations of soil solution nitrate, calcium, and magnesium coincident with *Allium* decomposition in early summer suggest that the synergism between mesofauna and microbial populations that has been initiated by *Allium* litter has fostered greater losses of nutrients than might otherwise occur in the absence of *Allium*. Wise and Schaefer (1994) also suggested that the rapid decomposition of other herbaceous species (*Anemone nemorosa* L. and *Mercurialis perennis* L.), due to high nitrogen concentration and low content of structural polysaccharides, served to enhance decomposition of overstory litter. Wolters (1999) suggested that the readily available source of carbon provided by *Allium ursinum* litter may partially overcome a deficit of carbon available to microbes in early summer, whereas Scheu (1997) argued that the potential for nutrient export is enhanced by the rapid decomposition of *Urtica dioica* L. litter in the fall when uptake by vegetation is low. Regardless of mechanism, phenological relationships are clearly important in understanding the role of herbaceous populations in ecosystem nutrient dynamics.

Much remains to be determined about the role of herbaceous species, in particular spring ephemerals, in nutrient cycling of forested ecosystems. The dynamics of nutrient acquisition by herbs and microbial populations requires more analysis to determine whether their relations are competitive (Zak et al. 1990) or additive. The possibility that spring ephemerals may facilitate nutrient loss raises new questions about the interactions of decomposition processes with herbaceous communities. It is important to recognize that there will be no single answer to any of these questions. Apparently competitive relations among herbs and microbes may become additive in more northerly ecosystems where continuous snow cover serves to separate phenology. Facilitated nutrient loss may occur only on those sites whose nutrient-rich qualities support highly productive spring herbaceous communities.

Summary

Although nutritional aspects of the herbaceous layer of deciduous forests have received considerably less attention than the overstory, available evidence suggests that this synusium is characterized by several unique features. Foliar concentrations of potassium are consistently 2–3 times higher among herbs than in woody overstory species. Similarly, magnesium concentrations can be up to 2 times higher than in the overstory. Although not true for all phenological groups, spring ephemerals appear to have significantly higher concentrations of nitrogen than either overstory species or herbaceous species from other phenological groups. It may be that the higher potassium concentrations in the foliage of herbaceous species play an important role in maintaining leaf structure through enhanced turgor pressure. Foliar nutrient concentrations, however, show surprisingly little variation among contrasting sites. Mechanisms accounting for this are unclear but may involve shifting carbon sinks from enhanced foliar growth on nutrient-rich sites or increased mycorrhizal support on nutrient poor sites. Seasonal patterns of nutrient concentration suggest that herbaceous species absorb nutrients during periods of high availability and that subsequent re-translocation supports growth during periods of active biomass accumulation. Thus, although herbaceous species do not exhibit luxury consumption in the traditional sense of site-to-site variation, they do exhibit seasonal patterns of luxury consumption that supports growth in seasons of limited nutrient availability.

As a component of forest ecosystems, the herbaceous layer contributes <5% of above-ground net primary productivity. However, several possible roles of the herbaceous layer in nutrient cycling have been proposed and debated. Some studies have indicated that herbaceous species are capable of absorbing nutrients from throughfall. Such a mechanism would reduce dependence on extensive root systems in a low-energy environment and, given the fairly extensive herbaceous leaf area in some forests, would also contribute to a conservation of nutrients at the ecosystem level. Other analyses have suggested that spring ephemerals play a particular role in nutrient conservation by absorbing nutrients during high soil moisture periods before the break of winter dormancy by deciduous trees and then releasing those nutrients through rapid decomposition during the early season of tree growth. The spring ephemerals, then, contribute to a vernal dam by retarding nutrient loss at a time of low biotic activity and high soil moisture movement. Further studies of microbial activity during the spring, however, have found that this effect is essentially swamped out by concurrent microbial growth. Thus, the vernal dam would appear to be mediated more by microbial populations than by the herbaceous layer. Yet other work has suggested that a rich spring herb assemblage may actually stimulate nutrient loss by serving as a nutrient-rich substrate for soil mesofauna. Enhanced fragmentation of overstory leaf litter by large mesofaunal populations was observed to stimulate mineralization by soil microbial populations, resulting in higher concentrations of nutrients in the soil solution. It is clear that the interactions among herbaceous populations, soil biota, and overstory species are undoubtedly more complex than currently recognized.

Acknowledgments I appreciate the facilities and support provided by the Santa Barbara Botanic Garden where this manuscript was completed during a recent sabbatical leave. This study (#01-09-89) is connected with a project of the Kentucky Agricultural Experiment Station and is published with approval of the director.

3

Ecophysiology of the Herbaceous Layer in Temperate Deciduous Forests

Howard S. Neufeld
Donald R. Young

Light availability is the primary limiting resource for understory plants of eastern deciduous forests, and their long-term persistence in these habitats depends on adaptations that enable photosynthesis sufficient to maintain a positive annual carbon balance (Pearcy et al. 1987). However, in addition to low light, understory plants may also be limited by water stress (Masarovičová and Eliáš 1986), nutrient deficiencies (chapter 2, this volume), and pollinator availability (chapter 5, this volume; Motten 1986). Release of CO_2 from the forest floor by decomposition processes and root respiration may elevate concentrations in the understory (Sparling and Alt 1965; Koyama and Kawano 1973; Garrett et al. 1978; Baldocchi et al. 1986; Bazzaz and Williams 1991) and improve the photosynthetic efficiency of understory plants (Lundegardh 1921), particularly during sunflecks (Naumburg and Ellsworth 2000).

Although plants in all habitats face some resource limitations, the limitations imposed on understory herbs have resulted in a suite of adaptations that distinguish them from plants that do not grow beneath forest canopies and from plants that are only temporary residents of the understory, such as the seedlings of canopy trees or shrubs. The flora of deciduous forest understories are highly adapted to these habitats, are distinctive in terms of their species composition, with close taxonomic affinities worldwide (Kawano 1970; Boufford and Spongberg 1983; Cheng 1983; Ying 1983), and express high degrees of evolutionary convergence toward similar physiologies and morphologies (Parkhurst and Loucks 1972; Givnish and Vermeij 1976; Shugart 1997). Yet at the same time, direct physiological comparisons between herbaceous understory plants and tree seedlings are few. Therefore, the conclusions of any such comparisons must be regarded as tentative at this time.

Adaptive responses to the environment can be both physiological and morphological (Givnish 1986) and can exist at multiple scales of organization ranging from the subcellular, to the leaf, to whole-plant and ecosystem levels (Ehleringer and Werk 1986; Givnish 1986; Küppers 1994; Pearcy and Sims 1994). This chapter focuses on the major ecomorphological and ecophysiological adaptations that characterize understory herbs and relates these features to the dynamic and limiting light environment in temperate deciduous forests. We begin our discussion by first characterizing the light microclimate within temperate deciduous forests and then move on to physiological and morphological adaptations at increasing scales of organization. Surprisingly, relatively few studies of herb physiology have been conducted in eastern North American forests, so we are drawing on the literature for temperate deciduous forests from other geographic regions in order to make global generalizations.

Our analytical approach is to use life-form analysis for scaling from the individual leaf or species to the ecosystem level. We use the concept of plant functional groups (T.M. Smith et al. 1997) to extrapolate general trends among species, using phenological strategies (Uemura 1994) as our grouping variable. As we will show, generalizations can be made concerning species functioning within specific phenological groups that pertain to temperate deciduous forests across the globe (Kawano 1970).

How woody seedlings cope with the understory environment as compared to understory herbs is not clearly understood. Certainly differences and similarities exist among these groups in terms of both their morphological and physiological attributes. For example, most (but not all) understory herbs are perennials, with large underground storage organs (Hicks and Chabot 1985) that are used to store carbohydrates for next year's growth and reproduction (Pfitsch and Pearcy 1992; Zimmerman and Whigham 1992; Tissue et al. 1995). Under conditions where rates of photosynthesis are low, such as during cloudy years or when substantial amounts of foliage are removed by herbivory, carbohydrates can be mobilized from below-ground reserves to maintain dormancy and to sustain leaf emergence in the spring (Lubbers and Lechowicz 1989). This often comes, however, at the expense of both vegetative and sexual reproduction (Pitelka et al. 1980, 1985; Lubbers and Lechowicz 1989; Pfitsch and Pearcy 1992). Tree seedlings can also mobilize below-ground reserves for maintenance and leaf emergence, but generally do not reproduce when they are in the herb strata; thus, they do not face the exact same carbon allocation dilemma. Both woody seedlings and herbs may idle for years in a light environment near their photosynthetic light compensation point until a gap in the canopy occurs and the light environment improves (Crawford 1989). Each year, however, the tree or shrub is putting on secondary growth in the above-ground stem, with the ultimate strategy of growing up and out of the herbaceous layer. Thus, a commitment to mechanical stability via secondary woody tissue is an allocation priority for these plants. In contrast, herbs have no secondary growth of the aboveground stem, and replacement each year comes at great cost in terms of carbon and nutrients.

The Understory Light Environment

Many authors have emphasized the dynamic nature of the understory light environment (M.C. Anderson 1964; Reifsnyder et al. 1971; Baldocchi et al. 1984; Chazdon 1988; Canham et al. 1990; Chazdon and Pearcy 1991; Baldocchi and Collineau 1994; Parker 1995; Küppers et al. 1996; Pearcy 1999; Vierling and Wessman 2000), and in particular the role of sunflecks with respect to plant carbon gain (Chazdon 1988; Chazdon and Pearcy 1991; Pfitsch and Pearcy 1995; Pearcy 1999). Although we focus here mainly on temperate deciduous forests, many of the conclusions we reach are common also to temperate and tropical rainforests, where much of this research has been concentrated (Becker and Smith 1990; McDonald and Norton 1992; Rich et al. 1993; Turnbull and Yates 1993; Clark et al. 1996; de Castro 2000).

Total Available Light in the Understory

In all forests, both the quantity (M.C. Anderson 1964; Reifsnyder et al. 1971; Hutchison and Matt 1977; Chazdon and Fetcher 1984; Brown and Parker 1994; Canham et al. 1994) and quality (Olesen 1992; Brown et al. 1994; Grant 1997) of light are modified by passage through the canopy of overstory trees and shrubs (Clinton 1995). Beneath forest canopies, the spatial and temporal patterns become highly variable and stochastic (G.C. Evans 1956; Reifsnyder et al. 1971; Chazdon and Pearcy 1991; Rich et al. 1993, Pearcy et al. 1994). This variability is caused by a number of factors, including overstory composition, leaf characteristics such as size, orientation, density, clumping, and fluttering (Hutchison et al. 1986), and gaps in the canopy due to branch or tree falls (Lawton 1990). In addition, there are penumbral influences (when canopy openings less than the diameter of the solar disk [0.5°] cause a shadow to develop along the edge of the sunfleck), topographical influences, changes in seasonal phenology, as well as daily and seasonal variations due to the weather and the solar path (Baldocchi and Collineau 1994).

The heterogeneity of the understory light environment means that an adequate sampling scheme for characterizing the photosynthetic photon flux density (PFD) of a particular locale necessarily involves a large number of sensors spread out over a substantial portion of the forest floor. Reifsnyder et al. (1971), working in Connecticut, estimated that it would take at least 18 sensors to adequately characterize the instantaneous, direct radiation patterns in a deciduous forest (to within a standard error of 7 W/m^2), but only 2 sensors to characterize the diffuse radiation pattern. However, just one or two sensors were adequate to characterize the daily or integrated radiation pattern, due to temporal averaging of the instantaneous values. The total daily PFD within a particular forest can be estimated from instantaneous measures of diffuse light, particularly on overcast days around noon, because the two are highly correlated in a diversity of forests (Koizumi and Oshima 1993; Parent and Messier 1996; Oshima et al. 1997; Machado and Reich 1999). Such measures can serve as good estimators of the relative amount of light available to understory plants over the course of a season, thus reducing the need for continuous measurements.

Site variation within a single forest stand is well illustrated by data from an oak forest in Tennessee, where correlations among light sensors became insignificant at distances >3.0 m, and about 70% of all sunflecks were <0.5 m in diameter (Baldocchi and Collineau 1994). Because canopy widths of most understory herbs are often much smaller than 0.5 m, plants spaced more than 3 m apart would almost never experience identical light environments. For example, Pfitsch and Pearcy (1992) reported that the number of sunflecks received per day by individuals of the understory herb *Adenocaulon bicolor* Hook. in a redwood forest ranged over two orders of magnitude, from <10 to >400. Similar analyses in eastern deciduous forests are lacking.

The concept of grain size has been useful in ecology for characterizing patchy habitats and for describing how organisms are adapted to dealing with different size patches (Levins 1968). The patchy distribution of herbs that is common in eastern forests (Bratton 1976, Rogers 1983) results not only from variations in soil characteristics such as nutrients, water contents, and soil depths, but also because understory herbs are responding to the heterogeneous distribution of light on the forest floor (Pitelka et al. 1985). As Pearcy (1999) has noted, if the patch size encompasses an entire plant (e.g., sun vs. shade habitats), the environment is perceived by the plant as coarse grained. Spending its entire life in just one patch, the plant must be well adapted to the local conditions, or else it will be extirpated. In contrast if the canopy of a plant extends beyond the edge of a patch or is subject to patches of varying quality, the environment is perceived by the plant as fine grained. In this case, acclimation to changing conditions (greater plasticity) would be most adaptive because the plant will not complete its growth entirely within a single patch. How a plant copes with a fine-grained environment depends on whether it has indeterminate or determinate growth. Plants with indeterminate growth can produce successive leaves differing in anatomy and physiology based on the conditions present at the time of formation, thereby increasing efficiency of resource utilization (Jurik and Chabot 1986). Plants with determinate growth do not have that option, and tolerance to a variable environment is limited by the degree of acclimation possible among existing leaves (Pearcy and Sims 1994; Rothstein and Zak 2001).

Background Levels of Diffuse PFD

When the leaf area index (LAI) of a temperate deciduous forest is at its maximum of about 6 m^2/m^2 in late summer (Hutchison and Matt 1977), light levels near the forest floor are very low, ranging from 0.7 to 7% of the PFD incident above the canopy, with the most commonly reported values centering around 1–3% (Chazdon and Pearcy 1991; Brown and Parker 1994; Canham et al. 1994; Pearcy et al. 1994; Constabel and Lieffers 1996). This range represents a 10-fold difference in PFD, and even though the absolute values are low, the variation is more than enough to influence species distribution and biomass (Reid 1964; Sparling 1967; Pitelka et al. 1985; Washitani and Tang 1991; Tang et al. 1992). Much higher light penetration values have been recorded for some forests, with values of 9–25% (see Baldocchi and Collineau 1994 and references therein), but most of these are from studies done before 1972, and forest conditions were often not specified. Truly higher PFDs (up to 20%), however, are commonly found in riparian forests (Oshima et al. 1997) because of abundant side light. These edge effects can have large impacts on the abundance of understory vegetation and may extend great distances even into nonriparian forests (Ruben et al. 1999; Gehlhausen et al. 2000). Forests with minimal edge-to-area ratios (square or circular in shape) have less side-light penetration than more irregularly shaped forests (Formann and Godron 1986).

Seasonal Patterns of Light Penetration

In temperate deciduous forests, light penetration to the herb layer is most strongly influenced by seasonality because of changing solar elevation and canopy phenology. The combination of these two factors results in several distinct phenoseasons (Hutchison and Matt 1977): winter/spring leafless, spring/summer leafing out, summer fully leafed, autumn partially leafed, and autumn/winter leafless. Many understory plants have phenological strategies that directly correspond to these demarcations (fig. 3.1) and can be grouped into several major categories (chapter 9, this volume; Robertson 1895; Seybold and Eagle 1937; Sparling 1964, 1967; Taylor and Pearcy 1976; Mahall and Bormann 1978; Chabot and Hicks 1982; Uemura 1994), and which are discussed in detail later in this chapter.

Solar radiation striking the forest floor is greatest in the spring leafless period (rather than in winter) because the solar angle is higher relative to the winter minimum (Hutchison and Matt 1976). At this time of year, diffuse radiation penetration can range from 10 to 60% of the radiation incident above the canopy (Hutchison and Matt 1977; Baldocchi et al. 1984; Curtis and Kincaid 1984). At the Walker Branch Forest in Oak Ridge, Tennessee, 90% of the incident irradiance was contributed by direct-beam radiation (Hutchison and Matt 1976, 1977). During this time of the year, PFDs may range up to full sun values of approximately 2000 $\mu mol\ m^{-2}\ s^{-1}$ and are

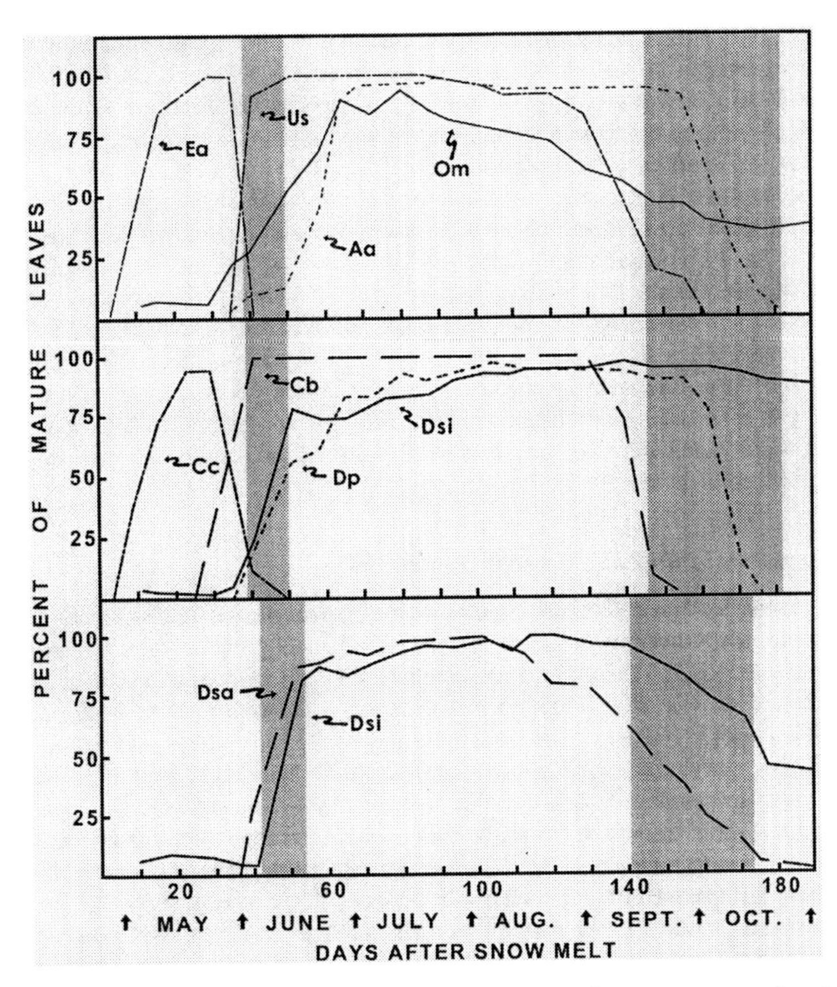

Figure 3.1. A comparison of the percentages of leaves in the mature stages for nine different taxa throughout a year. Shaded areas indicate periods of time when leaves are expanding or falling off the canopy. Species: Ea = *Erythronium americanum;* Us = *Uvularia sessilifolia*; Om = *Oxalis montana*; Aa = *Aster acuminatus*; Cc = *Claytonia caroliniana*; Cb = *Clintonia borealis*; Dp = *Dennstaedtia punctilobula*; Dsi = *Dryopteris spinulosa* var. *intermedia*; Dsa = *Dryopteris spinulosa* var. *americana*. Data from 664 m in Hubbard Brook, New Hampshire, except for the lower panel, which are from 782 m. Reprinted with permission from Mahall and Bormann (1978).

more than sufficient to allow herbs that are physiologically active in the leafless seasons to carry on some gas exchange (Hicks and Chabot 1985; Yoshie and Kawano 1986; Willmot 1989; McCarron 1995). As much as 20% of the annual irradiance can be received at the forest floor during just the spring months (Hutchison and Matt 1977).

As summer approaches, the solar angle continues to rise but understory PFD drops to a minimum because of attenuation by the fully leafed canopy (fig. 3.2a, b). For example, midday PFDs can be as low as 5–10 μmol m^{-2} s^{-1} and sometimes as low as 2–4 μmol m^{-2} s^{-1}. On an annual basis, only 13% of the total irradiance in a Tennessee oak-hickory forest falls in the 150 day period when the canopy is fully closed (Hutchison and Matt 1977). Once leaf fall begins in late summer, PFD begins to rise, even though the solar angle is now declining (Hutchison and Matt 1976). At equivalent solar angles, light penetration is greater in the spring than in the fall due to the persistence of leaves in the fall canopy.

Successional Influences on Light Penetration

It is often assumed that as forests age, light penetration to the forest floor monotonically decreases due to increasing LAI and greater bole and branch size (Horn 1971; Waring and Schlesinger 1985). Brown and Parker (1994), however, showed that, after an initial period of maximum light penetration before canopy closure, light transmittance peaked in 50-year-old stands of southern Maryland before dropping in older ones (fig. 3.3a). Light penetration in the younger stands was low because of high stem densities and foliage clumping. Over time, self-thinning reduces stem densities, and foliage density declines due to bole elongation, allowing greater light penetration (fig. 3.3b). Lower light penetration in older stands arises because of a greater preponderance of shade-tolerant canopy species, such as beech, sugar maple, and hemlock, all of which have deeper crowns and intercept more light than shade-intolerant species such as red oak and white ash (Canham et al. 1994).

Successional changes in light availability and other associated resources such as water and nutrients after disturbance have important implications for the distribution of understory plants (Whitney and Foster 1988; Meier et al. 1995; Singleton et al. 2001; but see Gilliam et al. 1995). If light were the sole, overriding limiting resource, we would expect greater diversity and biomass of herbs in mid-successional forests, rather than in early or late successional forests. Many eastern deciduous forests are young, second-growth stands (Whitney 1994), with relatively higher light levels compared to more mature stands, but herb abundances are actually reduced because of past disturbances (Meier et al. 1995; Singleton et al. 2001). Studies now suggest that recovery of the herb layer after logging, but particularly after conversion to agriculture, may take decades to centuries (chapter 13, this volume). Brown and Parker (1994) remarked on the near lack of an understory in the youngest stands they surveyed, but did not determine the reasons for this absence. Other studies suggest that recolonization of disturbed

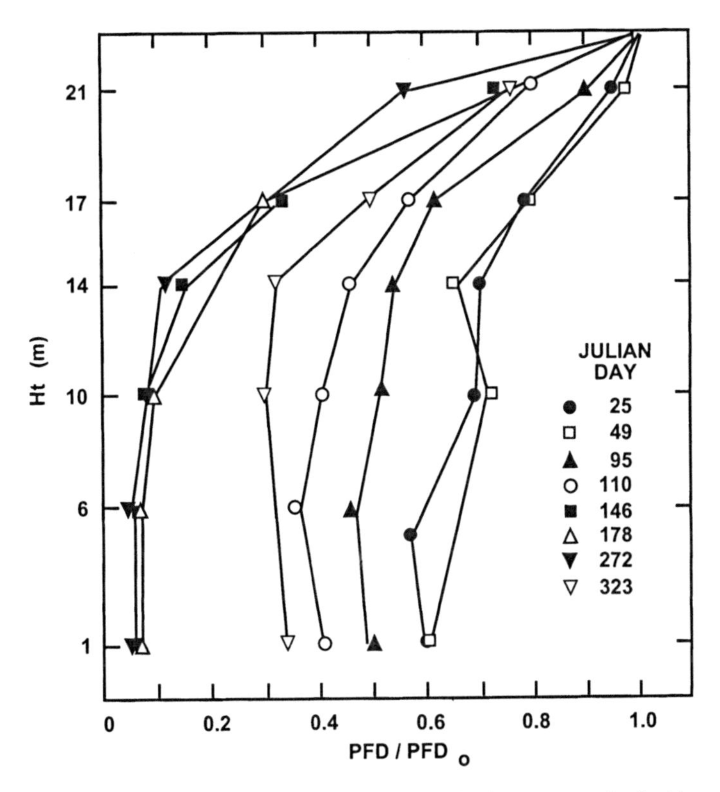

Figure 3.2. (Top) Hemispherical photographs through the canopy of a deciduous forest in North Carolina at different seasons and levels of openness (photos by Craig Brodersen). (Bottom) Seasonal variation in the vertical distribution of photon flux density (PFD) through the canopy normalized against that above the canopy (PFD_o). Reprinted with permission from Baldocchi et al. (1984).

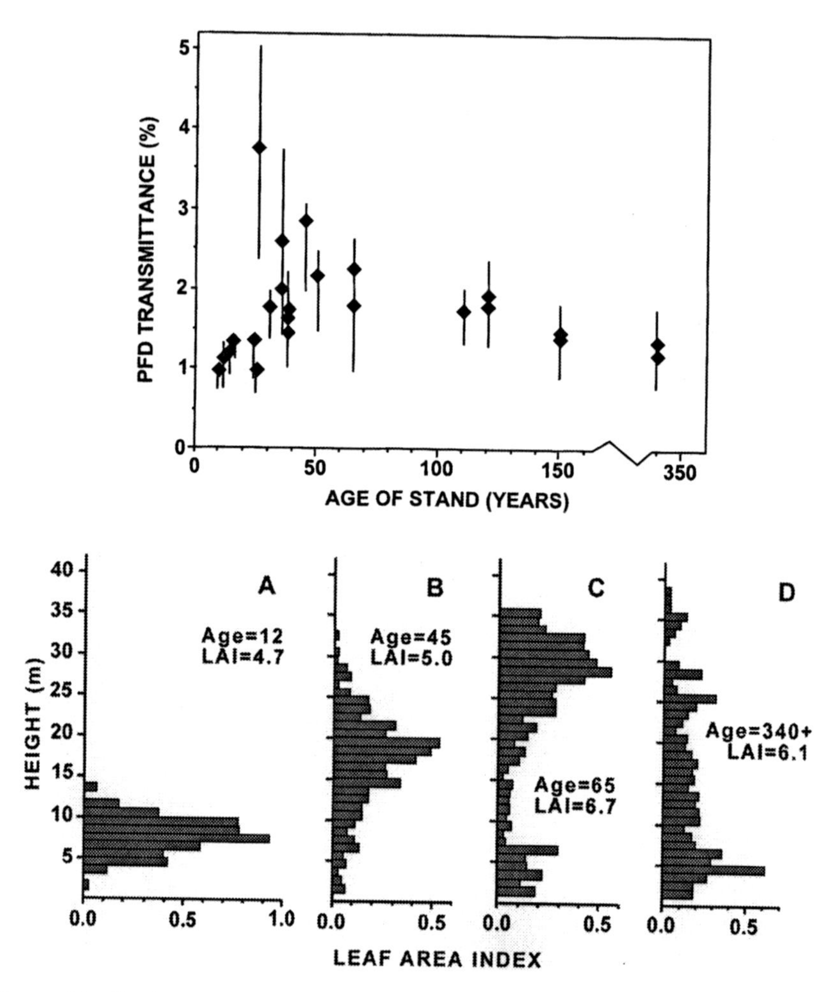

Figure 3.3. (Top) Percent transmittance of PFD through forest stands of various ages. (Bottom) Vertical distribution of leaf area index (LAI) for stands of various ages. Forests are in Maryland. Reprinted with permission from Brown and Parker (1994).

forests may take 75 or more years, due to a variety of factors such as prior use for agriculture (Singleton et al. 2001), dispersal limitations (Beattie and Culver 1981; Cain et al. 1988), and canopy influences, such as the abundance and distribution of safe sites (Beatty 1984). Other studies have found conflicting results for herb diversity over successional time. Bossuyt et al. (1999) found no differences in herb abundance or diversity between old and young forests in Belgium, although herbs associated with the older forests were poor colonizers of younger forests. Similar results for herb diversity were found for forests in West Virginia by Gilliam et al. (1995).

Most likely, successional trends in herb abundances and diversity interact with the type and extent of the disturbances initiating this process (chapter 13, this volume). As forests age and light levels decrease, colonization should result in increases in species diversity, but the counteracting effects of lower PFD might slow or reduce vegetative and reproductive success of some herbs, particularly the shade-adapted summer-greens, which grow and reproduce better in more open sites beneath the forest canopy (Pitelka et al. 1985).

Constabel and Lieffers (1996) noted that more open aspen stands had better developed shrub layers, which intercepted so much of the incoming PFD that percent penetration to the forest floor did not change as compared to more closed stands that had a less well-developed shrub layer. Oshima et al. (1997), working in a fragmented riparian forest in Japan, found that light availability in the spring above the herbaceous layer strongly correlated with light near the ground, but after canopy closure, the relationship became negative, which they attributed to more vigorous growth of herbaceous species in sunnier sites. These studies suggest that the amount of PFD available to low-growing herbs may be fixed for various forest types independent of whether light interception occurs in the tree canopy or lower in the shrub layer. However, even if the total PFD remains the same, a low canopy creates a substantially different light microclimate from a tall one, due to penumbral effects (W.K. Smith et al. 1989) and changes in the mean duration of sunflecks, which become shorter as canopy height decreases (Pearcy 1988; Pearcy et al. 1990). In undisturbed old-growth forests, maximum canopy heights have been estimated to be in the range of 20–37 m (Whitney 1994), while they may be only half that in second growth forests; thus, sunfleck dynamics may be strikingly different between these two types of forests.

Periodic disturbances, such as windstorms, tornados, and hurricanes, contribute to the formation of canopy gaps that temporarily raise light levels in forest stands (Canham et al. 1990; de Freitas and Enright 1995; Tang et al. 1999) but have a reoccurrence frequency of perhaps only once every 1000–2000 years (Whitney 1994). Tree and branch death due to disease, insects, or old age occur much more frequently; gaps created by these processes enhance the growth of herbs within old-growth stands (Tang et al. 1999), as well as contribute to their patchy distribution (Rogers 1982; Bratton 1976). Many perennial herbs live for decades (Antos and Zobel 1984), and some clonal plants may be hundreds of years old (Tamm 1956). Most will certainly encounter some form of canopy disturbance that substantially increases the incoming PFD during their lifetime and which will contribute to a period of rapid growth and enhanced reproduction. The average return rate for gaps in eastern forests is approximately once a century (Runkle 1982), which represents a spatially averaged maximum number of years that an understory herb has to endure the lowest light levels in a particular forest.

Topographic Influences on Light Penetration

A number of topographic variables can affect the amount of light received in the forest understory, including position with respect to surrounding topography, slope, aspect, latitude, longitude, and time of day (Cantlon 1953; Frank and Lee 1966; Nuñez 1980; Flint and Childs 1987). Forest community types may vary depending on topography (Cantlon 1953), which in turn can exert strong influences on the understory environment (Helvey et al. 1972). These complex interactions have important implications for the distribution of understory plants. In New Jersey, south-facing slopes have greater herb diversities and higher abundances than north-facing slopes. Certain species are primarily restricted to one slope or the other; *Asarum canadense* L., *Aralia nudicaulis* L., and *Cimicifuga racemosa* Nutt. are found on north-facing slopes, while *Viola palmata* L., *Lespedeza violacea* Ell. and *Eupatorium sessilifolium* L. occur on south-facing slopes (Cantlon 1953). On southwest-facing slopes in the southern Appalachians of Virginia, forests are dominated by evergreen pines, whereas deciduous oak forests prevail on northwest-facing slopes (Lipscomb 1986). Daily irradiance above the canopy is greater on southwest slopes, but interestingly, is nearly the same in the shrub layer beneath both canopies (Lipscomb and Nilsen 1990a), the result of differences in canopy structure between the pine and oak forests. However, there are differences in temperature between the two slopes, with maximum temperatures higher on southwest slopes, minimum temperatures higher than valley temperatures, and southwest ridge tops warmer than slope bottoms (Lipscomb and Nilsen 1990b). In other locations, north-facing slopes are generally cooler than south-facing slopes (Cantlon 1953, Stoutjesdijk 1974). Vapor pressure deficits are also lower on north-facing slopes, and the improved water status may favor ferns and bryophytes, which are more common on these slopes (Cantlon 1953).

Ridge-top forests tend to be shorter in stature, with more widely spaced trees and open canopies (Whittaker 1956; Hicks and Chabot 1985). They are highly prone to disturbances such as wind, ice, and defoliating insects (e.g., gypsy moths, *Lymantria dispar* L., prefer chestnut oak, *Quercus prinus* L., which is primarily a ridge-top species). If the surrounding hills are lower in elevation, then side light, particularly in the morning and afternoons, would be more abundant than in cove forests, or in forests in areas with little topographic relief. As a consequence, these forests would be expected to have higher understory PFDs. In some cases this could lead to greater herb abundances, while in other cases they could be lower because these forests might be situated on shallow, nutrient-poor soils (Whittaker 1956) that are more prone to drought stress in the summer. Much of the understory in these latter sites is composed of ericaceous shrubs and evergreen herbs, which are conservative in their water use (Hicks and Chabot 1985; Givnish 1986; McCarron 1995).

Effects of Overstory Composition

Many temperate forests contain a mixture of both deciduous and evergreen species (Braun 1950). Mixed conifer-hardwood forests often have intermediate understory PFDs during the summer compared to deciduous forests (Constabel and Lieffers 1996) but lower PFDs in the winter because the evergreen species retain their leaves all year. Mixed-species forests also have more patchy distributions of light because the conical geometries of the conifer crowns permit more light to penetrate between trees (Baldocchi and Collineau 1994).

Contribution of Sunflecks

No matter what type of forest or its successional stage, sunflecks are an important source of light for understory herbs. *Sunflecks* are transitory increases in PFD above the background diffuse PFD (Chazdon 1988; Pearcy et al. 1994) and are most often empirically defined as those periods of time when the PFD is > 50 or 100 μmol m^{-2} s^{-1}, which are typical background levels in temperate deciduous forests. However, as both Chazdon (1988) and Smith et al. (1989) point out, the criteria for defining a sunfleck are vague and imprecise. For example, Chazdon (1988) notes that the background PFD for a Wyoming forest (Young and Smith 1979) is nearly the same as that defined as a sunfleck for a tropical rainforest in Costa Rica (Chazdon and Fetcher 1984). Smith et al. (1989) proposed that openings in the canopy be categorized based on the duration and proportion of time each patch is in penumbra or umbra (direct beam radiation).

In temperate forests, most sunflecks last from seconds to minutes (Canham et al. 1990; Chazdon and Pearcy 1991; Horton and Neufeld 1998; Tang et al. 1999). Mean sunfleck durations for hardwood forests in the eastern United States are about 7 minutes (Canham et al. 1990), whereas in a closed stand in Cades Cove, Great Smoky Mountains National Park (fig. 3.4), nearly 80% of the sunflecks were 1 minute or less in length (Horton and Neufeld 1998). Sunfleck lengths in a temperate forest in Japan varied during the day, with the longest flecks around noon (12–18 minutes) and the shortest ones (6–10 minutes) at 10 AM and 2 PM (Koizumi and Oshima 1993).

Larger sunflecks usually have a higher PFD than smaller ones due to penumbral effects (Reifsnyder et al. 1971; Koizumi and Oshima 1993), which proportionally increase the amount of diffuse radiation, thereby lowering the total PFD. The proportion of a sunfleck in penumbra depends on both the height and size of the canopy opening (W. K. Smith et al. 1989). For a gap with a diameter-to-height ratio of 0.01, the entire sunfleck would be in penumbra, whereas if the ratio were 0.04, 80% of the sunfleck would be in umbra. In temperate forests, most short duration sunflecks are dominated by penumbra, and consequently have PFDs of 250 μmol m^{-2} s^{-1} or less (Chazdon 1988). In the Smoky Mountains, more than 60% of the sunflecks had PFDs < 200 μmol m^{-2} s^{-1} (Horton and Neufeld 1998).

Despite their transitory nature, sunflecks contribute a substantial amount to the total daily energy budget of forest understories. Values range from 21% in a Connecticut forest (Reifsnyder et al. 1971) to 45–68% in eastern North American and Japanese forests (Hutchison and Matt 1977; Weber et al. 1985; Koizumi and Oshima 1993). In the Smoky Mountains, nearly 80% of incoming PFD can be contributed by sunflecks (Horton and Neufeld 1998).

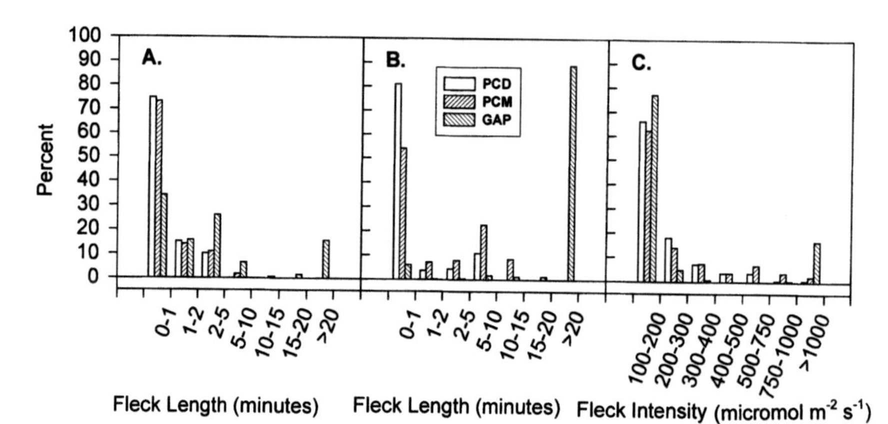

Figure 3.4. Frequency histograms for (A) sunfleck duration, (B) percentage of total daily PFD contributed by sunflecks of specific duration classes, and (C) sunfleck maximum intensity. Sites were sampled over 3 consecutive days and include a partially-closed canopy dry site (PCD), a partially-closed canopy moist site (PCM), and an open canopy gap site (GAP). Histograms represent combined data from 10 sensors per site. Reprinted with permission from Horton and Neufeld (1998).

Sunfleck contributions may be greater in the spring and fall months due to the lower LAIs at those times (Koizumi and Oshima 1993), but the proportional contributions of sunflecks are greater in forests with lower background diffuse PFD (Chazdon 1988).

Even though sunflecks make up a substantial portion of the radiation budget in an understory, their total cumulative duration is brief. Values ranging between 12 and 25% of the daylight hours are common (Weber et al. 1985; Koizumi and Oshima 1993). In the study by Koizumi and Oshima (1993), sunflecks were often more frequent in the morning and late afternoon and less common around solar noon due to the changing solar path through the canopy.

In mountainous areas, it might be expected that ridge-top forests would have substantial sunfleck activity when the solar elevation is low because there would be a shorter optical path into the understory and less obstruction by adjacent trees. In cove forests the opposite situation would predominate. Here, side light is almost entirely obliterated by slopes and other trees, and most sunflecks would occur near midday when the solar angle was high. North-facing slopes would have the shortest cumulative sunfleck durations because the sun drops below the horizon sooner than for a south-facing slope.

There are relatively few data on the temporal distribution of sunflecks. Because photosynthetic induction responses are dependent on closely spaced exposures to high light within a certain length of time (Chazdon 1988; Pearcy 1988, 1999, Pearcy et al. 1994), quantifying this parameter is important for modeling plant responses to sudden changes in PFD. Pearcy (1988) reported that 70% of sunflecks in a rainforest in Queensland occurred within 1 minute of the preceding sunfleck, and only 5% were preceded by low-light periods > 1 hour in length. However, there do not appear to be any similar analyses of sunfleck clumping in temperate deciduous forests.

Plant Strategies for Coping with the Varying Light Environment in the Understory

Temperate understory herbs use a combination of both physiological and morphological adaptations that give them a selective advantage in low light habitats. Besides general adaptations to the prevailing light conditions (typical sun/shade responses; see table 3.1), these herbs must also contend with cold temperatures for a large portion of the year. The combination of light and cold stress has resulted in the evolution of at least six distinct phenological strategies, whereas overstory trees have adopted primarily just the winter deciduous and evergreen habits (Uemura 1994), although other strategies do exist. For example, one shrub in Japan, *Daphne kamtschatica* Maxim. var. *jezoensis* (Maxim.) Ohwi is shade deciduous (Lei and Koike 1998) but maintains overwintering leaves. In the eastern United States, the understory tree *Aesculus sylvatica* Bartram behaves similarly to a spring ephemeral; that is, it leafs out early before the overstory does, begins to lose its leaves once the canopy closes, and is completely defoliated by the end of August (DePamphilis and Neufeld 1989). The fact that herbs with contrasting phenology coexist in the same communities, albeit to different degrees, suggests that there may be more than one evolutionarily stable strategy in these environments. In addition, by determining patterns of abundance among the different types of herbs, we may gain insight into the major selective factors that have resulted in the mix of herbs that populate deciduous forests.

Phenological Strategies of Understory Herbs

The six common phenological strategies of understory herbs (Uemura 1994) are (1) spring ephemerals, (2) summer-greens, (3) wintergreens, (4) heteroptics (Kikuzawa 1984) (i.e., species with two types of leaves: summer-green, and overwintering, each lasting less than 1 year, and two types of evergreen; (5) those whose leaves last just more than 1 year, called biennial-leaved by Uemura (1994), and (6) those whose leaves last for more than 2 years (Mahall and Bormann 1978; Chabot and Hicks 1982; Uemura 1994). Parasitic and saprophagous (often mycotrophic) herbs constitute two additional groups, but their phenological dependence on the canopy has not been well documented. Species that are hemiparasites can perform some photosynthesis, whereas holoparasites apparently have no photosynthetic activity (de-Pamphilis et al. 1997). Carbon for nonphotosynthetic herbs (such as *Monotropa uniflora* L.) probably comes from mycorrhizae that are connected to

Table 3.1. General differences between sun and shade leaves for leaves of plants adapted to high- and low-light intensities

Characteristic	Sun leaves	Shade leaves
Morphological features		
Leaf area	–	+
Leaf thickness	+	–
Palisade parenchyma thickness	+	–
Spongy parenchyma thickness	Similar	Similar
Specific leaf weight	+	–
Cell abundance	+	–
Cuticle thickness	+	–
Density of stomata	+	–
Cell ultrastructural features		
Cell size	–	+
Cell wall thickness	+	–
Chloroplasts per area	+	–
Chloroplast orientation	Vertical	Horizontal
Proportion of stacked membrane	–	+
Thylakoids per stroma volume	–	+
Thylakoids per granum	–	+
Starch grains in the chloroplasts	+	–
Chemical features		
Caloric content	+	–
Water content of fresh tissue	–	+
Cell sap concentration	+	–
Lipids	+	–
Anthocyanins, flavonoids, xanthophylls	+	–
Chlorophyll per chloroplast	–	+
Chlorophyll per area	Similar	Similar
Chlorophyll per unit dry mass	–	+
Chlorophyll a:b ratio	–	+
Light-harvesting complexes per area	–	+
Electron transport components per area	+	–
Coupling factor (ATPase) per area	+	–
RUBISCO per area	+	–
Nitrogen per area	+	–
Physiological functions		
Photosynthetic capacity per area	+	–
Dark and photorespiration per area	+	–
Photosynthetic capacity per dry mass	Similar	Similar
Dark and photorespiration per dry mass	Similar	Similar
Carboxylation capacity per area	+	–
Electron transport capacity per area	+	–
Quantum yield	– (Or similar)	+ (Or similar)
Light compensation point	+	–
Light saturation point	+	–
Photoinhibition likely	–	+
Transpiration	+	–

From Nilsen and Orcutt (1996) and Lambers et al. (1998), including generalizations from Boardman (1977).

other photosynthetically capable host plants, although one species of *Monotropa* appears to be able to perform non-photosynthetic carbon fixation via PEP carboxylase (Malik et al. 1983).

Spring ephemerals leaf out before canopy development when light levels are at their highest but exhibit little tolerance to the deep shade that develops later in the season. In more northerly latitudes, they may even leaf out in the winter or very early spring, when despite the lack of a canopy, total daily PFDs may be relatively low due to the low solar angle at these times of the year. Leaf senescence in this group begins once canopy closure occurs, and most species become dormant before midsummer (Kawano et al. 1978). Rothstein and Zak (2001) have termed these plants *shade avoiders*. In contrast, summer-green species leaf out during or after canopy closure (*shade tolerators*), but retain their leaves for less than 1 year. Senescence usually occurs in late summer or the fall. These species gain most of their carbon before the completion of canopy closure but are able to continue some carbon assimilation under the low light conditions that ensue later in the season. Wintergreen species, as defined by Uemura (1994), form overwintering leaves in late summer or fall, but these leaves detach early the next summer. The difference between heteroptic and wintergreen species is that the former have some type of leaf on the plant at all times during a year, whereas leaf life spans of true wintergreens are less than 1 year, and plants are often without leaves during the summer. Heteroptic plants can be considered a special case of evergreenness. True evergreen species retain their leaves for more than one year, and sometimes for up to 3 or 4 years (Chabot and Hicks 1982; Koizumi 1985, 1989; Koizumi and Oshima 1985; McCarron 1995; A.T. and B.G. Hallowell, pers. comm.).

In cool temperate forests in Japan (Uemura 1994), and most likely in all temperate deciduous forests, the majority of understory species are summer greens (69%), followed by evergreens (19%). Spring ephemerals constitute just 6% of the flora, wintergreens constitute 3%, and least abundant are parasitic and saprophagous herbs (2%) and heteroptic species (1%). Some of the more common species in each of these phenological categories and their physiological characteristics are listed in table 3.2.

Based on abundances only, it appears that evolution has favored those species that grow during the warmer portions of the season when temperatures are more favorable for photosynthesis and nutrient uptake. Evolving the ecophysiological mechanisms necessary for carbon gain when the canopy is leafless may be either more difficult, or the species pool from which the necessary adaptations could be obtained was limited, thereby curbing the number able to take advantage of these times of the year. In certain portions of northeastern England, though, where the winters are relatively mild and with little snow cover, a number of ephemeral species have evolved to leaf out in late winter such as *Ranunculus ficaria* L. and *Hyacinthoides non-scripta* L. Chouard (A. W. Davison, pers. comm.).

Deciduousness is most likely a derived condition and probably evolved in response to alternating periods of unfavorable conditions arising from either

Table 3.2. Summary of gas exchange characteristics for a sampling of understory herbs grouped by phenological class

Species	Light compensation point (μmol m^{-2} s^{-1})	Light saturation point (μmol m^{-2} s^{-1})	Apparent quantum yield (μmol CO_2 / μmol photons)	Light saturated photosynthesis (μmol CO_2 m^{-2} s^{-1})	Dark respiration (μmol CO_2 m^{-2} s^{-1})	Leaf life span (weeks)	References
Spring ephemerals							
Allium monanthum	5	700	0.041	6.4	1.1	11	Kawano et al. (1978)
Allium tricoccum	10; 22	400; 1800		13.1; 15.4	1.3; 1.4	10	Sparling (1967); Rothstein and Zak (2001)
Anemone raddeana	23	600	0.042	14.0	1.3	6	Yoshie and Yoshida (1987)
Claytonia caroliniana	36	300				11	Sparling (1967)
Dentaria diphylla	8	76				10	Sparling (1967)
Dicentra canadensis	36	600				9	Sparling (1967)
Decentra cucullaria						7	Harvey (1980)
Erythronium americanum	14–20	326–1000	0.046	20.0; 14.7	1.2	11	Taylor and Pearcy (1976); Hicks and Chabot (1985); Hull (2001)
Erythronium japonicum	20	>700[a]	0.051	6.9	0.7	9	Kawano et al. (1978)
Mertensia virginica						11	Harvey (1980)
Mean	20	592	0.045	11.8 (16.5)[b]	1.1	10	
Summer greens							
Adoxa Moschatellina	17	700	0.048	7.2	0.6	17	Kawano et al. (1978)
Allium Victorialis ssp. platyphyllum	6	200	0.124[c]	5.9	0.6	17	Kawano et al. (1978)
Aralia nudicaulis		185		6.0			Landhäusser et al. (1997)
Arisaema triphyllum	5	133	0.044	5.6	0.4		Hull (2001)
Asclepias exaltata	5	400	0.056	20.6	1.1	16	Souza and Neufeld (unpublished data)
Aster acuminatus	17–74	301–862	0.051	7.9–11.5	0.4–0.9		Pitelka and Curtis (1986)

Cardiocrinum cordatum	22	700	0.053	6.4	1.0	20	Kawano et al. (1978)
Caulophyllum thalictroides	20	600				20	Hicks and Chabot (1985)
Dryopteris filix-mas						32	Willmot (1989)
Dryopteris dilatata						32	Willmot (1989)
Dryopteris marginalis	4	60				30	Hicks and Chabot (1985)
Hydrophyllum appendiculatum	20	420		5.7			Hicks and Chabot (1985)
Jeffersonia diphylla						17	Harvey (1980)
Maianthemum canadense	2	50				22	Sparling (1987)
Mercurialis perennis	2–30	300	0.060	8.0	0.05–1.3		Kriebitzsch (1992a)
Microstegium vimineum	15	800	0.037–0.048	19.1	0.9	−28	Horton and Neufeld (1998)
Podophyllum peltatum	10–11	50–117	0.036	11.5		18	Hicks and Chabot (1985)
Rubus pubescens				4.6	0.8		Landhäusser et al. (1997)
Sanguinaria canadensis	4	100				20	Hicks and Chabot (1985)
Smilicina racemosa	4	212	0.041	8.5	0.9		Hull (2001)
Smyrnium perfoliatum				24.0			Oláh and Masarovičová (1997)
Solidago flexicaulis	10	240		11.0		28	Hicks and Chabot (1985)
Trientalis borealis				3.2			Adams (1970)
Trillium erectum	10	50				20	Hicks and Chabot (1985)
Trillium grandiflorum	10	280–460		10.3		11–22	Hicks and Chabot (1985)
Viola pubescens	8	700		12.1	0.8	23	Rothstein and Zak (2001)
Mean	12	340	0.056 (.048)	10.0	0.8	22	
Wintergreens							
Cornus canadensis				5.5			Landhäusser et al. (1997)
Dryopteris filix-mas						−52	Willmot (1989)
Dryopteris filatata						−52	Willmot (1989)
Heuchera americana			0.080			26	Skillman et al. (1996)
Tipularia discolor	10–20	550	0.040–0.059	8.5	0.8	−24	Tissue et al. (1995)
Mean	15.0	550	0.065	8.5	0.8	39	

(continued)

Table 3.2. *Continued*

Species	Light compensation point (μmol m^{-2} s^{-1})	Light saturation point (μmol m^{-2} s^{-1})	Apparent quantum yield (μmol CO_2 / μmol photons)	Light saturated photosynthesis (μmol CO_2 m^{-2} s^{-1})	Dark respiration (μmol CO_2 m^{-2} s^{-1})	Leaf life span (weeks)	References
Evergreens							
Dryopteris spinulosa	4	50				54	Hicks and Chabot (1985)
Fragaria vesca[a]	12–50	400–600		3.9		14–20	Hicks and Chabot (1985)
Galax urceolata	4	400	0.025	4.3	0.5	114	McCarron (1995)
Hexastylis arifolia	4–30	480	0.070	11.0		52	Hicks and Chabot (1985)
Mitchella repens	8	140				120	Hicks and Chabot (1985)
Pachysandra terminalis	8	200	0.023–0.080	9.0	0.8	182	Yoshie and Kawano (1986)
Polypodium virginianum	20–33	120–175	0.028–0.033	2.5	0.7–1.4		Gildner and Larson (1992)
Pyrola asarifolia		125		4.2			Landhäusser et al. (1997)
Pyrola elliptica	8	140				54	Hicks and Chabot (1985)
Tiarella cordifolia	9	330		6.8	0.3		Rothstein and Zak (2001)
Mean	10.7 (10.5)	291 (224)	0.044	6.0 (6.3)	0.7	85 (96)	

Calculation of means: where multiple values exist for a single parameter, the average of those values was used for that species. Maximum seasonal values reported for photosynthesis. Other parameters correspond to this same time for comparative purposes. Consult references for details regarding measurement conditions. Some values were obtained from potted plants, some from the field, and others from detached leaves.

[a]For determination of light saturation point, 700 μmol m^{-2} s^{-1} was used, even though the actual value is probably somewhat higher.

[b]Value in parentheses calculcated excluding the two values from Kawano et al. (1978), which for this group appear anomalously low. Measurements were made on detached leaves.

[c]Mean in parentheses does not include value reported by Kawano et al. (1978), which seems unusually high.

[d]Although this plant is technically evergreen, it differs from the others in this category in that its leaves do not last for more than 1 year. The value in parentheses for the mean of evergreens is the mean if this plant is deleted from the calculations.

cold or drought stress or, more likely, some combination of the two (Axelrod 1966; Stebbins 1974). Chabot and Hicks (1982) proposed several general hypotheses regarding the evolution of evergreen leaves that can be distilled down to two major themes: (1) evergreen leaves are more advantageous on low-nutrient soils and (2) evergreen leaves are an adaptation to drought or herbivory. Regarding the first hypothesis, Chabot and Hicks (1982) argued that longer leaf life spans lower the carbon construction costs per unit of nutrient taken up, resulting in higher nutrient-use efficiencies and a moderation of nutrient recycling rates (Monk 1966). To maintain functional leaves for more than 1 year, particularly through the harsh winters of temperate forests, these leaves need to be tough and resistant to injury. Such leaves are often sclerophyllous, with large amounts of lignin and other structural compounds that confer resistance to either abrasion or herbivory. These leaves also require a greater investment in carbon, which takes longer to pay back in terms of photosynthetic carbon gain. Evergreenness in herbs might also result from an inability to adapt to the low PFD during the summer, and hence their periods of peak photosynthetic activity are shifted to the colder portions of the year, when light is more available. In more northerly locations, where the growing season is much shorter, evergreenness confers an advantage by allowing species to begin carbon gain in the early spring as soon as temperatures become warm enough without incurring leaf construction costs that would delay the time when net photosynthesis becomes positive. In the fall, evergreen leaves could continue photosynthesizing until temperatures become too cold, thus extending the season of carbon gain as much as possible during times when the canopy is absent and light levels are highest. A similar argument applies to wintergreens and heteroptic species.

We also speculate that the proportion of evergreen herb species should increase toward the southern end of the distribution of temperate forests because in these locations temperatures are more favorable when the canopy is leafless than in more northern forests. For example, certain species are summergreen in colder areas, but evergreen in warmer areas (Sato and Sakai 1980). Landhäusser et al. (1997) reported that the evergreen *Pyrola asarifolia* Michx. has higher rates of photosynthesis in the summer and does not take advantage of the higher light in the spring and fall to raise assimilation rates. In fact, this plant behaves photosynthetically as a summer-green, with a lower compensation point than congeneric summer-green herbs. Yet in other studies of *Pyrola* species (*P. japonica*) from more southerly climates (southern Japan), plants acclimate to the high light periods in spring and fall and perform little photosynthesis in the summer when PFD is low (Koizumi 1989). We postulate that as the growing season shortens at more northerly latitudes, the windows of opportunity for appreciable carbon gain in the spring and fall shrink in length because of later frosts in the spring and earlier frosts in the fall, coupled with more rapid canopy leaf out in the spring (fig. 3.5). If so, there could be a shift in photosynthesis more toward the summer as latitude increases, until at the most northern locations, it becomes more

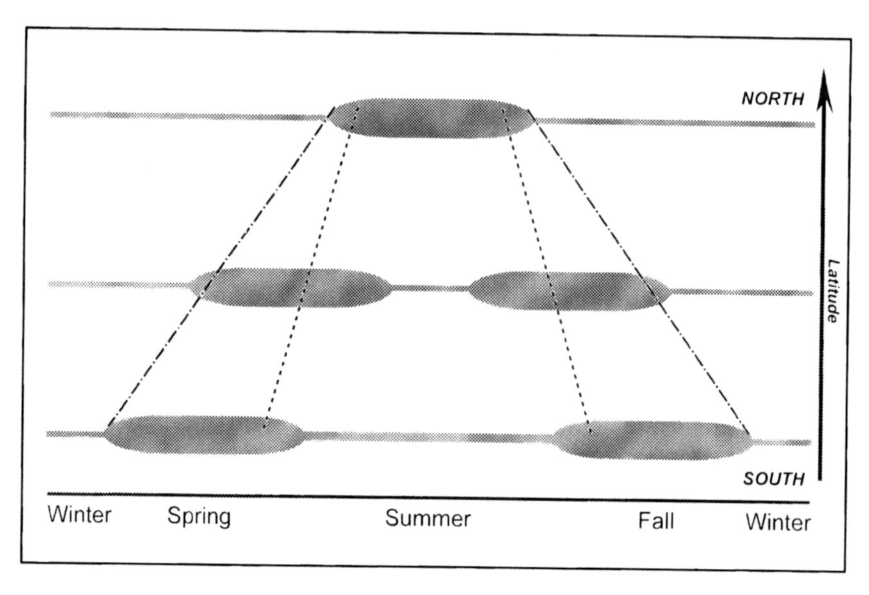

Figure 3.5. Model of phenology for an evergreen plant as a function of latitude. Dashed-and-dotted lines represent changes in date of last frost in spring and first frost in fall, short-dashed lines represent dates when canopies are fully leafed out or fully leafless. Gray areas represent photosynthetic activity of evergreen herbs; width of gray area is proportional to activity. As latitude increases, the period for favorable carbon gain is shifted more toward the summer, and time available to take advantage of high light in spring and fall is greatly reduced.

advantageous to gain carbon in the summer when temperatures are at their peak even if PFD is low.

Finally, phenology may also be a function of plant size. *Dryopteris filix-mas* (L.) Schott, an understory species characteristic of British woodlands, switches from being wintergreen when small to summer-green when large, presumably because smaller plants need more light during the leafless seasons to bolster their carbon gain (Willmot 1989).

Leaf Morphology of Understory Herbs

Light strongly alters the leaf anatomy of a plant. Plants in low light environments tend to have larger but thinner leaves, with reduced palisade layer development, or in some cases, an absence of palisade tissue altogether (Boardman 1977; table 3.1). The two understory herbs *Hydrophyllum canadense* L. and *Asarum canadense* L. each have only a single layer of photosynthetic cells (DeLucia et al. 1991). Shade leaves are most commonly hypostomatous, with a greater preponderance of stomata on the abaxial surface. This helps distribute CO_2 within the leaf to those tissues receiving the most light, thus maximizing photosynthetic efficiencies when PFD is limiting (J. R.

Evans 1996, 1999; W. K. Smith et al. 1997). Shade leaves also produce more widely spaced stomata than sun leaves (Salisbury 1928), which may reflect lower mean leaf temperatures and transpirational demands in the understory. However, there is a suggestion that spring ephemerals may also have low stomatal densities (<160 stomata/mm^2), and these plants may be either hypo- or amphistomatous (Eliáš 1981).

Another strategy for increasing light absorption is for leaves to increase internal light scattering and as a consequence reduce light transmission through the leaf. One way to do this is to produce more spongy mesophyll cells because these irregularly shaped cells scatter light more than do palisade cells. When leaves of *H. canadense* or *A. canadense* are infiltrated with mineral oil (which nearly eliminates refraction from cell wall–air interfaces), light absorption is reduced nearly twofold, potentially lowering photosynthesis by 20% (DeLucia et al. 1996).

Shade leaves frequently have more reflective lower surfaces (W. K. Smith et al. 1997) which direct light back into the leaf that otherwise would not be captured, but less reflective upper surfaces due to fewer intercellular spaces in the palisade layer (Lee and Graham 1986). Other species contain a deeply pigmented abaxial epidermis that may enhance absorption of PAR, particularly among tropical understory plants (Lee and Graham 1986). This latter phenomenon appears to be less prominent in temperate forests, perhaps because light levels are not as low as in tropical forests (Baldocchi et al. 1994), but *Oxalis* and *Glechoma* species are notable exceptions.

Leaves of a number of evergreen herbs, (e.g., *Galax urceolata* (Poir.) Brummitt [formerly *G. aphylla* L.]) accumulate anthocyanins when exposed to high PFD and turn bright red in winter (McCarron 1995). The anthocyanin buildup is localized in the adaxial epidermis, and leaves that are shaded remain green. The adaptive value of this reddening is not well understood (Chalker-Scott 1999; Gould and Quinn 1999; Gould et al. 2000) but may involve protection against photoinhibition (Powles 1984) caused by sunflecks during cold periods.

Some understory plants have glabrous leaves and lack the trichomes and hairs that might reflect light that could otherwise be used for photosynthesis (Ehleringer and Mooney 1978). Cells in the adaxial surface are often shaped as concave lenses that may direct light more efficiently to those tissues within the leaf that contain the chloroplasts (Haberlandt 1914; Vogelmann 1993; Vogelmann et al. 1996). Palisade cells also act as light pipes because the chloroplasts are often appressed against the anticlinal walls, which allows light to penetrate deeper into the leaf, reducing self-shading effects. However, palisade light piping does not appear to work in the diffuse light conditions most typical in the understory, but it may be of primary use during sunflecks, when direct beam radiation strikes the leaf surface (Pearcy 1998). If epidermal cells in shade plants are more spherical than in sun plants, light would focus more shallowly within the leaf and light intensities would peak directly within the palisade and spongy mesophyll cell layers where most of the chlorophyll resides. Using microfiber optic probes, light intensities as high as 3 times full sunlight have been measured at the focal points inside leaves (Vogelmann 1993; Vogelmann et al. 1996), with concomitant reductions in transmitted light and increases in light use efficiencies.

Understory Plant Forms

Approximately 94% of the understory herbs in deciduous forests in North America are perennials; nearly two-thirds of these are hemicryptophytes (with perennial buds or shoots close to ground level), and slightly less than one third are cryptophytes (with buds below ground on bulb or rhizome) (Buell and Wilbur 1948; Cain 1950; Struik and Curtis 1962). Only 6% are annuals (Hicks and Chabot 1982). Many of the perennial herbs are clonal, and production of ramets that explore the forest floor is an efficient means for vegetatively reproducing as well as buffering individual ramets against localized resource deprivation. Several studies have shown that various resources, including carbon (Silva 1978; Ashmun et al. 1982; Alpert and Mooney 1986; Stuefer et al. 1994), water (Lau and Young 1988; de Kroon et al. 1996), and nutrients (Birch and Hutchings 1994) are transferred among ramets and that severing connections among ramets can be detrimental to their survival (Lau and Young 1988).

The degree of integration among ramets and the length of time they remain functionally integrated varies by species (Ashmun et al. 1982) and may depend on the life span of the rhizome. For example, *Aster acuminatus* Michx. rhizomes decay within 2 years of formation, whereas those of *Clintonia borealis* (Aiton) Raf. last many years. There is little or no inter-ramet translocation of ^{14}C in *A. acuminatus*, but significant transfer in *C. borealis* (Ashmun et al. 1982). *Clintonia borealis*, which produces leaves early in the season before to canopy closure, may be subjected to large extremes in resource availability at this time of year (Ashmun et al. 1982), and selection pressures to maintain physiological integration may be more intense than that for the later growing *A. acuminatus*. In sites with extreme microhabitat heterogeneity, selection for maintenance of physiological integration may be favored, whereas in more homogeneous sites little translocation of resources may occur, even though ramets may remain connected (Silva 1978). Disturbances that disrupt resource availability may result in reestablishment of physiological integration among connected ramets, improving survival of the genet as a whole (Ashmun et al. 1982; Lau and Young 1988; de Kroon et al. 1996).

Height of Forest Herbs

Tall plants are better competitors for light than short plants (Jahnke and Lawrence 1965; Horn 1975), and competition between herbs and tree seedlings depends, in part, on the maximum height that herbs can attain. Here we show that there are clear patterns of herb height that correlate well with conditions in the understory and phenological strategy.

Figure 3.6 shows maximum heights for a variety of herbs of differing

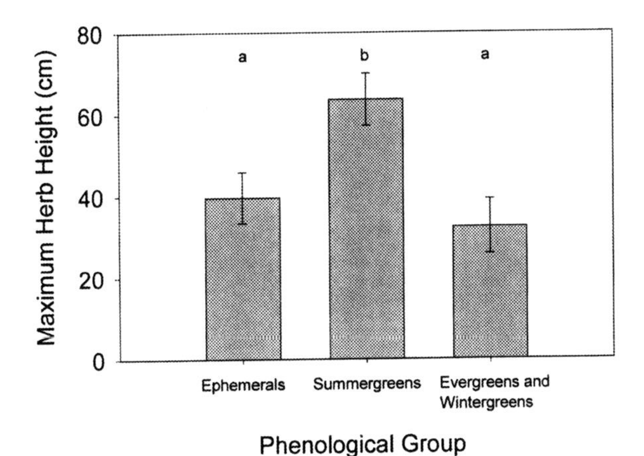

Figure 3.6. Mean maximum heights (± SE) of herbs as a function of phenological class. Means with the same letter are not statistically different at $p = .05$ ($N = 11–15$ among the groups). Data from Radford et al. (1968).

phenology. Using data obtained from the *Manual of the Flora of the Carolinas* (Radford et al. 1968), we show that spring ephemerals and evergreens are the shortest, and summer-greens tend to be the tallest ($p < .05$) forest herbs. Because the former two groups are active in those portions of the year when light availability is highest and other resources such as nutrients and water are less limiting, there may be fewer selective pressures to grow tall (Reid 1964; Givnish 1982). In the summer, competition for light becomes more intense as leaf area coverage by summer-green herbs and woody plants increases. As a consequence, selection for height growth is intensified to maximize light interception (Givnish 1982, 1986).

The maximum heights that herbs achieve represent a compromise between allocation to support structures and allocation to leaves for photosynthetic carbon gain (Givnish 1982, 1986). For each unit increase in height, there is a proportionally greater investment in support rather than leaf tissue due to the high costs of maintaining mechanical integrity. For competing plants, there is some maximum height at which the advantage in terms of carbon gained no longer warrants any further increase in height. Givnish (1986) has shown that in a Virginia forest, each 7% increase in leaf coverage roughly results in a doubling of leaf height (fig. 3.7). The consequences of this for tree regeneration are great because the habitats with the greatest light availability are often the most conducive to seedling growth but are also those in which competition for light will be greatest.

Givnish (1982) suggested that evergreen species are restricted to infertile, open sites because of the conflict between maintaining warm leaves in the winter and competition for light in the summer. Species active in the colder portions of the year may position their leaves close to, or directly on, the ground to obtain heat thermally reradiated from the soil. Leaf temperatures will be elevated above cold air temperatures, and significant photosynthesis may occur in the winter months. This would exclude them from more

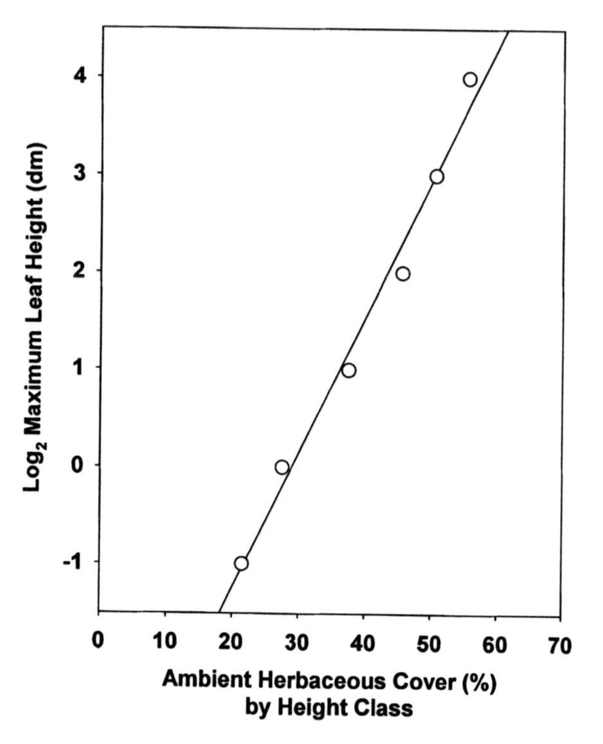

Figure 3.7. Maximum leaf height versus ambient cover for all species or seasonal morphs falling into a given height class. The least mean squares regression line is $y = -4.03 + 0.138x$, indicating that leaf height roughly doubles with each 7% increase in herbaceous cover. Redrawn from Givnish (1982) with permission.

resource-rich sites, where they would be easily overtopped. However, we know of no leaf energy-budget studies on spring ephemerals, and the only one on an evergreen appears to be an exception to the above rule. Christmas fern (*Polystichum acrostichoides* (Michx.) Schott) is an evergreen that grows on north-facing, nutrient-rich slopes instead of more xeric, sterile sites and presents an interesting strategy for dealing with the seasonal demands for light. In spring, the fronds are produced in an erect posture that is maintained throughout the summer. But after the first frosts, the fronds reorient to lie flat on the ground. Givnish (1982, 1986) showed that prostrate leaves can be up to 5°C warmer when in contact with the ground compared to fronds raised to their summer-erect position. This magnitude of temperature increase would theoretically enhance photosynthetic rates by 22%, although no actual measurements of photosynthesis have been done in the field.

Crown Architecture of Forest Herbs

Plants in shady habitats tend to form single-layered crowns, with minimal leaf overlap (Horn 1971; Valladares 1999). Light interception is maximized by maintaining large, horizontally oriented leaves (Pons 1977; DeLucia et al. 1991). The shadier the habitat, the more horizontal the leaves (Menges

1987; Muraoka et al. 1998; Leach and Givnish 1999), which tends to maximize light interception. If there are openings in the canopy from gaps or trails where side light can enter the forest, leaves will often orient toward the direction of higher diffuse irradiance (Björkman and Demmig-Adams 1995; Muraoka et al. 1998). Under closed canopies, however, most of the light on the forest floor comes from within 20–40° of vertical (Canham et al. 1990); thus, near-horizontal leaf orientations will maximize light interception.

A horizontal leaf orientation can pose a problem during periods of prolonged high light such as in long sunflecks or canopy gaps after a disturbance. During these times high amounts of radiation can cause overheating, increased transpiration rates, and photoinhibition (Young and Smith 1979; Oberhuber and Bauer 1991; Pearcy and Sims 1994; Björkman and Demmig-Adams 1995; Muraoka et al. 1998). To avoid overexposure, some plants (mainly in the families Fabaceae and Oxalidaceae) actively reorient their leaves parallel to the direction of incoming radiation (Björkman and Demmig-Adams 1995), which reduces the flux density of radiation by the cosine of the angle of incidence (Kriedeman et al. 1964). Leaves of *Oxalis oregana* Nutt. reorient in just a few minutes after exposure to high PFD. This lowers the incident PFD to levels that do not cause photoinhibition while still permitting substantial carbon assimilation. Immobilization of leaves causes photoinhibition within a short period of time (Björkman and Demmig-Adams 1995). *Arisaema heterophyllum* Blume folds its leaves along the main vein when growing in high light situations (Muraoka et al. 1998). Other species passively reorient their leaves through changes in turgor pressure brought about by water loss when exposed to high PFD. Leaves of *Impatiens capensis* Meerb., *I. pallida* Nutt., and *Rudbeckia laciniata* L., to name but a few, wilt rapidly when exposed to sunflecks, causing the leaves to droop to a more vertical position. Reducing incident PFD will lower leaf temperatures, which in turn decreases the vapor pressure deficit and subsequent water loss rates. It must be noted though, that the majority of plants in the understory show no active leaf movements. Raven (1989) suggested that when sunflecks are of long duration and separated by an interval long enough to allow for any repair of photodamage, avoiding incident PFD by reorienting leaves (pulvinar leaf folders) is energetically favored over simply repairing the injury (repairers, no reorienting of leaves). However, if flecks are of short duration and close together, leaf folding is the more costly option. Because the majority of plants in understories do not possess the pulvinar leaf-folding mechanism, responses to sunflecks may generally favor repairers over leaf folders (Raven 1989), or there may be some phylogenetic constraint limiting the number of species that are leaf folders.

Phyllotaxy can also be adjusted to maximize light interception and to reduce self-shading. Distichous and spiral leaf arrangements ensure that no leaf is fully shaded by an upper one, and these arrangements are common among plants in shady habitats (Leach and Givnish 1999; Valladares 1999).

In a tropical forest understory in Panama, computer simulations showed that such arrangements could limit mutual shading to approximately 10% (Pearcy and Yang 1996).

Topographic relief may also influence crown architecture. Some herbs, such as *Smilicina racemosa* (L.) Desf., *Disporum lanuginosum* (Michx.) Nicholson, and *Polygonatum biflorum* (Willd.) Pursh, have stems with distichous leaves that arch down-slope. In the Smoky Mountains, the compass orientation of these species matches that of the slope on which they are growing (Givnish 1986), so that they arch out over other species lower down the slope. Although this architectural strategy is more expensive (in terms of the carbon costs for gaining height) than those incurred by building an erect umbrella crown, the marginal gain in height on steep slopes presumably gives these species a competitive edge in these habitats (Givnish 1982, 1986). The mechanism by which these plants sense the appropriate orientation, if in fact they do, has not been investigated.

Physiological Adaptations of Understory Herbs

Photosynthetic Pathways of Understory Herbs

Most understory herb species use the C_3 pathway of photosynthesis, although there are a few representatives from the C_4 and CAM pathways (Leach and Givnish 1999). *Microstegium vimineum* (Trin.) A. Camus and *Muhlenbergia sobolifera* (Muhl. ex Willd.) Trinius are shade-tolerant C_4 grasses that can become prevalent in forest understories (Winter et al. 1982; M. Smith and Stocker 1992; M. Smith and Wu 1994; L. D. Williams 1998). *M. vimineum* is an annual exotic that has spread throughout the eastern United States and may be displacing native forest herbs (L. D. Williams 1998). *Sedum ternatum* Michx. grows on rocky soils and shaded outcrops in southern Appalachian forests. It performs C_3 photosynthesis when well watered, but switches to CAM cycling if water stressed (Gravatt and Martin 1992), which may serve to moderate photoinhibitory effects (Luettge 2000). Although plants with the C_4 and CAM pathways have the potential to adapt to low-light conditions (Pearcy 1983; Sage et al. 1999), their success in shaded habitats may ultimately be constrained by their inherently higher ATP requirements (von Caemmerer and Furbank 1999).

Carbon Gain by Herbs Before and After Canopy Closure: Carbon Gain and Leaf Life Span

Data on the relationship between mean maximum photosynthetic rate and leaf life span are shown in table 3.2. Leaves of spring ephemerals live on average only 10 weeks, while leaves of some evergreen species may last for up to 120 weeks, a 12-fold difference. Mean maximum assimilation rates (±

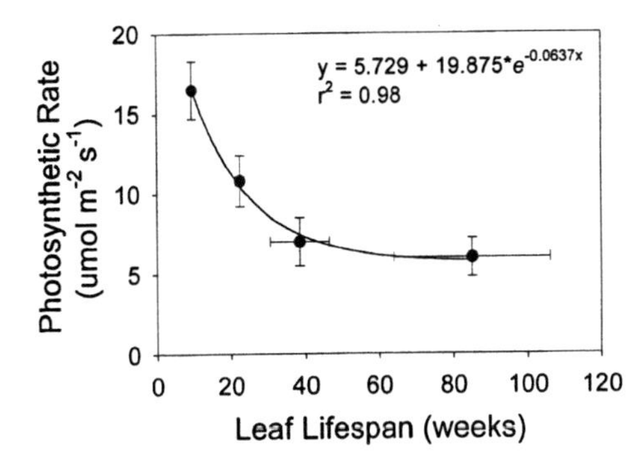

Figure 3.8. Mean maximum photosynthetic rates (± SE) as a function of mean leaf life span (± SE) for herbs from various phenological classes. Data from table 3.1.

standard error) vary threefold among the phenological groups, ranging from 16.5 ± 1.8 μmol m^{-2} s^{-1} among spring ephemerals, to 6.0 ± 1.2 μmol m^{-2} s^{-1} for evergreens. When mean assimilation rates are plotted against mean leaf durations (from table 3.2), a negative but highly significant nonlinear relationship is found (fig. 3.8). The curve shows that as leaf life spans increase beyond one growing season, there are diminishing costs in terms of decreased photosynthetic rates. Extending leaf life spans from the mean for spring ephemerals to that for summer greens (a 13-week increase) results in a decrease in A_{max} of 0.44 μmol m^{-2} s^{-1} per week of extended life span, whereas going from summer greens to evergreens, A_{max} decreases only 0.08 μmol m^{-2} s^{-1} per week of extended life span over a mean of 63 more weeks. Extended life spans are necessary to pay back the costs of construction when rates of photosynthesis are low. Construction costs on a per-unit mass basis are similar between sun and shade leaves, but higher for sclerophylls due to their higher lignin contents (Poorter 1994; Pearcy 1999).

The variety of leaf phenologies among understory herbs can be understood within the context of cost–benefit models (Chabot and Hicks 1982; K. Williams et al. 1989; Kikuzawa 1991; Kikuzawa and Ackerly 1999). According to Kikuzawa's (1991) econometric model, leaf longevities are short when initial photosynthetic rates are high, the decline of photosynthesis over time is rapid, and construction costs are low. If the costs for maintaining a leaf through an unfavorable period exceed the benefits of the preceding favorable period, the leaf should be dropped (Fig. 3.9). The optimal time for a leaf to be replaced is when its net carbon gain per unit time over its life span is maximum (i.e., when the marginal gain is maximized). The utility of this model is that it takes into consideration temporal and spatial variations in resource availability to explain patterns of leaf longevity among disparate groups of plants. In high-resource sites, where nitrogen, water, and light are freely available, rates of photosynthesis can be high, and, as a consequence, leaf life spans are short. In low-resource sites, the acquisition of these resources takes more time; hence, leaf life spans are longer.

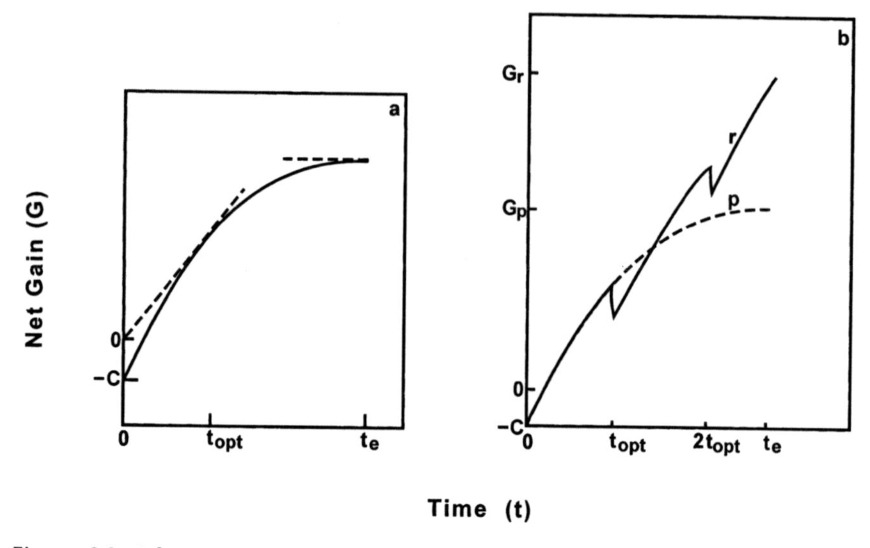

Figure 3.9. Schematic representation of net gain of carbon per leaf (G) to time (t) curve. (a) Net gain at time zero is minus construction cost (−C) and increases rapidly at first and then gradually because of decrease in photosynthetic rate with time due to aging. Net gain is maximized by replacing leaves when tangent line from origin touches curve (t_{opt}). Waiting until net gain is zero does not maximize gain. (b) Comparison of net gain for a plant that continually replaces leaves (r) and one that has persisting leaves. The net gain for the plant that replaces leaves (G_r) at time t_{opt} exceeds that of one with persisting leaves (G_p) that retains leaves to time t_e. Reprinted with permission from Kikuzawa (1991).

Carbon Gain by Herbs Before and After Canopy Closure: Spring Ephemerals

Table 3.2 shows gas exchange characteristics and leaf life spans among a large number of forest understory herbs. Spring ephemerals, such as *Erythronium americanum* Ker, *Allium tricoccum* Aiton, *Anemone flaccida* Fr. Schmidt, and *A. raddeana* Regel, are the first deciduous herbs to leaf out after winter and complete leaf growth either before or just after the canopy develops (Kawano et al. 1978; Koizumi and Oshima 1985). Average leaf life spans last only about 10–11 weeks but can be as short as 6 weeks in more northerly latitudes (Yoshie and Yoshida 1987). In *A. flaccida*, leaf senescence begins before full canopy closure, suggesting that leaf life span may be genetically fixed (Koizumi and Oshima 1985). Taylor and Pearcy (1976) noted that shaded leaves of *E. americanum* senesced earlier than unshaded leaves, presumably because of a lack of carbon gain. For all the species in this group, senescence is complete by the time the canopy is fully developed.

Many spring ephemerals produce typical sun-type leaves, with high maximum rates of light-saturated photosynthesis (A_{max}) on a per-unit area basis, high-light compensation points (LCP), high light saturation points for photosynthesis (L_{sat}), and high dark respiration (R_d) rates (Taylor and Pearcy 1976; Boardman 1977; Kawano et al. 1978; Björkman 1981; Koizumi and Oshima 1985; see table 3.2). Spring ephemerals also have thicker leaves than shade-tolerant summer-green herbs, a better developed palisade mesophyll with one or possibly two layers of cells, higher specific leaf mass (SLM, g/m^2), and higher chlorophyll a:b ratios (Sparling 1967; Taylor and Pearcy 1976; Rothstein and Zak 2001). Photosynthetic rates peak at or near full leaf expansion, but remain high for only 1–2 weeks before declining sharply (Taylor and Pearcy 1976; Kawano et al. 1978; Koizumi and Oshima 1985). Maximum rates of assimilation average nearly 17.0 μmol m^{-2} s^{-1}, but individual values can range between 13 and 20 μmol m^{-2} s^{-1} (table 3.2). The rates reported by Kawano et al. (1978) appear anomalously low, perhaps because they were done on detached leaves. However, if they are accurate, then these species have unusually low rates for plants in this phenological group.

High rates of assimilation in spring ephemerals result from a large investment in electron transport capacity (ET) and in enzymes associated with photosynthetic carbon reduction (PCR), particularly ribulose-1,5-bisphosphate carboxylase/oxygenase (RUBISCO) (Taylor and Pearcy 1976; Rothstein and Zak 2001). These high fixation rates require substantial amounts of leaf nitrogen because much of the leaf-protein nitrogen is allocated directly to RUBISCO (Poorter and Evans 1998). Additional nitrogen fractions are associated with the light-harvesting complexes and processes related to nitrogen reduction (J.R. Evans 1986; Pearcy 1999). Although data are limited, it appears that spring ephemerals have the highest leaf nitrogen contents (nearly 4% in *Erythronium americanum*; chapter 2, this volume) among the various phenological groups of herbs. Leaf nitrogen concentrations of summer-green herbs are generally lower (~2.5%) and those of evergreens (1.4%) even lower than nitrogen concentration of spring ephemerals (McCarron 1995). Nitrogen is more readily available for a short period in spring because of mineralization due to increasing soil temperatures and plentiful soil moisture, whereas later in the season, nitrogen becomes more limiting due to uptake and immobilization. With their high rates of photosynthesis, and generally greater metabolic activity, spring ephemerals are able to accumulate high concentrations of nitrogen, whereas the slower growing evergreens cannot, even though they, too, are active in the early spring. The dichotomy in response can best be explained by viewing spring ephemerals as ruderal-type species in the sense of Grime (1979) and evergreen species more as stress-tolerators, with inherently slower growth rates and a lowered ability to take up nutrients even when supplied in excess.

As is typical of many shade-intolerant plants, spring ephemerals exhibit little photosynthetic plasticity with respect to lower prevailing PFD, as occurs once the canopy closes (Pearcy and Sims 1994). Adaptations that confer a significant advantage in high light appear to be incompatible with persistence under low-light conditions. For example, greater leaf thickness means more light attenuation within the leaf tissues, and cells near the abaxial surface may be at or below their LCP at the PFDs beneath a closed canopy. High investments in leaf nitrogen and enzymatic processes require substantial amounts of ATP and carbon that are not available once the canopy closes, and it becomes a liability to maintain these sun-type leaves under low-light conditions.

Some studies have suggested that the absorbed quantum yield (ϕ) of shade plants (μmol CO_2/μmol absorbed photon) is higher under low PFD than it is for sun plants (Nilsen and Orcutt 1996; but see Björkman 1981). However, more recent studies have shown that the more commonly measured apparent ϕ (μmol CO_2/μmol incident photon) is highly variable and influenced by physiological and environmental conditions. Apparent ϕ among four understory herbs is only slightly lower in the spring ephemeral *Erythronium americanum* (0.045 μmol CO_2/μmol photons) than that for several shade-tolerant summer-green herbs (0.048 μmol CO_2/μmol photons), while in other studies the apparent ϕ depends strongly on A_{max} (Yoshie and Kawano 1986; Yoshie and Yoshida 1987; Hull 2002). In *Anemone raddeana*, apparent ϕ varies seasonally in concert with A_{max}, ranging from ~0.020 to ~0.040 μmol CO_2/μmol photons. Spring ephemerals, for example, have both the lowest and highest values (0.029 μmol CO_2/μmol photons in *Allium ursinum* L. to a maximum of 0.071 μmol CO_2/μmol photons in *Arum maculatum*) among the herbs surveyed by Kriebitzsch (1992a). The complexity of responses and the large number of factors affecting ϕ make generalizations among phenological groups suspect, and it is not possible at this time to conclude that under field conditions any significant differences exist among the functional groups (table 3.2).

Spring ephemerals achieve high rates of photosynthesis early in the season because of their ability to carry on gas exchange when temperatures are still low. This suggests that these plants might have a lower temperature optima (T_{opt}) than summer greens. Wintergreens and evergreens, in contrast, might adjust their T_{opt} to the prevailing temperatures over the year. However, the patterns found in the field are more complex and less distinguishing than hypothesized. Kriebitzsch (1992b) defined the T_{opt} as the range of temperature over which photosynthesis is 90% of A_{max}. Working in Germany, he found no large differences among the phenological groups, despite large differences in prevailing air temperatures when the species were photosynthetically active. In spring, the T_{opt} for six species spread among most of the phenological groups ranged from 8–10°C at the lower end to 16–26°C at the upper end, with a spring ephemeral (*Arum maculatum* L.) having the highest T_{opt}. The anomalously high T_{opt} for this species may be related to its taxonomic status as a member of the Araceae because species in this family are known to express high rates of the alternative respiratory pathway, a heat-generating process. Over the season and across species, the T_{opt} shifted upward only about 2–4°C before going back down toward spring values in the fall (Kriebitzsch 1992b). Over most of the season, the T_{opt} ranges were broad and averaged 8–10°C for all species. For those species persisting into the summer, the T_{opt} tended to shift upward as the canopy closed, probably because of lowered respiratory costs. The T_{opt} range may also be a function of the biophysical characteristics inherent in variations of leaf size, pigmentation, and pubescence.

Carbon Gain by Herbs Before and After Canopy Closure: Wintergreen and Heteroptic Species

Other species active in the early spring are the wintergreen/evergreen and heteroptic species, plus summer-greens that produce leaves before canopy closure. Although wintergreens and evergreens have leaves present at the same time as spring ephemerals, they have a much lower A_{max} (table 3.2), with values more similar to summer-greens after canopy closure. The highest A_{max} reported in this group is 11 μmol m^{-2} s^{-1} for *Hexastylis arifolia* (Michx.) Small in North Carolina (Gonzalez 1972), but nearly half of the reported values for these species are < 6 μmol m^{-2} s^{-1}. Although there may not be any inherent physiological reason these plants could not attain higher rates of photosynthesis (Chabot and Hicks 1982), factors associated with evergreenness, such as lignification, defense against herbivores, and perhaps drier and more nutrient-deficient soils, appear to constrain photosynthesis in this group to lower mean values.

Wintergreen species, such as *Tipularia discolor* (Pursh) Nutt., take particular advantage of the spring and the fall seasons. *T. discolor* produces a single new leaf in the fall which then overwinters before senescing in late spring. Photosynthetic rates are highest in the fall and continually decline through the winter until just before senescence, when A_{max} is only 39% of that in the fall (Tissue et al. 1995). Despite this dramatic decrease in photosynthetic potential, actual assimilation rates in the field decline by only 30% from fall to spring because of compensating changes in light and temperature. Because the T_{opt} for this species is invariant and remains at about 26°C throughout the year, warm air temperatures and high light in the spring contribute to higher than expected rates of carbon uptake (Tissue et al. 1995). This differentiates *T. discolor* physiologically from another wintergreen orchid, *Aplectrum hyemale* (Muhl. ex Willd.) Torrey, which can change T_{opt} by nearly 10°C in response to changing air temperatures (Adams 1970). In essence, *T. discolor* behaves much like a spring ephemeral by making use of those periods when the canopy is absent, while avoiding the deep shade of the summer months by senescing its leaves. In contrast, *A. hyemale* has higher assimilation rates when preconditioned at low temperatures (T_{opt} is 15°C). For example, plants are able to obtain 75% of their A_{max} at 5°C and nearly 90% at 10°C. Below 5°C photosynthesis declines markedly, suggesting that little carbon uptake occurs during the winter, especially when plants are covered with snow.

Heteroptic species produce specialized leaves that maximize carbon gain during those times of the year when each leaf type is present. These species are both developmentally and physiologically more plastic than evergreen species that only produce leaves at one time during the year (Skillman et al.

1996). *Heuchera americana* L. produces leaves in both the fall and spring, each of which only last about 6 months. Based on laboratory measurements, leaves produced in the winter have a much higher A_{max} (by 250%) than summer leaves, while the T_{opt} for summer leaves is higher by about 8°C. A_{max} increases over the winter for leaves produced in the fall such that they are able to take immediate advantage of the high light and favorable temperatures in the spring before senescing as the canopy closes (Skillman et al. 1996). Given the higher prevailing air temperatures in summer, the higher T_{opt} for summer leaves allows for reasonable carbon gain, even though the PFD is low because of the tree canopy overhead.

Shifts in the way in which nitrogen is apportioned within the leaves appear to be responsible for the different assimilation abilities of the two types of leaves. Nitrogen per unit leaf area does not differ between summer and winter leaves but is much higher on a per unit chlorophyll basis in the winter for leaves produced in the fall. The ratio of A_{max} to nitrogen is low for this species, but it is within the range found for shade-tolerant, tropical understory plants (Skillman et al. 1996). This suggests that much of the nitrogen has been diverted away from the photosynthetic apparatus to defensive compounds, although no systematic studies have been done that relate this ratio to secondary chemical concentrations in leaves of understory herbs. There appears to be little or no chronic photoinhibition (as measured by fluorescence) in *H. americana* leaves over the winter. Summer leaves allocate approximately 31% of their leaf nitrogen to the light-harvesting complexes in the thylakoid membranes, compared to only 17% for overwintering leaves, which enables this species to gain substantial carbon in the summer, even though the PFD can be quite low (Skillman et al. 1996).

Carbon Gain by Herbs Before and After Canopy Closure: Evergreen Species

True evergreen species (leaf life spans > 1 year) use both the fall and spring leafless periods to obtain much of their carbon gain (Koizumi 1985; Koizumi and Oshima 1985; Graves 1990; McCarron 1995; Rothstein and Zak, 2001). They may also gain additional carbon during the summer, in contrast to wintergreens that are dormant at this time of the year. Many evergreen species have leaves with some characteristics typical of shade plants, such as low photosynthetic rates and low light-compensation points (table 3.2). When exposed to high light and favorable temperatures, A_{max} remains low and in fact is sometimes inhibited by prolonged exposure to high light, as in *Galax urceolata* (McCarron 1995) and the summer-adapted (shade) leaves of *Pachysandra terminalis* Siebold & Zucc. (Yoshie and Kawano 1986). Koizumi (1989) showed that *Pyrola japonica* Klenze was able to acclimate to rising temperatures in the summer but was not able to lower the T_{opt} to match air temperatures when they fell below 15°C, indicating only partial temperature acclimation. As a group, evergreens appear to gain most of their carbon in the spring and the fall, when light levels are the highest. The highest rates of assimilation for both *G. urceolata* and *P. terminalis* are found in the spring and fall when temperatures are moderate, and the lowest rates are found in the summer, due to extreme shade, and in the winter, when air temperatures are low and soils are frozen (Yoshie and Kawano 1986; McCarron 1995). There is some indication that populations at more northern latitudes have a higher A_{max} than more southerly populations; presumably this is a form of compensation for shorter growing seasons in northern forests (Yoshie and Kawano 1986).

Rates of photosynthesis in evergreens also progressively decline with leaf age (Koizumi and Oshima 1985; McCarron 1995). A_{max} in *Pyrola japonica* drops from near 2.3 μmol m^{-2} s^{-1} in current year leaves to 1.6 μmol m^{-2} s^{-1} in 1-year-old leaves, down to only 0.6 μmol m^{-2} s^{-1} in 2 year old leaves. Similar decreases in A_{max} have been found in *Pachysandra terminalis* and *Galax urceolata* also (Yoshie and Kawano 1986; McCarron 1995). These decreases represent losses of photosynthetic capacity and are not due to greater stomatal limitations on the diffusion of CO_2. Although stomatal conductances (g_s) decrease as leaves age, they do so in concert with decreases in A_{max} such that the ratio of internal to external CO_2 remains constant throughout the season (Koizumi and Oshima 1985). Losses in photosynthetic capacity with aging result from either natural processes of senescence, redistribution of resources (Tissue et al. 1995), or a combination of these events, but the actual biochemical/physiological factors causing age-related declines in A_{max} are not well understood. As occurs in the woody understory shrub *Rhododendron maximum*, loss in A_{max} may be due to progressive deterioration of thylakoid membranes in the photosynthetic apparatus after prolonged exposure to high light during the winter (Bao and Nilsen 1988; Nilsen et al. 1988).

Respiratory Demands of Spring Ephemerals

Spring ephemerals as a group possess high rates of R_d (Sparling 1967; Taylor and Pearcy 1976; Koizumi and Oshima 1985; Hull 2001; Rothstein and Zak 2001), which may be due to the large metabolic costs of sustaining and repairing PCR enzymes, maintaining the protein integrity of the light-harvesting complexes, and processing carbohydrates that are present in higher concentrations than in shade leaves (Lambers 1985). Rates are dramatically higher during leaf formation when construction costs are high, but then drop significantly afterward. In *Anemone raddeana*, R_d was 1.5 μmol m^{-2} s^{-1} during leaf expansion but just 0.8 μmol m^{-2} s^{-1} once leaves matured (Yoshie and Yoshida 1987). The result is that spring ephemerals as a group have a 40% higher R_d, an LCP nearly twice as high, and an L_{sat} of photosynthesis 250 μmol m^{-2} s^{-1} higher than more shade-tolerant summer-green herbs (table 3.2).

One of the major unanswered questions concerning spring ephemerals is how much assimilated carbon is translocated below ground and then lost via respiration by tissues such as roots and storage structures. Knowledge of these fluxes, in addition to understanding processes within individual leaves, is critical for understanding the carbon balance of forest herbs. High belowground respiratory demands may result in a negative whole-plant carbon budget if leaves do not photosynthesize adequately at low PFD (Kikuzawa and Ackerly 1999). Because leaves of spring ephemerals, and sun leaves in general, appear to be unable to downregulate their photosynthetic apparatus in response to lower light conditions (Pearcy and Sims 1994), and because they produce only a single flush of leaves per year, they may not be able to maintain a positive whole-plant carbon balance at the low PFDs typically found beneath a forested canopy. Thus, the only solution is to avoid shade stress by senescing leaves and remaining dormant until the next spring.

Respiratory Demands of Wintergreen and Evergreen Species

There are some methodological problems associated with making comparisons among phenological groups in terms of whether they differ in R_d. For example, most measurements of R_d are derived from light-response curves at 0 PFD; but these measurements may be inappropriate for making comparisons among species that grow in habitats that differ greatly in the background levels of PFD. As Pearcy (1999) cautions, measurements of R_d for sun plants obtained in this manner may overestimate the actual rates in the shade because of the dependency of R_d on carbohydrate processing. R_d is typically a function of the previous day's net photosynthesis (Syvertsen and Cunningham 1979) and is lowest just before dawn, by which time most carbohydrate reserves have been consumed. Thus, distinctions in R_d among the different phenological groupings as reported in most of the literature may reflect the different rates of carbohydrate processing more than inherent differences in maintenance R_d among sun and shade-type leaves. More accurate comparisons could be obtained if measurements of R_d were made after a standardized period in the dark to eliminate the influence of carbohydrate concentrations. This being said, R_d for wintergreen and evergreen herbs, taken at the time of A_{max}, do not differ greatly from those of summer-green herbs (table 3.2).

Physiological Adaptations to the Light Environment After Canopy Closure

Strategies to Maximize Light Use Efficiency

The persistence of herbs in the understory after the canopy has leafed out requires physiological, biochemical, and morphological changes in leaves that maximize photosynthetic efficiency under light-limited conditions. Most plants adopt a strategy for coping with decreasing PFD by shifting resources away from the processes associated with fixation to those involved with light harvesting (Pearcy and Sims 1994; Rothstein and Zak 2001). This is accomplished in a variety of ways, including maximizing light absorption and the ϕ of photosynthesis and by minimizing R_d. Not all species are equally effective in adjusting photosynthetic parameters to decreasing light. In particular, spring ephemerals and evergreen species appear incapable of much adjustment. For example, evergreens have trouble adjusting ϕ in response to low PFD beneath a closed canopy, which may reflect preadaptation to the high light conditions of the leafless seasons. Quantum efficiencies in 1-year-old *Pachysandra terminalis* leaves are highest in the spring and fall (0.077 μmol CO_2/μmol photons) but much lower in the summer, suggesting little acclimation to the shade (Yoshie and Kawano 1986). Insignificant change is observed for the evergreen *Asarum maculatum* in Europe during the spring-to-summer transition (Kriebitzsch 1992a). In contrast, the ϕ in summer-green species increases from spring to summer and is positively correlated with increases in total chlorophyll (Kriebitzsch 1992a). Together these two changes contribute to more efficient photosynthesis by summer-green herbs at low PFD. The failure of the evergreen species to adjust suggests they are preadapted to the periods of high light before and after canopy closure but will do a small amount of photosynthesis after canopy closure.

In addition to light, temperature strongly influences ϕ (Kriebitzsch 1992a). In the spring, ϕ is positively related to temperature when temperatures are below 15°C, but above this point ϕ begins to decline sharply with rising temperature. This pattern is consistent among species from different phenological groupings and there is no suggestion that one group or another has a higher or lower ϕ.

Changes in Nitrogen Use Efficiency in Low Light

On a per-unit mass basis, amounts of chlorophyll in sun and shade leaves are reported to be similar (Björkman 1981), but shade leaves do shift chlorophyll away from the light-harvesting complexes of PSI to those associated with PSII core proteins (LHCII). The result is a lowering of the chlorophyll a:b ratio, because chlorophyll b is primarily found only in LHCII, whereas the core protein complex contains only chlorophyll a. This reallocation increases light harvesting at the expense of ET, but raises the nitrogen-use efficiency of photosynthesis because there is less nitrogen per unit chlorophyll–protein complex in LHCII (J. R. Evans 1986). Such a shift in chlorophyll a:b ratios has been documented in a number of European herbs, including spring ephemerals and summer-greens (Kriebitzsch 1984; Eliáš and Masarovĉová 1986a, b). During canopy closure, total chlorophyll (a+b) amounts in the summer-green *Mercurialis perennis* L. rise by nearly 2.5 times, while the chlorophyll a:b ratio drops by about 3 times, the result of a substantial (~5 times) increase in chlorophyll b (Kriebitzsch 1984). A similar, albeit smaller response in total chlorophyll (1.5 times) occurs for *Viola pubescens* Ait. in a hardwood forest in Michigan (Rothstein and Zak 2001), but no changes in total chlorophyll were observed for the evergreens *Tiarella cordifolia* L. or *Galax urceolata* (McCarron 1995). In contrast to these trends, Harvey

(1980) found that the evergreen *Hepatica acutiloba* DC actually increases its chlorophyll a:b ratio after canopy closure by 25%, mainly the result of large (84%) increases in chlorophyll a production. Thus, although some evergreens may make adjustments to the low PFD of summer, some may not, instead remaining ready to take advantage of increases in PFD in the fall and spring.

Changes in Dark Respiration During and After Canopy Closure

Respiration is consistently lower in shade leaves and, in fact, may be the predominant factor that determines shade tolerance in plants (Pearcy 1999). Low rates of R_d result in a lower LCP and a more positive balance between photosynthesis and respiration, and hence a greater ability to photosynthesize at the PFDs typically found in forest understories. On a seasonal basis, R_d decreases after canopy closure because of lowered construction and maintenance costs because by this time most leaf growth has been completed and there is a decrease in activity of the photosynthetic carbon reduction cycle. Large decreases in R_d are characteristic of all herbs that make the transition from spring to summer (Kawano et al. 1978; Kriebitzsch 1992a; Rothstein and Zak 2001). The decrease in rates can span an order of magnitude (from 39 to 400%), but absolute differences are much smaller for wintergreen/evergreen herbs because of their overall lower gas exchange rates (Kriebitzsch 1992a; Rothstein and Zak 2001). After canopy closure, rates of R_d in the shade are so low that differences among groups of species can no longer be distinguished (table 3.2).

Because of the temperature dependency of R_d, static measurements in the laboratory do not provide a clear picture of the dynamic nature of respiratory processes in field situations. Kriebitzsch (1992a) made an extensive series of measurements of the temperature dependence of R_d in a variety of understory herbs. He showed that not only did R_d increase with temperature, but the response to temperature varied seasonally, with greatest sensitivity in the period of transition from high to low light. As a result, the LCPs increased with increasing temperature. Despite this strong temperature dependency (nearly tripling every 10°C in the spring, but only doubling in summer and fall), carbon losses in the summer were greatly moderated because canopy closure reduced the range of understory air temperatures. Similar results were obtained for the evergreen herb *Pyrola japonica* (Koizumi 1989). Respiration rates were highest in leaves and flower organs in early spring and low and constant the rest of the year. Rhizome respiration peaked in midsummer and then declined the rest of the year. Only new buds showed an increasing trend into the fall and winter, presumably due to construction costs associated with maturation for the next growing season (Koizumi 1989). One consequence of these trends is that, across a group of species representing spring ephemerals to evergreens, maintenance R_d values in the field are within a few percent of each other most of the season, particularly when the canopy is closed (Kriebitzsch 1992a). Similar field studies of R_d in North American woodland herbs have not yet been done.

Seasonal Timing of Carbon Gain with Respect to Canopy Closure

Both summer-green and evergreen/wintergreen herbs often produce leaves before canopy closure and use this time of high light availability to gain much of their annual carbon (Blackman and Rutter 1946; Koizumi 1985; Koizumi and Oshima 1985; Pitelka et al. 1985; Graves 1990; Chazdon and Pearcy 1991; Kriebitzsch 1992c; Rothstein and Zak, 2001). There is usually a short window of time before canopy closure in which temperatures rise enough to permit appreciable carbon gain (Graves 1990). In Michigan, *Viola pubescens* accumulates 66% of its annual biomass increment in the 6 weeks before canopy closure, while the remaining 34% takes another 20 weeks (Rothstein and Zak 2001). If we assume that sunflecks can enhance carbon gain after canopy closure by an average of 30–40% (Weber et al. 1985) but occupy only about 10% of that time (Chazdon and Pearcy 1991), then *V. pubescens* could potentially accumulate 80% of its annual carbon in only 30% of the season (~10% per week). Over the remaining 18 weeks, accumulation would average only 1% per week, suggesting that for much of this time the plant is barely above its LCP. Kriebitzsch (1992c) calculated that *Mercurialis perennis* gained 73% of its yearly carbon gain in just 44 days, and the remaining 27% took an additional 168 days. In some stands in Europe, total shoot carbon balances are actually negative for much of the summer after canopy closure for both *M. perennis* and *Geum urbanum* L. (Graves 1990).

Changes in A_{max} After Canopy Closure

Once the canopy is fully leafed out, A_{max} drops sharply in a wide spectrum of summer-green species, with rates declining by nearly half over this period (Taylor and Pearcy 1976; Koizumi and Oshima 1985; Graves 1990; Kriebitzsch 1992a; Rothstein and Zak 2001). In contrast, little or no acclimation is found in boreal hardwood forests in Canada for either *Rubus pubescens* Raf. or *Aralia nudicaulis* L. (Landhäusser et al. 1997). These species may not be able to take advantage of the higher light in the spring because of the cold temperatures. Instead they wait for the warmer temperatures of summer to gain carbon, obviating the need to downregulate A_{max}.

Declines in A_{max} result primarily from reduced mesophyll conductances to CO_2 and not from greater limitations on the diffusion of CO_2 because of lowered stomatal conductances (Taylor and Pearcy 1976; Yoshie and Kawano 1986; Yoshie and Yoshida 1987). Lower mesophyll conductances result from a loss of RUBISCO (Taylor and Pearcy 1976; Rothstein and Zak 2001) and decreases in activation state of this and associated enzymes (Pearcy 1999). These processes bring fixation processes in line with reduced ET capabilities. The mechanism by which these changes are initiated is still not fully understood, but it may involve changes in redox states and pH gradients within the chloroplast membranes that trigger subsequent coordinated changes in photosynthetic capacity under low PFD (Pearcy and Sims 1994).

Many summer-green herbs, such as *Asclepias exaltata* L. and *Fragaria virginiana* Duchesne, produce cohorts of leaves well after canopy closure. Strawberry (*F. virginiana*) continually replaces leaves throughout the season with those more suited to the prevailing environmental conditions (Jurik and Chabot 1986), thus maximizing whole-plant carbon gain over the season. The biennial *Hydrophyllum appendiculatum* Michx. produces sun leaves before canopy closure, followed by shade leaves beneath a closed canopy (Morgan 1968). *Viola pubescens*, a summer-green that produces leaves before canopy closure, has the ability to decrease specific leaf mass during the transition to low light to a degree not observed in other species (Rothstein and Zak 2001). This morphological adjustment, which occurs presumably through continued leaf expansion after canopy closure, allows the leaf to take on characteristics more typical of shade leaves and contributes to shade acclimation.

Photosynthetic Responses to Sunflecks

Once the canopy closes, sunflecks represent the main supply of PFD and may provide up to 80% of total daily PFD (Horton and Neufeld 1998). Herbs depend on these brief periods of high light to gain the bulk of their carbon after canopy closure, which may constitute up to 60% of their total carbon gain in some cases (Chazdon 1988). The importance of sunflecks for the growth of understory herbs varies, however, depending on factors that include prevailing forest conditions such as background levels of diffuse PFD, intensity and duration of sunflecks, temporal patterns of sunflecks (morning or afternoon, in rapid succession, or spaced widely apart), and interactions of sunflecks with vapor pressure, temperature, and drought. Together the myriad combinations make it difficult to determine how a species will respond to sunflecks unless detailed modeling and field measurements are made.

Among the studies of understory herbs that we reviewed, only two, one by Pfitsch and Pearcy (1989a, 1989b, 1992) and another by Koizumi and Oshima (1993) have attempted to determine the contribution of sunflecks to carbon gain in temperate understory herbs. Using shadow bands designed to eliminate sunflecks, Pfitsch and Pearcy (1992) found that *Adenocaulon bicolor*, a common herb in redwood forests, became smaller and had less reproductive biomass. Plants receiving sunflecks did not increase in size, perhaps due to other effects associated with sunflecks, such as water stress and high leaf temperatures. However, the shadow bands also reduced the incoming diffuse radiation, which may have confounded interpretation of whether the growth effects were due solely to removal of the sunflecks, even though this factor was accounted for statistically. Measurements in the field suggested that in sites with abundant sunflecks nearly 70% of the carbon gain was due to sunflecks.

Koizumi and Oshima (1993) compared sunfleck responses of a summer-green herb (*Syneilesis palmata* Maxim) to an evergreen herb (*Pyrola japonica*) in a warm temperate forest in Japan. Using steady-state light curves of photosynthesis combined with detailed analyses of the sunfleck dynamics at their site, they concluded that sunflecks contribute at most 7–10% of the carbon from May through July for *S. palmata*, and only 2–3% for *P. japonica*. The summer-green *S. palmata* is able to respond to sunflecks more dynamically because of its higher L_{sat} and A_{max}. In reality, these estimated contributions due to sunflecks may be too high. Calculations derived from steady-state responses of photosynthesis to PFD most likely overestimate expected carbon gain due to the failure to take into account the dynamics of photosynthetic induction and de-induction (Pearcy 1990; Sims and Pearcy 1993; Pearcy et al. 1994), as well as possible photoinhibition under prolonged sunflecks. Furthermore, in the work reported by Koizumi and Oshima (1993), high background levels of diffuse PFD may have reduced the relative impact of the sunflecks (Chazdon 1986).

In a study of tropical understory herbs, Watling et al. (1997) showed enhanced growth for two Australian understory species under fluctuating versus constant light of similar daily total PFD. In some cases, biomass was 2–3 times greater in fluctuating light. Aside from these studies, we know of no other attempts to quantify the importance of sunflecks for the growth of understory herbs. Whether differences exist among the phenological groups in their response to sunflecks is still a open question. However, given the much lower rates of photosynthesis and the potentially longer induction times of evergreen leaves (see below), we suspect that evergreen/wintergreen species may respond less to sunflecks than either spring ephemerals or summer greens.

Factors Affecting Utilization of Sunflecks: Induction and De-Induction Times

The *induction state* of a leaf is its photosynthetic capacity upon sudden exposure to high PFD after a prolonged period of low PFD, relative to steady-state rates at saturating PFD (Chazdon and Pearcy 1986a, 1986b). Induction involves light-induced increases in g_s, increases in amounts of enzymes responsible for ribulose bisphosphate regeneration, and RUBISCO activation (Chazdon and Pearcy 1986a; Sassenrath-Cole and Pearcy 1992; Sage 1993; Sage and Seemann 1993). Usually, changes in g_s lag behind those of photosynthesis because activation times for RUBISCO and other enzymes can be on the order of just 1–2 minutes (Pearcy 1999), while changes in stomatal aperture may require 5–20 minutes. In very short flecks, g_s may even reach a maximum after the fleck has passed (Pearcy 1987).

The induction state is affected by prior environmental conditions, such as the length of the preceding low-light period and the temporal sequence and intensity of those sunflecks (Chazdon and Pearcy 1986a, 1986b; Tang et al. 1994; Pearcy 1999). Induction velocity, or change in induction per unit time after being exposed to a sunfleck, also depends on sunfleck intensity and the physiological status of the plant, such as whether it is drought stressed or suffering from temperature limitations.

A number of researchers have suggested that shade-tolerant plants induce faster and maintain induction longer than shade-intolerant plants after a return to low light (Paliwal et al. 1994; Küppers et al. 1996). In a survey of both woody and herbaceous plants from shady and sunny habitats, Ögren and Sundin (1996) found that the mean induction time to reach 90% of A_{max} for shade plants was only about 7 minutes, whereas fast-growing and slow-growing sun plants took 18 and 32 minutes, respectively. Yanhong et al. (1994) found faster induction in leaves of *Quercus serrata* Thunb. seedlings grown in low light (50 μmol m^{-2} s^{-1}) compared to leaves grown in high light (500 μmol m^{-2} s^{-1}). A more recent review of rates in woody species, however, found no clear patterns among sun or shade plants, except that gymnosperms take longer to induce (Naumburg and Ellsworth 2000). In the C_4 grass *Microstegium vimineum* leaves showed no induction differences when grown in either 25% or 50% light, with leaves from both treatments reaching 90% maximum induction in 10–13 minutes (Horton and Neufeld 1998). Failure of the low-light leaves to induce faster may have been because leaves were already partially induced at the lower light level.

Hull (2001) found that the spring ephemeral *Erythronium americanum* and the summer-green species *Podophyllum peltatum* L. both required higher light to maintain induction than did the summergreen species *Smilicina racemosa* and *Arisaema triphyllum* L. Schott. The spring ephemerals were more clearly delineated from the summer-green herbs by the minimum PFD necessary for maintaining induction, rather than induction velocity itself. Because typical background diffuse PFD in temperate forests is similar to the levels required by the summer-green species, they are likely to be in a moderate to high induction state throughout the day and primed to take advantage of an initial series of sunflecks (Hull 2001).

Among a group of tropical understory plants, Kursar and Coley (1993) observed that plants with shorter-lived leaves (~1 year) induced more quickly than plants with longer-lived (> 4 years) leaves. Induction times to 90% A_{max} were only 3–6 minutes for plants with short-lived leaves compared to 11–36 minutes for plants with long-lived leaves. Using oxygen electrodes, they found that induction of O_2 production was also faster in short-lived leaves, suggesting that slower induction times in longer-lived leaves were caused by slower activation of RUBISCO rather than a delay in the induction of ET. Thus, mean induction times may be longer for evergreen/wintergreen herbs; however, if there is seasonal adjustment to changing light conditions, activation times may vary accordingly.

Loss of photosynthetic induction begins once the PFD drops below the minimum level necessary to maintain an induced state. This occurs in two phases: an initial fast component, thought to be associated with deactivation of PCR enzymes, especially RUBISCO (Sassenrath-Cole and Pearcy 1992), followed by a progressive loss of g_s, which occurs more slowly (Chazdon

1988). Typical time constants for deactivation of associated enzymes are in the range 3–5 minutes, whereas g_s may continue to change over longer periods of up to an hour or more. Species that maintain induction for longer periods of time will be better able to utilize subsequent sunflecks, including those that occur widely spaced apart than species that de-induce rapidly. At present, there are no clear patterns with regard to which types of species maintain induction for longer periods of time. Poorter and Oberbauer (1993) found that an understory tree maintained induction longer than a pioneer species, but Hull (2001) found no correlation of induction loss rate with phenological grouping among the species he studied. Induction losses may be more rapid in plants grown in high light (Poorter and Oberbauer 1993; Tinoco-Ojanguren and Pearcy 1993a, 1993b; Horton and Neufeld 1998), suggesting that under these conditions the emphasis is on maintaining water-use efficiency at the expense of carbon assimilation. In addition, time of day may influence induction losses. Limitations on g_s can be brought about by increasing vapor pressure deficits in the afternoon, but also possibly because biochemical limitations change from morning to afternoon (Poorter and Oberbauer 1993; Tinoco-Ojanguren and Pearcy 1993a).

Direct comparative studies of induction/de-induction times between trees and herbs have not been done. There is not enough information to conclude that major differences exist between these two functional groups. Both groups make use of sunflecks to bolster carbon gain in light-limited environments, but it has not yet been demonstrated that one species group might gain a competitive advantage because of more efficient utilization of sunflecks. The temporal sequence and distribution of sunflecks need to be better characterized to determine induction states of herbs and tree seedlings at PFD levels typical of closed canopies. Only in this way can relative utilization efficiencies be calculated for assessing the benefits of sunflecks to these two groups of plants.

Photoinhibition and Sunflecks

In a fraction of a second, a sunfleck can expose the leaf of an understory herb to more than 100 times the background PFD (Chazdon and Pearcy 1991). This sudden rise in PFD is often used effectively for photosynthesis, but when fixation processes are overwhelmed, the leaf is unable to dissipate this excess energy without harming the photosynthetic reaction centers, resulting in photoinhibition (Björkman and Demmig-Adams 1995). Leaves adapted to shady habitats, and with low rates of photosynthesis, have a reduced capacity to process excess light. Although sunflecks contribute substantially to carbon gain beneath the closed canopy, there is also a potential for sunflecks to be injurious if they are too intense or last too long (Pearcy et al. 1994).

Photoinhibition in temperate herbs can occur when high light coincides with low temperatures, as in the spring, fall, or winter, or in the summer when shade-tolerant herbs are suddenly exposed to long sunflecks with high

PFD. The dissipation of this excess energy can be accomplished by conversion into chemical compounds through fixation processes or other metabolic pathways, by reemission through fluorescence, or by conversion to heat in the pigment beds (thermal dissipation). Once metabolic pathways are saturated, however, the only two viable alternatives are fluorescence and thermal dissipation. The latter is thought to be more important because fluorescence can dissipate at most only about 3–4% of the energy (Björkman and Demmig-Adams 1995). Current research suggests that development of a trans-thylakoid pH gradient in the presence of reduced ascorbate optimizes the conversion of the carotenoids violaxanthin and antheraxanthin to zeaxanthin, which plays a major role in the thermal dissipation of excess light energy away from the harvesting antennae (Demmig-Adams et al. 1996; Watling et al. 1997). The result is that the excess energy is liberated as heat, preventing the deleterious formation of triplet-excited chlorophyll and singlet oxygen. Sun leaves are known to accumulate greater amounts of xanthophyll-cycle intermediates than shade leaves (Königer et al. 1995), and leaves of plants exposed to sunflecks also show enhancement of the pigment concentrations (Watling et al. 1997; Schiefthaler et al. 1999). Low temperatures can also stimulate the formation of large amounts of zeaxanthin, which may prepare a leaf for subsequent exposure to high light in order to reduce the amount of potential photoinhibition that might occur (Oberhuber and Bauer 1991).

Low-Temperature Photoinhibition

Spring ephemerals, early-leafing summer greens, and wintergreen/evergreens are most at risk from low-temperature photoinhibition, but relatively few studies have been done on these plants. Germino and Smith (2000) found small amounts of low-temperature photoinhibition in *Erythronium grandiflorum* Pursh (a close relative of the spring ephemeral *E. americanum*) on sunny days after nights with frost, but photosynthetic capacity recovered by sunset of the same day. Appreciable carbon gain was possible even though temperatures were quite low at this time of year, in large part because low-temperature photoinhibition was not a serious problem. It is likely that other spring ephemerals, such as the closely related *E. americanum*, use analogous mechanisms to avoid or minimize low-temperature photoinhibition. However, there are no other studies on this group that we know of at this time.

The situation with regard to evergreen/wintergreen species is more problematic. These species maintain functional leaves for much longer and are exposed to the greatest temperature variations on a seasonal basis. They can potentially be subject to both low- and high-temperature photoinhibition. Skillman et al. (1996) suggest that trade-offs exist between the frequency of leaf production and the balance between photosynthetic acclimation and photoinhibition. The wintergreen *Hexastylis arifolia*, which produces one flush of leaves per year, suffers greater low-temperature photoinhibition than the heteroptic *Heuchera americana*, which produces new leaves in the spring and fall. In northern Japan, overwintered leaves of *Pachysandra terminalis* do not show any low-temperature photoinhibition in early spring, although they do show high temperature photoinhibition later in the summer when exposed to prolonged high light (Yoshie and Kawano 1986). In Europe, east-facing leaves of the evergreen vine *Hedera helix* L. show substantial photoinhibition in the fall after being exposed to high light, but north-facing leaves do not (Oberhuber and Bauer 1991). In conclusion, low-temperature photoinhibition is probably not a major problem for spring ephemerals, but it can cause significant depressions of A_{max} in some, but not all, evergreens. Heteroptic species with specialized overwintering leaves may be less prone to such injury.

High-Temperature Photoinhibition

Summer-green herbs growing in the shade beneath a closed canopy will suffer photoinhibition if exposed to high light for a prolonged time (Critchley 1998), and an interesting question is whether they do so when exposed to natural sunflecks. The negative impacts of photoinhibition may be ameliorated by either avoiding or tolerating high PFDs. Avoidance is best illustrated by *Oxalis oregana* Nutt., which grows in very shady habitats below redwood trees (background PFD ~ 3–4 μmol m^{-2} s^{-1}). During sunflecks the PFD can quickly rise to 1800 μmol m^{-2} s^{-1}, and within 6 minutes, leaves reorient nearly parallel to the solar beam, lowering the intercepted radiation by 90%. Immobilizing the leaves results in severe photoinhibition within 30 minutes (Björkman and Powles 1981). Banner (1998) has shown a similar phenomenon in *O. acetosella* L., but leaf folding only proceeds to 60°, which attenuates PFD just 50%, still above L_{sat} for this species. However, when sunfleck durations were short, little photoinhibition was detected. We suspect that *O. grandis* Small, which is common in rich woods and beneath hemlocks in the eastern United States, may use this same strategy, but we are not aware of any published studies. Muraoka et al. (1998) have shown a similar behavior in *Arisaema heterophyllum*. This species displays leaves horizontally in intact forest understories, but in disturbed forests with higher PFD, it folds leaves along the main vein so that the blades are parallel to the incoming light. This apparently prevents photoinhibition in this species. We do not know of any similar studies on active leaf orientation changes in tree seedlings, although it is well known that leaf angles decline with depth through the canopy for overstory trees (Kinerson 1979; McMillen and McClendon 1979; Pickett and Kempf 1980).

Studies of photoinhibition in temperate forest herbs under natural sunfleck conditions are scarce, and the majority of studies done have been on tropical understory species (Königer et al. 1995; Logan et al. 1997; Watling et al. 1997, Schiefthaler et al. 1999). Pearcy et al. (1994) concluded that most sunflecks have too low PFD or are too short to cause lasting photoinhibition, but sensitivity does increase at higher temperatures. For example, in *Alocasia macrorrhiza* Schott, injury does not occur at a leaf temperature of 40°C under a nearly saturating PFD of 375 μmol m^{-2} s^{-1}, but injury does occur at a similar PFD once leaf temperatures pass this critical point (Pearcy et al. 1994). Banner (1998) similarly found little evidence for photoinhibition in *Oxalis acetosella* after lightflecks of up to 1000 μmol m^{-2} s^{-1}, but some inhibition was recorded after flecks of 2000 μmol m^{-2} s^{-1}. Even so, nearly complete recovery occurred after about 30 minutes, suggesting that lasting effects were minimal. However, prolonged exposure to high light resulted in photodamage and leaf loss in this species, which may explain its restriction to shady sites.

Leaves exposed to a series of high-light flecks may show a capacity for acclimation through increased resistance to further photoinhibition, most likely by increasing the pools of xanthophyll-cycle intermediates which are used to thermally dissipate the excess energy. This may be particularly important for herbs with evergreen or wintergreen leaves, which are exposed to many more episodes of high light. For summer-greens that are suddenly exposed to high light through formation of a gap in the canopy, increased resistance to photoinhibition may occur by production of new leaves better adapted to high-light conditions (Pearcy et al. 1994). Finally, it should be noted that increased resistance to photoinhibition is not necessarily associated with increases in A_{max}. In fact, A_{max} is often reduced by sudden exposure to high light. Rather, increased resistance may be a mechanism to maintain some carbon gain in response to the change in the light environment until either new leaves are produced or the older ones have a chance to acclimate, which may take several days in some species (Pearcy et al. 1994).

Water Relations of Understory Herbs

Although there is no pronounced drought season in temperate deciduous forests, understory herbs may be limited by insufficient water at times, particularly in the summer, but also in the winter when soils are frozen. Competition for soil water between herbs and trees is evidenced by the large increases in biomass in the understory seen after trenching experiments (see extensive review by Coomes and Grubb 2000). However, despite the importance of water for the productivity of understory plants, less attention has been paid to the water relations of understory herbs than to photosynthesis and respiration.

Stomatal Conductance Patterns

Stomatal conductance will vary among understory herbs because of differences in stomatal densities, sizes, and sensitivity to vapor pressure deficit (Δw) and responses to PFD. In most studies, g_s responds strongly to PFD, increasing sharply in the morning as PFD rises, and then dropping in the afternoon as PFD decreases. Eliáš (1983) found that g_s for a variety of understory herbs peaked between 9 and 11 AM, but that sunflecks had a strong influence on rates, even among leaves on the same plant. For *Mercurialis perennis*, PFD and Δw were the two variables most responsible for variations in g_s (Krie-

Table 3.3. Maximum values for stomatal conductance (g_s) among phenological groups of herbs

Phenological grouping	Maximum g_s (mol H_2O m^{-2} s^{-1})	Reference
Spring ephemerals		
Allium ursinum	0.100	Kriebitzsch (1992b)[a]
Allium tricoccum	0.204	Taylor and Pearcy (1976)[b,c]
Anemone raddeana	0.310	Yoshie and Yoshida (1987)[c]
Arum maculatum	0.166	Kriebitzsch (1992b)
Erythronium americanum	0.187	Taylor and Pearcy (1976)
Mean	0.193	
Summer greens		
Asclepias exaltata	0.600	Davison et al. (unpublished data)[a]
Aralia nudicaulis	0.160	Landhäusser et al. (1997)[c]
Galeobdolon luteum	0.092	Eliáš (1983)[a,b]
Geum urbanum	0.058	Eliáš (1983)
Glechoma hirsuta	0.067	Eliáš (1983)
Hordelymus europaeus	0.060	Kriebitzsch (1992b)
Melica uniflora	0.069	Kriebitzsch (1992b)
Mercurialis perennis	0.083; 0.105	Eliáš (1983); Kriebitzsch (1992b)
Microstegium vimineum	0.100	Horton and Neufeld (1998)[c]
Podophyllum peltatum	0.250	Taylor and Pearcy (1976)
Polygonatum latifolium	0.075	Eliáš (1983)
Rubus pubescens	0.075	Landhäusser et al. (1997)
Solidago flexicaulis	0.129	Taylor and Pearcy (1976)
Trillium grandiflorum	0.092	Taylor and Pearcy (1976)
Viola mirabilis	0.100	Eliáš (1983)
Mean	0.132	
Wintergreens/evergreens		
Asarum europaeum	0.100	Kriebitzsch (1992b)
Cornus canadensis	0.100	Landhäusser et al. (1997)
Galax urceolata	0.140	McCarron (1995)[c]
Pachysandra terminalis	0.160	Yoshie and Kawano (1986)[c]
Pyrola asarifolia	0.153	Landhäusser et al. (1997)
Mean	0.131	

[a]Values from field measurements.
[b]Values converted from cm s^{-1} or mm s^{-1} using a correction factor of 1 mm s^{-1} = 0.0416 mol m^{-2} s^{-1}.
[c]Values from laboratory measurements.

bitzsch 1984). Because PFD beneath the canopy drops substantially by early afternoon, g_s often declines earlier in the forest understory than outside (Neufeld and Young pers. obs.).

Maximum values for g_s among understory herbs are generally low (table 3.3), ranging between 0.050 and 0.250 mol H_2O m^{-2} s^{-1} (Körner et al. 1979; Eliáš 1983; Kriebitzsch 1992b) but occasionally higher (Davison et al. unpublished data; 0.600 mol H_2O m^{-2} s^{-1} for *Asclepias exaltata*). Because g_s often scales with A_{max} in a variety of plants (Yoshie and Kawano 1986; Pearcy et al. 1987; Yoshie and Yoshida 1987) spring ephemerals should have the highest g_s, followed by summer greens and then wintergreens/evergreens. Although spring ephemerals do appear to have somewhat higher g_s table 3.3 shows nearly identical values for g_s among summer greens and wintergreens/evergreens. Additional measurements need to be collected before more conclusive statements about the patterns of maximum g_s among the phenological groups can be made.

Kriebitzsch (1992b) showed a moderate dependency of stomatal conductance on vapor pressure deficit among several species of herbs. The steepest declines in g_s in response to increasing Δw were found in the spring ephemerals, with weaker dependencies among the other species. *Mercurialis perennis* showed little response to Δw in summer, leading to high transpiration rates and low water potentials. Across groups, one might expect that species resistant to cavitation (most likely the evergreens) to show the least sensitivity to Δw, as there would be less of an incentive to close stomata in response to increased drought stress.

Water Potentials of Understory Herbs

Few studies have followed diurnal and seasonal patterns of leaf water potential (ψ_w) in understory herbs. Early studies by Eliáš (1981) showed a range of minimum ψ_w from -0.32 to -0.63 MPa among several spring ephemerals and summer greens, with an occasional reading as low as -0.90 MPa. Masarovičová and Eliáš (1986) showed that water saturation deficits (a surrogate for ψ_w) increased when plants were exposed to sunflecks. Kriebitzsch (1988) followed changes in minimum ψ_w over two growing season among six herbs in a beech forest in Germany. The two spring ephemerals (*Allium ursinum* and *Arum maculatum*) showed high ψ_w in both years due to low temperatures and Δw and the fact that soil water contents were high in the spring. The summer-green herbs and grasses developed low ψ_w by midsummer, ranging between -2.0 and -3.0 MPa, respectively. Because understory herbs have the majority of their roots concentrated between 5 and 15 cm in depth (Plašilová 1970), where soil water deficits can rapidly develop, the occurrence of low ψ_w in these herbs is not unexpected, even at the low irradiance levels typical beneath the canopy. In the evergreen *Galax urceolata*, McCarron (1995) found that ψ_w were highest in midsummer (minimum ψ_w of -0.40 MPa), most likely because of the low PFD then, and lowest in winter (minimum ψ_w of -0.80 to -1.2 MPa). On one particular date in February, the air warmed considerably while the soils remained frozen, and PFD was quite high because of the lack of a canopy. On this date ψ_w dropped as low as -1.6 MPa. In summary, it is most likely that spring ephemerals will exhibit the least water stress, summer greens should exhibit greater water stress in midsummer, and wintergreens/evergreens will have their lowest ψ_w in winter, when cold soils prevent the movement of water from roots to leaves.

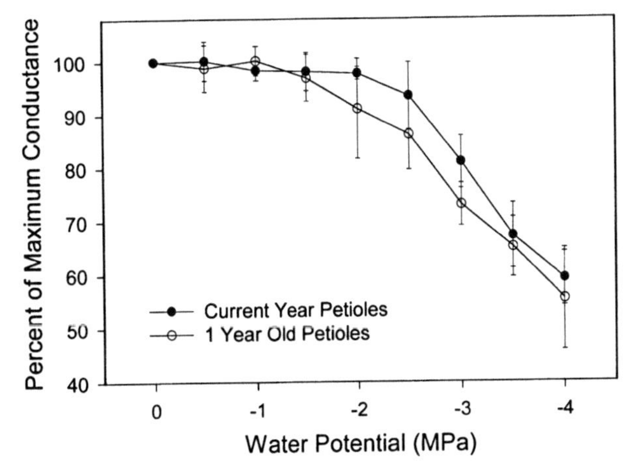

Figure 3.10. Vulnerability curve for mean hydraulic conductance (± SE) in *Galax urceolata* petioles. There were no statistically significant ($p < .05$) differences between petiole age classes ($n = 5$ for each age class). A pressure sleeve was used to induce embolisms, and conductances were measured using a Sperry apparatus (Sperry et al. 1988). Data from Sobieraj (2002).

Hydraulic Conductances in Understory Herbs

The ability to conduct water from the roots to the leaves is a major determining factor of the water balance of plants. Almost nothing is known concerning the hydraulic conductivities (K_h) of understory herbs. Sobieraj (2002) have worked on characteristics of K_h in petioles of the evergreen *Galax urceolata.* Xylem elements in this species are small (80% of diameters < 30 mm) and may enhance tolerance of cold-induced embolisms, an important feature for an evergreen. *G. urceolata* is also extremely drought tolerant. Embolisms in the petioles do not appear until water potentials drop below −2.0 MPa, and even at water potentials as low as −4.0 MPa, K_h is still near 50% of maximum (fig. 3.10). This pattern is more typical of that found in desert plants (Pockman and Sperry 2000) and suggests that *G. urceolata* periodically experiences severe water stress, perhaps because it is shallowly rooted in soils that can substantially dry out between rainfalls. Since the lowest ψ_w found in the field was −1.6 MPa, *G. urceolata* may rarely embolize in response to typical droughts. But soil water potential values as low as −2.5 MPa can occur during severe droughts in the upper soil layers of deciduous forests in Tennessee (S.D. Wullschleger, pers. comm.) and are low enough to possibly induce at least a minimal number of embolisms. Maximum K_h values in *G. urceolata* are also quite low (Sobieraj 2002) and may restrict g_s in this species to low values, which in turn prevents water deficits from building up rapidly. We suspect that other evergreen understory herbs may display similar hydraulic properties. Spring ephemerals should have higher K_h values associated with their higher g_s, but some aspects of their morphology, such as a lack of an above-ground stem, make determination of K_h

in this group technically difficult. As a group, there is some evidence that herbs might have higher K_h values than woody plants, due in part to the differences in physical structural of the xylem between these two groups (Camacho-B et al. 1974).

Community-Level Carbon and Water Exchange

Carbon Exchange

Herbs generally make up less than 3% of the total biomass of temperate forests and contribute nearly the same proportion toward above-ground net primary productivity (ANPP). In the northern hardwood forests at Hubbard Brook, New Hampshire, ANPP is 2.7% of the ecosystem total (Bormann and Likens 1979); evergreens and summer greens account for 2.2% of that amount, with spring ephemerals (mainly *Erythronium americanum*) contributing the remaining 0.5%. In a Tennessee forest, herbs constitute 3.1% of the standing crop, but take up 7% of the net photosynthetic carbon gain (Harris et al. 1975). Early estimates of the actual values for ANPP by understory herbs ranged between 10 and 100 g m^{-2} $year^{-1}$ (Ovington 1962), but more recent studies by Bazzaz and Bliss (1971) and Kriebitzsch (1992c) have extended the upper limit to between 135 and 175 g m^{-2} $year^{-1}$, although values fluctuate from year to year, depending on complex interactions among light, temperature, and drought.

The relative contribution of herbs to ANPP does not always correlate directly with leaf-level measures of carbon exchange. For example, despite high assimilation rates, grasses contribute relatively little to total understory ANPP in beech forests in Germany, mainly because in these communities grasses have a very low LAI. In contrast, *Mercurialis perennis* and *Allium ursinum*, with more moderate gas exchange rates but high LAI, are the two dominant species in terms of ANPP (Kriebitzsch 1992c).

How much of the carbon assimilated above ground is lost to below-ground processes is poorly understood. Schulze (1972) estimated that below-ground respiration in *Oxalis acetosella* constituted 28% of above-ground assimilation, and Kriebitzsch and Regel (1982) estimated losses of 26% for *Mercurialis perennis*. After reviewing the literature, Kriebitzsch (1992c) concluded that the best estimates ranged between 20% and 30% of above-ground assimilation. In the spring and fall, when there is still considerable above-ground activity by herbs, below-ground respiration rates will be limited by cold soil temperatures. Appreciable soil warming can occur during the day because of the lack of a canopy, which could stimulate respiration rates (Kawano 1970). Once the canopy closes in the spring, and once a snow cover is established in the winter, diurnal variations in soil temperature are greatly reduced, and estimates of below-ground respiration are easier to model.

The translocation of carbon from above-ground to below-ground storage organs has been shown repeatedly to be important for growth and reproduction of understory herbs. Tissue et al. (1995) showed that 62–76% of [^{14}C]CO_2 given to plants was stored in the corms, then later used to support the growth of reproductive structures and new leaves, roots, and corms. This confirmed earlier studies showing the high dependency of growth and reproduction on stored reserves (Whigham 1990; Zimmerman and Whigham 1992). A substantial portion of the translocated carbon was assimilated late in the season, even as photosynthetic rates declined and leaves began to senesce. Similar findings were reported by Hutchings and Barkham (1976), who suggested that carbon assimilated in the shade of summer was then transported to rhizomes in *Mercurialis perennis*. Muller (1979) showed that cormlet weights of *E. albidum* increased more than 100% during leaf senescence due to translocation of carbohydrates from above- to below-ground tissues. Considering the substantial proportion of biomass belowground for understory herbs, the importance of these processes in determining carbon mass budgets for ecosystems cannot be underestimated.

Water Exchange

Among the species Kriebitzsch (1993) tested, a spring ephemeral, *Arum maculatum*, had the highest transpiration rates, although the maximum values obtained for the grasses (*Hordelymus europaeus* Jessen ex Harz and *Melica uniflora* Retz.) and the herb *Mercurialis perennis* were similar. The evergreen *Asarum europaeum* and the spring ephemeral *Allium ursinum* both had the lowest water-loss rates. The single most important factor in determining transpiration rates was PFD, whereas Δw played an important secondary role. For example, 1983 was a sunnier year than 1982, and transpiration was enhanced in 1983 between 60 and 200% over that in 1982. In concordance with the community-wide patterns of ANPP, Kriebitzsch (1993) calculated that *M. perennis* and *A. ursinum* achieved the highest water loss rates per unit ground area because of their higher LAIs. The grasses and *A. maculatum* had much lower loss rates due to their lower LAIs.

Kriebitzsch (1993) integrated water loss rates across the species groups in his extensive studies in Germany over two seasons. Using empirical models to predict leaf-level water loss over the season, he scaled up to the community level by multiplying those rates by the LAI of each species. He then compared these water loss rates to those for the entire forest and calculated that understory herbs could transpire between 20 and 40 mm of water, which comes to about 10% of the total water lost from the system over the course of a year. If we assume a similar biomass distribution in North America, then herbs in North America would also be disproportionately critical components of ecosystem water fluxes, even though they grow in severely light-limited environments. Because stomata of understory herbs are sensitive to low light and open at relatively low PFD, it is not surprising that substantial water loss can occur from the understory.

More recently, carbon isotope discrimination has been used to determine ecosystem-wide patterns of water use among life forms (Brooks et al. 1997). In boreal aspen forests, deciduous forbs showed the most discrimination against ^{13}C, followed by evergreen forbs, shrubs, and mosses, with the least discrimination found among the overstory trees. This occurred in both a dry and wet year, suggesting that these patterns are diagnostic characteristics of these ecosystems. Differences among life forms (tree, shrub, or forb) disappeared when plant stature and leaf longevity were factored out. Within a life form, evergreen plants had lower discrimination than deciduous leaves, no matter whether a tree or forb, and, most interesting, the difference between evergreen and deciduous forbs was greater than that between evergreen and deciduous trees. Less discrimination by evergreens may reflect their inherently higher water use efficiencies and low A_{max} and g_s (Chabot and Hicks 1982).

Global Change and Influences on the Ecophysiology of Understory Herbs

Global changes in atmospheric CO_2 concentrations, along with those of air and soil temperature, may pose significant problems for understory herbs in the future. Although higher CO_2 may be beneficial for photosynthesis, particularly in a light-limited environment, it may bring about concomitant changes in overstory canopy characteristics, such as increased LAI, that may further reduce PFD in the understory. Higher rates of photosynthesis would demand higher rates of nutrient extraction, particularly for phosphorus and nitrogen, and there could be greater competition for these nutrients between woody plants and herbs. It is interesting to note that one of the more responsive understory plants to increased CO_2 in the Free Air CO_2 Enrichment system at Duke University, North Carolina, has been poison ivy (*Rhus toxicodendron* L.; J. Mohan, pers. comm.).

Changes in global air and soil temperatures may disrupt the concordance between day length and phenology for both overstory and understory plants. Because the ability to gain sufficient carbon for spring ephemerals and summer greens is closely tied to the timing of canopy phenology, any changes, such as earlier leafing of the overstory, could be detrimental. Higher soil temperatures may alter the dormancy requirements for herbs, which could then sprout at inopportune times, when they may become subject to early or late-season frosts. If flowering times are also disrupted, pollinator services could be reduced. Higher soil temperatures would also result in higher belowground respiration rates, which could lead to carbohydrate depletion, perhaps decreased frost tolerance, and lowered reproduction, not to mention reduced vigor in the spring when new leaves need to be produced. Low carbohydrate reserves might also alter disease resistance in herbs, as well as their ability to withstand pollutants such as ozone and/or acidic deposition.

Summary

The presence of herbs in the understory of temperate deciduous forests depends in large part on the ability to grow in a light environment that is highly heterogeneous on both temporal and spatial scales. Light penetration to the forest understory herb layer is most pronounced during the spring overstory leafless period. This is followed by the most light-limited time of the year during the summer when the overstory is fully developed. Variations in irradiance can drive differences in other microclimatic factors, especially temperature, humidity, and water availability. Adaptive responses to this dynamic environment can be both physiological and morphological and exist at multiple scales from subcellular to leaf, whole plant, and ecosystem levels.

The presence of seasonality, coupled with distinct periods when the overstory canopy is present or absent, results in a suite of phenological strategies among herbs, including spring ephemerals, summer greens, wintergreens, evergreens, heteroptics, and parasitic and saprophytic plants. It appears that evolution has favored species that are physiologically active during the warmer portions of the years when temperatures are more favorable for photosynthesis and nutrient uptake. The proportion of evergreen herb species in the understory should increase toward the southern latitudinal limits of the distribution of temperate forest due to more favorable temperatures when the canopy is leafless.

Strategies to increase light absorption in the forest understory include a well-developed spongy mesophyll to scatter light, reflective lower leaf surface to direct light back into the leaf, and adaxial surface cells that are concavely shaped to direct light toward cells containing chloroplasts. Selection for height growth and a monolayer canopy are intensified to maximize light interception during the light-limiting summer months.

Spring ephemerals have photosynthetic characteristics similar to sun-adapted plants and maximize carbon gain before development of the overstory canopy. These species are also well adapted to low temperatures. In contrast, wintergreen and heteroptic herbs have much lower photosynthetic rates but are also much more plastic in response to seasonal variations in both temperature and light. Although evergreens are similar, these herbs obtain most of their carbon during the fall and spring leafless periods. There is a strong inverse relationship between photosynthetic rate and leaf life span; evergreens have the lowest photosynthetic rate and the greatest leaf life span.

For herbs that persist after canopy closure, physiological, biochemical, and morphological changes are required to maximize photosynthetic efficiency under light-limited conditions. Resources are shifted from processes associated with carbon assimilation to those involved with light harvesting. Once the canopy has closed, sunflecks represent the primary source of light. Many species use light-induced increases in stomatal conductance, RuBP regeneration, and RUBISCO to enhance sunfleck-related carbon gain. However, the high-intensity irradiance of sunflecks may also lead to photoinhibition, especially for species adapted to extremely shaded environments, although the durations of most sunflecks are not long enough to cause lasting injury. The majority of sunfleck research has been done in tropical forests, and understory herb responses in temperate deciduous forests have not been well documented.

Clearly, the morphological, biochemical, and physiological processes involved in the response of herbs to the dynamic light environment of the deciduous forest understory remain to be thoroughly investigated. The water relations of understory herbs have not been comparatively examined across the phenological strategies, but there is a suggestion that maximum hydraulic conductivities are higher, as a group, in herbs as compared to woody seedlings. In addition, the contribution of understory herbs to forest community-level carbon gain, production, and both CO_2 and H_2O exchange with the atmosphere have not been adequately addressed. Existing data suggest that herbs can be disproportionally important to these fluxes despite their relatively small contribution to the biomass of an ecosystem. These ecophysiological problems varying in scales of time, space, and biodiversity are essential to understanding the response of both understory herbs and the forest community to changing global climate.

Appendix 3.1. Symbols Used in Chapter 3

Abbreviation or Symbol	Definition	Units
ϕ	Quantum yield	μmol CO_2 / μmol photons
A_{max}	Maximum rate of light saturated photosynthesis	μmol CO_2 m^{-2} s^{-1}
ANPP	Annual net primary productivity	g m^{-2} $year^{-1}$
ET	Electron transport capacity	—
g_s	Stomatal conductance	mol H_2O m^{-2} s^{-1}
K_h	Hydraulic conductance	mol H_2O s^{-1} m^{-1} MPa^{-1}
LAI	Leaf area index	m^2 leaves per m^2 ground area
LCP	Light-compensation point	μmol photons m^{-2} s^{-1}
LHCII	Light harvesting complex of photosystem II	—
L_{sat}	Light saturation point of photosynthesis	μmol photons m^{-2} s^{-1}
PCR	Photosynthetic carbon reduction cycle	—
PFD	Photosynthetic photon flux density	μmol photons m^{-2} s^{-1}
R_d	Dark respiration rates	μmol CO_2 m^{-2} s^{-1}
RUBISCO	Ribulose-1,5-bisphosphate carboxylase/oxygenase	—
SLM	Specific leaf mass	g/m^2
T_{opt}	Temperature optimum of photosynthesis	°C
Δw	Vapor pressure deficit	kPa

Acknowledgments We acknowledge the guidance and assistance of the editors, Mark Roberts and Frank Gilliam, toward improving this chapter. In addition, several colleagues reviewed the manuscript, including Alan Davison, Jonathon Horton, and Erik Nilsen, which helped greatly to improve it, and to whom we express our most sincere thanks. Beverly Moser translated several articles from the original German, for which we are grateful. Authorities for scientific names came from either Radford et al. (1968) or the *International Plant Name Index* (www.ipni.org) from Kew Gardens.

4

Interactions of Nutrient Effects with Other Biotic Factors in the Herbaceous Layer

Wendy B. Anderson

The intent of this chapter is to elucidate both the importance of nutrients in forest herb physiology and community interactions and the importance of forest herbs in regulating the nutrient environment of their ecosystems. Although plants in the herbaceous layer account for only a small proportion of deciduous forest biomass, they perform many important functions in intrasystem nutrient cycling. Understanding the physiological processes that occur in individuals provides insight into how herbaceous species interact with other species and how they function in the environment as individual species and as a functional guild. Most physiological studies of forest herbs have focused on responses to light as the primary limiting resource (chapter 3, this volume); however, nutrients are also potentially limiting and may be as important as light in determining physiological, growth, and reproductive parameters for forest herbs.

The herbaceous species of eastern deciduous forests experience a spatially and temporally dynamic nutrient environment (chapter 9, this volume). Nutrient availability changes seasonally, peaking in the early spring and declining throughout the summer, and the phenology and physiology of the various herbaceous species reflect these patterns. Nutrient availability is also spatially heterogeneous, varying across spatial scales from centimeters to kilometers, and can alter physiological and community interactions within and among species. I focus first on the physiological responses that plants in the herbaceous layer of deciduous forests exhibit in response to variability in their nutrient environment. Then I address the effects of nutrient availability on community-level interactions between herbaceous species and their mycorrhizal associates and between herbs and their herbivores. Finally, I summarize the possible ecosystem-level responses that these physiological and community processes may induce in a changing environment.

Physiological Responses of Forest Herbs to Nutrient Availability

Interactions Between Light and Nutrients

Many of the physiological studies addressing the relationship between nutrients and the herbaceous layer focus on the interaction between light and nutrients. As a general pattern across biomes, shading lessens the demand for soil nutrients via reduced photosynthesis rates and biomass accumulation (Cui and Caldwell 1997). For the herbaceous layer of forests, this pattern may explain variation in carbon acquisition rates between spring and summer herbs. In this section, I highlight some representative studies from deciduous forests that show the relationships between light and nutrients for forest herb physiology.

In a classic study by Taylor and Pearcy (1976), spring ephemeral herbs that actively photosynthesized before canopy closure exhibited high light-saturated photosynthesis rates, high respiration rates, and high-light compensation points, followed by negative net photosynthesis rates and ultimately senescence upon canopy closure. This time period is also associated with extremely high nutrient availability as freeze–thaw cycles lyse microbes that formerly immobilized nutrients (Zak et al. 1990; DeLuca et al. 1992), nutrients continue to leach from warming, leaf litter decomposes, and competition with woody plant species is minimized (Bormann and Likens 1979). These available nutrients facilitate the high metabolic rates of spring ephemeral herbs. In contrast, summer herbs from the same deciduous forest site in New York exhibited low light-saturated photosynthesis rates, low respiration rates, and low compensation points but could acclimate to increased irradiance in gaps and from sunflecks with slight increases in those rates (Taylor and Pearcy 1976). Nutrients are much less available in the summer as microbes and woody plants immobilize greater quantities; hence the shade plants of forest floor in the summer must maintain low metabolic rates, even when light is occasionally available.

Two experimental studies of *Claytonia virginica* L. further illustrated the importance of light–nutrient interactions for spring ephemerals in a deciduous forest in Tennessee (Eickmeier and Schlusser 1993; W. B. Anderson and Eickmeier 1998). In the earlier study (Eickmeier and Schlusser 1993), additions of an NPK fertilizer enhanced carboxylation by ribulose-1,5-bisphosphate carboxylase/oxygenase (RUBISCO) and increased vegetative and reproductive biomass of plants growing in full sunlight. However, plants growing under shade treatments did not exhibit the same increases. In the latter study (W. B. Anderson and Eickmeier 1998), both shaded and unshaded plants responded to nutrient additions with increased biomass, but shaded plants did not respond to the same degree as unshaded plants (fig. 4.1). Further, root:shoot

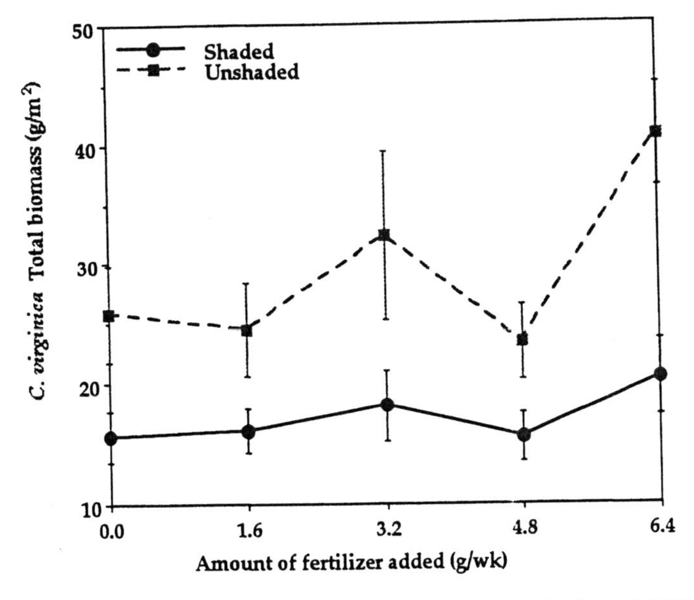

Figure 4.1. Total biomass of *Claytonia virginica* in response to shade and NPK fertilizer addition treatments. Data from 3 years are combined and presented as means ± SE (n = 12 samples per treatment combination). Reprinted from W.B. Anderson and Eickmeier (1998), with permission from *Canadian Journal of Botany*.

biomass ratios decreased with nutrient additions when plants were in full sun but did not respond to nutrient additions when shaded (fig. 4.1).

In addition to the growth responses described above, nutrient (nitrogen and phosphorus) concentrations of *C. virginica* increased with increasing nutrient additions up to a point where the plants became saturated with nutrients (W. B. Anderson and Eickmeier 1998). The shaded plants reached this saturation point at a lower nutrient-addition level than the plants in full sun. The shaded plants maintained internal nutrient concentrations that were slightly higher than those of plants in full sunlight because of their lower total biomass. However, the nitrogen and phosphorus content (concentration × biomass) in shaded plants was significantly lower than in unshaded plants (fig. 4.2; W. B. Anderson and Eickmeier 1998). These studies show that shaded plants have a lower demand for nutrients and may not be able to respond to higher nutrient availability when they are shaded.

Nutrient Availability and Nutrient Resorption

Nutrient availability in deciduous forests can also affect herbaceous species' ability to reabsorb nutrients before leaf abscission. Nutrient resorption is a metabolically expensive process that breaks down nutrient-rich organic molecules in senescing leaves and transports the nutrients into perennating or-

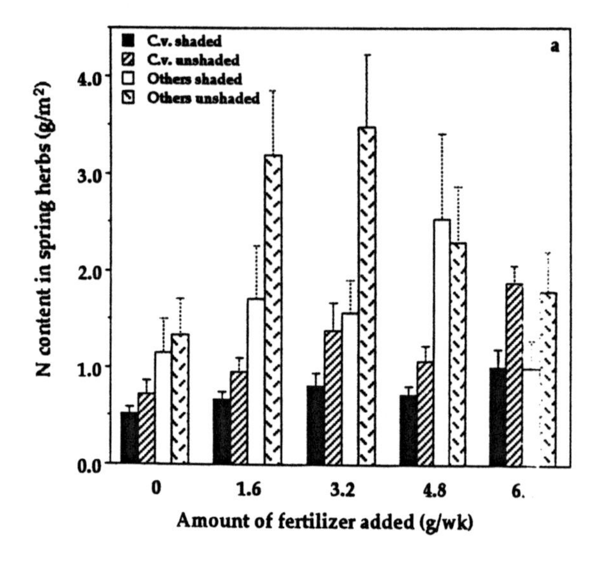

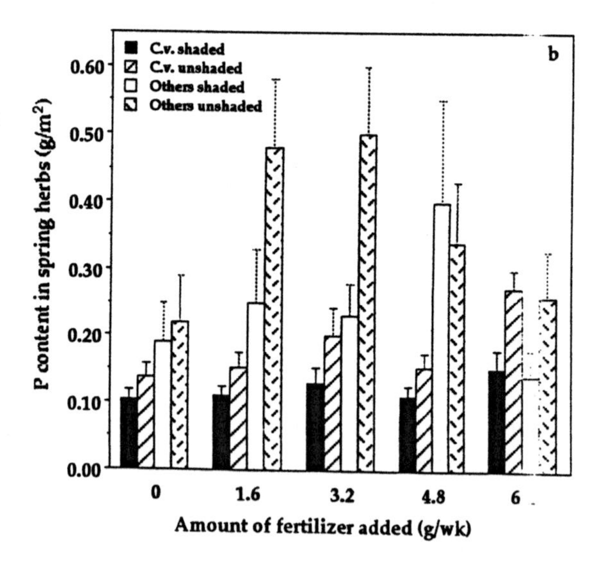

Figure 4.2. Total nutrient pool responses to shade and NPK fertilizer additions treatments for (top) nitrogen in *Claytonia virginica* (C.v.) and other herbs and (bottom) phosphorus in *C. virginica* and other herbs. Data for 3 years are combined and presented as means ± SE (n = 12 samples per treatment combination). Reprinted from W.B. Anderson and Eickmeier (1998), with permission from *Canadian Journal of Botany*.

gans to be stored until the next growing season. As a general pattern, plants growing in less fertile sites resorb greater proportions of nutrients than plants growing in more fertile sites, although this may not apply to herbaceous plants (Chapin 1980; Vitousek 1982; Shaver and Melillo 1984; but see Bridgham et al. 1995, Pastor and Bridgham 1999 for conceptual and empirical alternatives). Measuring nutrient resorption rates in herbaceous plants provides information on the length of time nutrients reside in that compartment of the ecosystem and on what proportions are stored in that compartment versus returned to the soil (W. B. Anderson and Eickmeier 2000).

In an Ohio forest, *Smilacina racemosa* L., a summer herb; *Cardamine concatenata* (Michx.) O. Schwarz, an early spring herb; and *Trillium flexipes* Raf., a late spring herb, each resorbed greater proportions of phosphorus at lowland sites with low phosphorus availability than at upland sites with higher phosphorus availability (DeMars and Boerner 1997). The resorption patterns for nitrogen in these species were less clear, in that all species on slope sites with intermediate nitrogen resorbed less nitrogen than in upland or lowland sites where nitrogen was more or less available, respectively. An interaction between nutrient availability and soil moisture might have confounded that resorption pattern (fig. 4.3).

DeMars and Boerner (1997) also made comparisons among summer and spring species to address resorption in terms of temporal variability in nutrient availability. They found that both *S. racemosa* and *T. flexipes* resorbed proportionally less phosphorus and nitrogen than did *C. concatenata* (fig. 4.3). They provided two possible explanations for this. First, although nutrients may be most available during peak months of *C. concatenata* and least available during those for *S. racemosa* and *T. flexipes*, the shade-tolerant physiology of the later spring and summer species may not allow for the high energy expenditure required by the process of resorption. In contrast, high photosynthetic rates of *C. concatenata* in full sun may afford that species the requisite energy for high proportions of resorption. Second, *C. concatenata*, like most early spring ephemerals, maintains relatively high green-leaf nutrient concentrations, so even with high proportional resorption rates it may still maintain relatively high concentrations in its senescent tissues.

W. B. Anderson and Eickmeier (2000) found that nutrient resorption in the spring ephemeral herb *C. virginica* was less efficient when plants were grown with higher nutrient availability. This was particularly true for nitrogen resorption and less so for phosphorus resorption. *C. virginica* growing under ambient nutrient conditions resorbed up to 80% of green-leaf nitrogen and 86% of phosphorus, whereas *C. virginica* growing in fertilized conditions resorbed 47% of nitrogen and 56% of phosphorus, which is in the normal range of nutrient resorption efficiencies of many other spring ephemeral species (Muller 1978; DeMars and Boerner 1997) and other species across several biomes (Aerts 1996).

Another way to consider nutrient resorption is nutrient resorption proficiency, which is the lowest concentration to which a plant draws down foliar nutrients during senescence (Killingbeck 1996). It reflects the adaptation of species to specific nutrient environments such that species adapted to typically high-nutrient environments tend to retain higher nutrient concentrations in senescing tissues (low proficiency) than do species adapted to low-nutrient environments (high proficiency). According to Killingbeck (1996), complete resorption would be reducing senescent tissue concentrations to 0.3% for nitrogen and 0.01% for phosphorus. In *C. virginica*, resorption proficiency

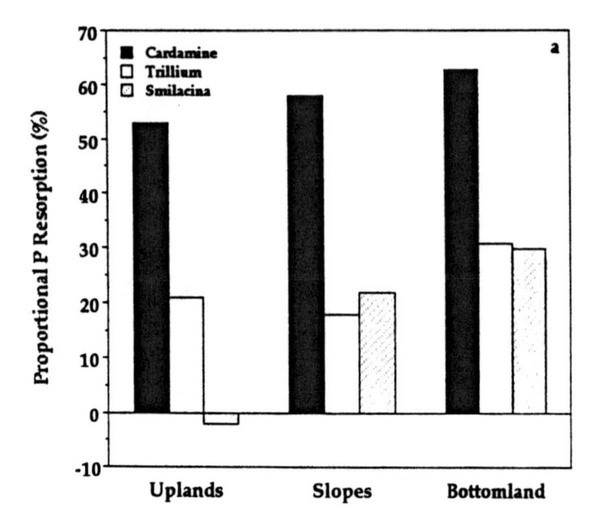

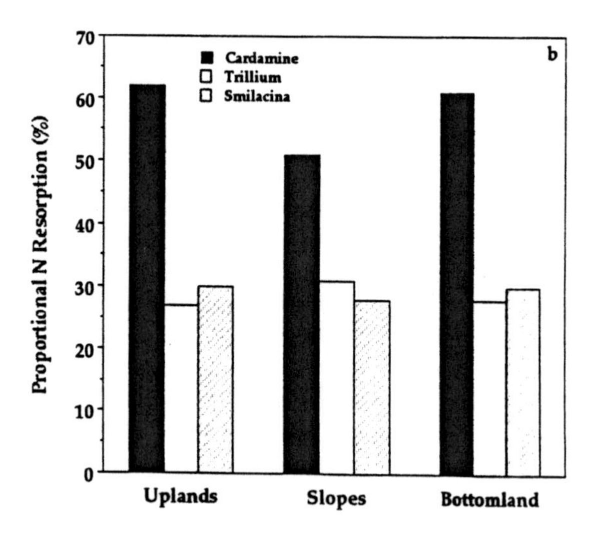

Figure 4.3. Overall proportional (top) phosphorus and (bottom) nitrogen resorption by species and topographic position. Each bar represents mean phosphorus or nitrogen for all years and sites. Data were pooled because phosphorus or nitrogen resorption did not differ significantly among years or sites for any topographic position. Reprinted from DeMars and Boerner (1997), with permission from *Castanea.*

was relatively low, particularly when fertilized (W. B. Anderson and Eickmeier 2000). For *C. virginica,* resorption efficiency could be average or even high while proficiency was low because green-leaf tissues contained extremely high concentrations of nitrogen (3.5–6%), and their senescent tissues, although 47–80% lower, still had 2.0–4.2% nitrogen. Although efficiency and proficiency results initially suggest conflicting conclusions about resorption, they show that *C. virginica,* which is the dominant spring herb species at many study sites, can take up large quantities of nutrients during its short growing season but will also return to the soil a large quantity of nutrients from its senescent leaves.

Physiological Effects of Pollution on Forest Herbs

Nutrient-based pollutants exert a range of effects on herbaceous species in deciduous forests. Although low levels of many pollutants can stimulate many plant-growth responses, moderate levels can inhibit growth responses, and high levels can be fatal (W. H. Smith 1981). A series of fumigation studies in Europe demonstrated this for forest herbs by focusing on the physiological responses of herbs exposed to increased nutrients and acid deposition (Fangmeier 1989; Steubing et al. 1989). These studies, summarized below, provide models of how herbaceous species in eastern deciduous forests of North America might respond to similar pollutants.

Fumigation with SO_2, NO_2, and O_3, either individually or in combination, reduced the leaf area index, chlorophyll contents, and glutamate dehydrogenase (GDH) activity of *Allium* L., *Anemone* L., *Viola* L., and *Arum* L. species, but not of *Melica* L., *Milium* L., *Hedera* L., *Lamium* L., *Oxalis* L., or *Galium* L. However, responses varied across forest sites and across years, and some species, but not all, were more sensitive to combinations of pollutants than to a single source (Fangmeier 1989). Single or mixed fumigation also corresponded to a breakdown of cuticular waxes in all species and lower transpiration and photosynthesis rates in most species, particularly at ambient PFD (see Appendix 3.1). Combinations of gases also increased starch content of leaves (Steubing et al. 1989). Antagonistic interactions among the pollutants were also common. For example, sulfur content of some leaves reflected the fumigation levels of SO_2 when it was applied alone, but when it was combined with O_3, leaf sulfur content decreased as O_3 leached sulfates from plant tissues. Finally, fumigation also increased the numbers of soil fungi and decreased the bacteria/fungi ratios of the soil (Steubing et al. 1989). These results indicate that herbaceous species are particularly sensitive to inputs of airborne pollutants, and these inputs can be as important as the light environment in altering photosynthesis rates and other growth-and nutrient-related responses.

The Influence of Nutrients on the Relationships Between Forest Herbs and Mycorrhizal Fungi and Herbivores

Mycorrhizal Associations and Nutrients

The nutrient environment can have profound effects not only on the physiology of species in the herbaceous layer, but also on the relationships that herbaceous species maintain (or avoid) with other organisms in their environment. As a general pattern, the frequency of vesicular–arbuscular (V-A) mycorrhizal associations in plants varies among sites depending on availability of nutrients, particularly phosphorus (Mosse and Phillips 1971; Menge et al. 1978; Chapin 1980). Individuals or species that grow in sites with higher phosphorus availability are usually less likely to facilitate mycorrhizal associations than those in lower phosphorus areas and thus can maintain higher levels of carbon for growth and reproduction (Snellgrove et al. 1982). However, the lack of mycorrhizal associations could inadvertently increase water stress and possibly limit nitrogen uptake for those individuals or species.

For example, mycorrhizal associations are quite expensive for the carbohydrate reserves of *Erythronium americanum* Ker Gawler, but they may alleviate water stress during the spring growing season and enhance nutrient uptake in the autumn (Lapointe and Molard 1997). Although soils in the spring are usually not as dry as in the summer, water limitation could still occur in spring herbs because they maintain high light-saturation points and photosynthesis rates (Taylor and Pearcy 1976), which would require that their stomates remain open longer, thus allowing plants to potentially become water stressed. Summer herbs growing in the shade may be less likely to experience water stress but may also be more likely to form mycorrhizal associations because of the increased nutrient limitation.

Mycorrhizal associations in two summer herbs, *Geranium maculatum* L. and *Polygonatum pubescens* (Willd.) Pursh. in an Ohio forest exhibited the general pattern of higher V-A infection rates in sites with lower phosphorus availability (Boerner 1986). Foliar phosphorus concentrations were positively correlated with V-A infection rates; however, proportional nutrient resorption was lower in individuals with high infection rates. Although general patterns (described earlier in this chapter) suggest that nutrient resorption should be higher in less fertile sites, this may not be true for species with mycorrhizal associations. The energy expenditure to maintain the fungal association may preclude the expense of resorbing nutrients before leaf abscission. However, Boerner (1986) suggested that shade-tolerant individuals with mycorrhizal associations growing in canopy gaps might be able to accumulate carbon at a high enough rate to support both the fungal symbiont and the energy-demanding process of resorption.

Herbivory and Nutrients

Invertebrate and mammalian herbivory of both spring and summer herbs occurs frequently and can substantially impact individuals and populations of forest herbs (Lubbers and Lechowicz 1989; chapter 5, this volume). In the spring herbs *Jeffersonia diphylla* (L.) Pers., *Sanguinaria canadensis* L., *E. americanum*, and *Trillium sessile* L., experimental or natural defoliation > 50% reduced reproductive performance, but often not until the following year (Rockwood and Lobstein 1994). Whigham (1990) showed that defoliation affected both reproduction and below-ground storage in *Tipularia discolor* (Pursh) Nutt., a wintergreen perennial. Corm biomass dwindled each year

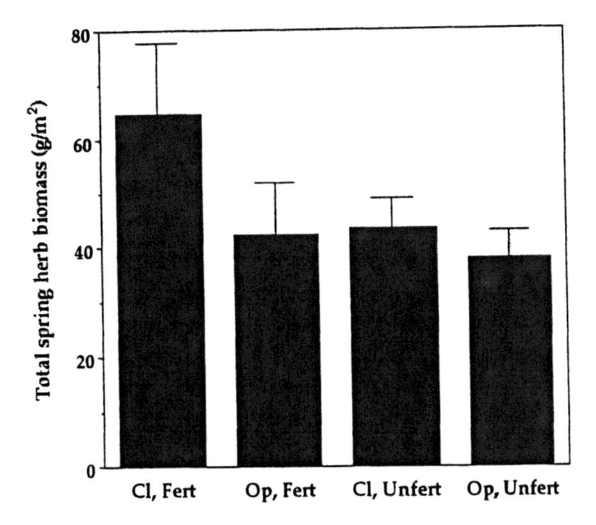

Figure 4.4. Biomass of spring herbs at Radnor Lake State Natural Area, Davidson County, Tennessee, subjected to herbivore exclosures (Cl) or open, control plots (Op) and 6.4 g/week NPK fertilizer (Fert) or no fertilizer (Unfert).

after 50% and 100% annual defoliations over a period of 3 years. However, reproduction was diminished only in the second and third year of the 100% defoliation treatment, and only in the third year of the 50% defoliation. Apparently, translocation of stored carbon and nutrients from the corm could replace lost above-ground tissue for only 1 or 2 years.

The nutrient environment may influence the severity of herbivory and the ability of herbaceous species to recover from herbivory. In a boreal forest in the southern Yukon, John and Turkington (1995) found that many herbaceous plants responded to both fertilization and protection from herbivores. *Festuca altaica* Trin. and *Mertensia paniculata* (Aiton) G. Don. increased in biomass in response to nutrients, whereas *Anemone parviflora* Michx. and *Arctostaphylos uva-ursi* (L.) Sprengel. decreased in abundance when fertilized. *F. altaica*, which is highly preferred by snowshoe hares, the predominant herbivore, was less abundant in fertilized plots open to herbivory than in hare exclosures. In contrast, *A. uva-ursi* was less preferred by the hares and performed better in open plots. Overall, herbivory had less of an impact on the total abundance of herbaceous species than fertilizer additions in this nutrient-limited environment. However, the interaction between nutrients and herbivory can alter the community structure of the herbaceous layer by increasing abundance or susceptibility of individual species.

An interaction between nutrient availability and herbivory was also found in a community of spring herbs in a Tennessee forest (W. B. Anderson 1998). Total biomass of all herbaceous species was greater in herbivore exclosures than in open control plots, but these differences were much more striking in fertilized treatments than unfertilized treatments (fig. 4.4). This suggested that fertilized plants not protected by an exclosure were more susceptible to herbivory than unfertilized plants that were equally unprotected. However, this was not true for every species in the community. *Claytonia virginica*, which had lower concentrations of nitrogen and phosphorus than other species in all treatment combinations, had greater biomass in fertilized control plots than in fertilized exclosures, possibly because it was not preferred by herbivores or because it was released from interspecific competition for light.

Application to Ecosystem Processes

Species in the herbaceous layer of deciduous forests play an important role in nutrient cycling at the ecosystem level, particularly in the spring when many other species are still dormant (Muller 1978; Bormann and Likens 1979; chapter 2, this volume). Anthropogenic alterations of the nutrient environment (e.g., atmospheric deposition, burning and harvesting regimes) could have significant impacts not only on the physiological and community dynamics of the herbaceous layer, but on all forest species that interact directly or indirectly with the herbaceous layer. These effects could translate into changes in how nutrients cycle within deciduous forests.

The ability of spring herbs, which are sun plants (Taylor and Pearcy 1976), to photosynthesize, assimilate biomass, create large, thick sun leaves, and take up large quantities of nutrients in proportion to their biomass depends on both the light and nutrient environment (W. B. Anderson and Eickmeier 1998). Increased nutrient availability increases photosynthesis rates, biomass, and nutrient uptake by herbs, but only in full sun, and only up to a certain growth threshold and nutrient saturation point (Taylor and Pearcy 1976; W. B. Anderson and Eickmeier 1998). In the shade, these thresholds are lowered. If forests are invaded by evergreen or semi-evergreen shrub species (e.g. *Juniperus* or exotic species of *Lonicera*; chapter 12, this volume) or if the timing of canopy closure is sped up by global climate change, light availability will be reduced, which will reduce the ability of spring herbs to grow, store nutrients, and reproduce. Summer-green species may have inherently lower growth thresholds and nutrient saturation points than spring species because of their lower photosynthesis rates. In both spring and summer herbs, either light or nutrient limitation on growth will reduce the total amount of nutrients that flow through the herbaceous layer and increase the probably of nutrients leaching into streams (W. B. Anderson and Eickmeier 1998).

Increased nutrients also reduce nutrient resorption rates before senescence and potentially increase decomposition rates of the senescing tissues (Shaver and Melillo 1984). Senescence under these conditions could release more nutrients for uptake by other plant or microbial species. These mineralized nutrients might also be susceptible to being leached from the system. Higher levels of nutrients could also alter interactions between several herbaceous species and their mycorrhizal associates and could increase some herbaceous species' susceptibility to herbivory. In both cases, these interactions could alter the competitive relationships among herbaceous species and ultimately alter the species composition and diversity of the entire forest community.

Physiological and community-level responses of herbaceous species to the nutrient environment vary among species, sites, and seasons. Overall, though, an increase in nutrient availability could increase short-term nutrient uptake and growth, decrease resorption and mycorrhizal associations, and increase herbivory in the herbaceous layer, which would all increase the rate at which nutrients cycle through that layer and decrease the time that nutrients actually reside in that compartment of deciduous forest ecosystems.

Summary

Although the nutrient environment of the herbaceous layer of deciduous forests may be relatively rich, light–nutrient interactions may affect herbaceous species' ability to exploit available nutrients. Shading may lessen the demand for soil nutrients, as the uptake of nutrients may be limited by the biomass accumulation of individual plants and their specific nutrient saturation points. Nutrient resorption may also occur less efficiently or proficiently in species or individuals that experience relatively high nutrient availability. Physiological responses of herbaceous species to fumigation by various pollutants vary depending on the species and the combinations of pollutants.

Nutrient availability may mediate interactions between herbaceous species and members of other kingdoms. Mycorrhizal associations are more common when nutrients are limited and may indirectly mediate water stress for some species, although the associations have a high carbon cost. Nutrients may also increase susceptibility of herbaceous species to invertebrate or vertebrate herbivores. Finally, natural or anthropogenically altered nutrient environments may shift physiological responses and community patterns in such a way that nutrients reside in the herbaceous layer for shorter periods of time and possibly become more susceptible to being leached from forest ecosystems.

II

Population Dynamics of the Herbaceous Layer

5

Populations of and Threats to Rare Plants of the Herb Layer

More Challenges and Opportunities for Conservation Biologists

Claudia L. Jolls

Nature provides exceptions to every rule.
—Margaret Fuller (1810–1850)

Nature goes on her way, and all that to us seems an exception is really according to order.
—Johann Wolfgang von Goethe (1749–1832)

In 1982, Paulette Bierzychudek presented a seminal review on population biology of shade-tolerant herbs of temperate deciduous forests. Her focus was largely on life histories, demography, and processes occurring at the population level, inspired, in part, by Harper's (1977) treatise. In her opinion, our knowledge of herbs of particular forests was incomplete, particularly for coniferous forests, floodplains, rocky ledges, forested dunes, savannahs, and subtropical forests. Particular taxa were underrepresented in the literature as well. For example, at the time, no life-history study had yet been done on many shade-tolerant herbs that flower in later summer (Newell and Tramer 1978; Bierzychudek 1982a). Bierzychudek concluded that inadequate information existed for making generalizations about life histories of deciduous forest herbs, particularly their population growth rates or the temporal stability of their population sizes or structures.

Threats to and Conservation of Herbs of Forested Eastern North America

Our ability to understand populations of herbs of forests of eastern North America is complicated by their declining abundances and increasing risk of extinctions, particularly for rare taxa. Major causes for global declines in biodiversity have been attributed to the so-called Four Horsemen of the Environmental Apocalypse or "the evil quartet" (Diamond 1984; Pimm and Gilpin 1989): habitat loss, introductions of alien species, overexploitation, and secondary (or indirect) effects resulting from impacts of other taxa. In addition, global change has the potential of becoming yet another major cause of the decline in biodiversity (Vitousek et al. 1996). I use this framework to first review some major challenges to the conservation of populations of herbs of wooded eastern North America. I then attempt to review our knowledge of the population biology and dynamics of plants of the forest herb layer, relative to their conservation, with particular emphasis on taxa listed as endangered or threatened in Canada or the United States. Using Bierzychudek's (1982a) original paper as a model, I consider the progress we have made since that time in our knowledge of the plant population dynamics of this flora, particularly of listed taxa, and consider some recent approaches to their population biology.

To limit the volume of literature to be reviewed, I follow Bierzychudek's lead and define a member of the herb layer as any vascular herbaceous taxon (forb, grass, grasslike plant, e.g., sedge) growing under a wooded (largely forested or rarely shrubby) overstory. I primarily include the temperate deciduous and boreal coniferous forests east of the Mississippi River and in the associated Canadian provinces. I consider some taxa of subtropical forests and of riverine and savannah ecosystems to a lesser extent, given their treatment in the literature. In keeping with the book's focus on the understory of forests of eastern North America, floating or emergent aquatics and woody vines are excluded. Nor do I include species associated with salt water coastlines, brackish marshes, freshwater lacustrine shorelines, mountain lakes, prairie, rocky outcrops, limestone barrens, tundralike vegetation of boreal Nova Scotia, bogs, fens, calcareous meadows, old-fields, dunes, interdunal slacks, or sandy ridges.

Habitat Loss

Anthropogenic influences are responsible for major loss of habitat for herbs of the forest understory. In addition to direct habitat destruction, human activities in forests and other ecosystems can dramatically alter habitat, including natural disturbance regimes. Although some taxa are dependent on natural disturbance for persistence (e.g., *Aster acuminatus* Michx., Hughes et al. 1988, *Pedicularis furbishiae* S. Wats., Menges 1990), current forest management practices are impacting populations (chapter 14, this volume). Fire-dependent communities such as those of the southeastern coastal plain are affected by human suppression of natural disturbance. Rare habitats are at even greater risk. At least 30 to 60% of xeric vegetation (scrub, scrubby flatwood, southern ridge sandhills) has been lost in some Florida counties since settlement (Peroni and Abrahamson 1985).

Forest herbs may be more at risk than other co-occurring growth forms or life histories; our methodologies may underestimate their extinction risk. Quintana-Ascencio and Menges (1996) used a metapopulation model to estimate colonization and extinction rates of 25 plant species of Florida scrub. Their analyses suggest that herbs suffer higher risk of extinction than do woody plants as patch size decreases.

Some forest herbs may be more sensitive than other taxa to human-induced disturbance, such as logging (Gilliam and Roberts 1995; Meier et al. 1995; chapter 13, this volume). For example, vernal herbs associated with more mesic habitats of the Susquehanna River corridor in Pennsylvania occupied a more limited range of habitats and were less tolerant of disturbance than upland taxa (Bratton et al. 1994). Thus, the optimal sites for spring wildflowers may be most vulnerable, if forest is disturbed, particularly with canopy removal. Intensive human development has nearly extirpated these most sensitive vernal herbs from the Susquehanna tributaries. Bratton et al. (1994) call for preservation of "an appropriate matrix of microhabitats," not only along the riverine corridor, but also at the mouths of creeks.

Habitat Fragmentation

Management practices directly destroy habitat and fragment what remains. Such fragmentation can increase risk of extinction through its impact on small populations and on sensitive phases of the life cycle. Fragmentation requires that conservation efforts be based on a firm knowledge of the dynamics of local populations and of the linkages among them throughout the landscape, particularly if they function as a metapopulation. Studying local-scale processes may provide insights into landscape-level patterns, even in the absence of "an explicitly spatial broad-scale map" (Ives et al. 1998, p. 35). Ives et al. (1998) used a simple simulation model for a perennial plant on a landscape of suitable and unsuitable habitat. They demonstrated that when suitable habitat sites are rare and randomly distributed rather than clumped, small populations can go extinct due to demographic stochasticity.

The impact of fragmentation, specifically edge effects, may be specific to particular phases of the life cycle. For example, edge effects reduced recruitment of *Trillium ovatum* Pursh. of coniferous forests of western North America through impact on some but not all phases of the life cycle. Seed production decreased due to changes in pollination and increased seed predation by rodents. In contrast, flowering phenology, resource limitation of seed set, seed dispersal, germination, herbivory, or survivorship of established plants showed no significant relationship with distance of populations to a forest edge, eliminating them as possible mechanisms of reduced recruitment on population growth (Jules and Rathcke 1999).

Fragmentation can have genetic consequences, both among and within populations. Habitat fragmentation in any part of the species range may impose a potential reproductive bottleneck by reducing genetic diversity, although spatial isolation of plant populations may not lead to losses of all types of genetic variation (Young et al. 1996). Genetic diversity of plants with limited ranges is low; decreases in population size can be associated with reduced diversity (see Godt et al. 1996 for four rare Appalachian perennials of high montane rock outcrops). In *Linnaea borealis* L., extreme reproductive failure in an isolated population was reversed by experimental hand pollination with viable pollen imported from plants from another population. That is not to say, however, that all rare species have limited genetic variability. *Marshallia mohrii* Headle & F. E. Boynton had the greatest total genetic variation of three species in its complex, with greater variation within than among populations (Watson et al. 1991). Godt and Hamrick (1998a) found relatively high genetic diversity for *Sarracenia rubra* ssp. *alabamensis* (Case & Case) D. E. Schnell, suggesting maintenance of its evolutionary potential if population sizes are maintained or increased. In contrast, genetic diversity for *S. oreophila* Wherry, *Helonias bullata* L., and *Schwalbea americana* L. was extremely low (Godt et al. 1995; Godt and Hamrick 1996; 1998b). Management and restoration will need to include reintroduction of compatible mating partners, specifically, translocation of different genotypes from different populations (Montalvo et al. 1997, Wilcock and Jennings 1999). Increasing fragmentation of forest herb habitat will necessitate even greater knowledge of genetic differences among populations.

Corridors have been proposed as ways to restore connectedness among fragments; however, the spatial configuration of remaining habitat may be unable to compensate for overall loss of habitat and fragmentation (Harrison and Bruna 1999). Van Dorp et al. (1997) demonstrated that linear landscape elements are not effective corridors in fragmented landscapes for taxa with limited dispersal, given low migration rates (< 5 m/year), variable availability of high-quality patches, and wide separation between refugia and suitable habitat patches. They argued that seed and propagule transplant, rather than the promise of dispersal through corridors, may be the best way to conserve endangered plants. Similarly, woodland herbs may be infrequent in fencerows and forest edges that can serve as corridors (Fritz and Merriam 1994). Thus, corridors may not provide effective links between forest fragments due to limited dispersal of many forest herbs, particularly rare taxa (Whitney and Foster 1988; Cain et al. 1998).

Alien Species Introductions

Herbs of forested eastern North America are threatened by alien invaders (chapter 12, this volume). Such invasives can alter competitive regimes (chapter 11, this volume), compete with native taxa for regeneration opportunities in response to disturbance (see Horvitz et al. 1998 for tropical hardwood hammocks of southern Florida), and alter environmental conditions, producing impacts at the population, community, and ecosystem levels in forests. *Robinia pseudoacacia* L., black locust, is a nitrogen-fixing tree, invading savannah and woodland areas of Indiana Dunes National Lakeshore. Peloquin and Hiebert (1999) found that shade from black locust trees limited the native herbaceous species. Increased soil nitrogen from the black locust also facilitates a nonindigenous grass, *Bromus tectorum* L. The phenology of invasives relative to that of native taxa also can affect their impact. For example, the exotic shrub *Lonicera maackii* (Rupr.) Maxim reduced survival and fecundity of the native annuals, *Galium aparine* L., *Impatiens pallida* Nutt., and *Pilea pumila* A. Gray, of eastern temperate forests (Gould and Gorchov 2000). This invasive honeysuckle has early season expansion of long-lived leaves that particularly affect those natives that are shade intolerant or photosynthesize only in the early spring. *Agalinis skinneriana* (Alph. Wood) Britton, *Chamaesyce deltoidea* ssp. *deltoidea* (Engleman ex Chapman) Small, *Jacquemontia reclinata* House ex Small, *Sarracenia rubra* spp. *alabamensis*, *Silene polypetala* (Walter) Fernald & Schubert, *Symphyotrichum sericeum* (Vent.) G. L. Nesom, and *Trillium reliquum* J. D. Freeman are some of the threatened and endangered taxa of forests of eastern North America noted to be at risk from invasive plants.

Damage from the introduced aphidlike hemlock woolly adelgid (HWA) from Asia, *Adelges tsugae* Annand, is creating canopy gaps, increasing light availability, and promoting opportunistic understory herbaceous species such as *Erechtites hieracifolia* Rafin. ex DC., *Phytolacca americana* L. and exotics *Ailanthus altissima* Swingle and *Microstegium vimineum* A. Camus (Orwig and Foster 1998). Orwig and Foster (1998) predict dramatic declines in eastern hemlock because of invasion by the HWA, presumably with associated indirect effects resulting in changes in the composition of the understory. Similarly, fast-growing, shade-intolerant forbs have increased rapidly after tornado damage in upland and lowland forests in southeastern Missouri (Peterson and Rebertus 1997). Chapter 13 of this volume provides further examples of changes in the herbaceous layer after disturbance by insects.

Plant invaders such as *Microstegium vimineum* may be met with increasing use of biological control agents for management; the impacts on forest herbs, particularly rare herbs, are largely unknown. McEvoy and Coombs (1999) call for a new approach that deemphasizes top-down control of communities and limits new introductions of biological control agents. Specifically, they promote a combination of herbivore and resource limitation on plant growth, targeting sensitive transitions between plant life-cycle stages, illustrating these approaches with the weedy herb *Senecio jacobaea* L.

Unfortunately, our richest sites for forest herbs may be at greatest risk from alien species (chapter 12, this volume). Although there is some suggestion that areas of low species richness are more readily invaded than areas of high plant-species richness, Stohlgren et al. (1999) concluded that sites high in herbaceous cover, soil fertility, and hot spots of plant biodiversity were prone to alien introductions at many scales (e.g., 1 m^2 and 1000 m^2 plots in grasslands of Colorado, Wyoming, South Dakota and Minnesota). They suggested that this pattern of high risk of alien invasion in hot spots might reflect resource availability rather than species richness. Either way, the herbaceous understory faces substantial threats from invasions by exotic plants and animals.

Overexploitation

Shortsighted harvest of declining populations results in increased risk of extinction, in addition to other indirect effects. Nantel et al. (1996) used previously published transition matrices (Charron and Gagnon 1991) and data from confiscated harvests to compute λ, extinction thresholds, and minimum viable population sizes for American ginseng (*Panax quinquefolium* L.) and wild leek (*Allium tricoccum* Ait.). Illegal harvests of 3600 wild leek bulbs per person and estimates as high as 25,000 ramets total per park have occurred. Even relatively modest harvest of large American ginseng plants in Canada was sufficient to bring population growth rates below equilibrium (Nantel et al. 1996).

Overexploitation of forest canopy resources, notably clearcutting, promotes early successional, shade-intolerant species (chapter 13, this volume). Ground flora diversity declined in cove-hardwoods and mixed-oak hardwoods 16 years after cutting, with increases in *Erechtites, Solidago, Eupatorium, Panicum*, and *Aster* spp. Late successional, shade-tolerant understory taxa such as *Viola, Galium, Sanguinaria, Uvularia*, and *Veratrum* spp. also declined after clearcutting (Elliott et al. 1997). Such a response, however, can be highly site dependent (chapter 13, this volume).

Secondary Effects and Population Regulation

Impacts on native herbs may be mediated by indirect effects on other taxa, e.g., loss of keystone species or alteration of competitive effects through extinction of one population or species, such as predators of generalized herbivores or pollinators. Loss of pollinators through habitat loss, fragmentation, or competitive displacement from species introductions can impact seed set and genetic diversity of forest herbs. Yet some plants with specialized floral morphology may survive outside the range of specialist pollinators with the help of opportunistic flower foragers (e.g., *Corydalis cava* Schweigg. & Kort. in Kansas; Olesen 1996).

Changes in interactions between herbivores and their food plants are a major influence on changes in forest composition in eastern North America (chapter 4, this volume)—notably, extreme browsing by overabundant populations of native herbivores. *Eurbytia divaricata* (L.) G. L. Nesom (= *Aster divaricatus* L.), threatened in Canada and without status in the United States, is preferentially browsed by white-tailed deer (*Odocoileus virginianus* Zimmerman) and is proposed as a possible indicator of browsing intensity (C. E. Williams et al. 2000). Rooney and Dress (1997a, 1997b) observed declines in species richness because of direct and indirect effects of herbivory by white-tailed deer in old-growth hemlock-northern hardwood and riparian zone forests of Pennsylvania. Artificially high densities of herbivores have greater impact on plant species composition and abundance; however, at more typical levels of herbivory, productivity of the forest may play a larger role in determination of understory in boreal forests (John and Turkington 1995).

The impact of herbivores is highly variable and can be secondary to internal factors such as demography, population structure, and population dynamics. For *Lathyrus vernus* Bernh., the effect of external factors, such as seed predation, on population growth rate largely depended on the demographic transition rates in the investigated population (Ehrlén 1996). Some have argued that our understanding of the role of herbivores on population-level processes is inadequate relative to our knowledge at the level of the individual and community (e.g., for subarctic and arctic forests; Mulder 1999).

Global Change

Global change, particularly that mediated by anthropogenic effects (e.g., acid deposition, ozone effects, climate change), will significantly impact the herbaceous forest understory (chapter 3, this volume). Dramatic declines in surface soil pH on the north shore of Long Island, New York, since 1922 are associated with decreasing species diversity and a shift to acid-tolerant flora, such as the common forest herb *Maianthemum canadense* Desf., and *Eurybia divaricata* (=*Aster divaricatus*), listed in Canada but common in the United States (Greller et al. 1990). Forest herbs will undoubtedly differ in their responses to global change. *A. divaricatus* was less sensitive to ozone damage that was *Rudbeckia laciniata* L. in Great Smoky Mountains National Park (L. S. Evans et al. 1996a, 1996b).

Global warming could alter phenology of forest herbs more so than other growth forms of other habitats. Artificial warming of plots in a temperate deciduous forest did not affect relative abundances of 16 common understory herbs. However, this simulation of global change did affect plant densities early in the growing season (e.g., increases in *Maianthemum canadense* and *Uvularia sessilifolia* L). Of 12 herbaceous, 6 shrub, and 8 tree species, herbs were most sensitive to soil warming, with potential influences on carbon and nutrient acquisition dynamics (Farnsworth et al. 1995).

Emphasis on Rare Flora and Conservation

Threats to plant populations and declining biodiversity make studies of the population biology of plants of forested eastern North America timely and critical. Of particular concern are those taxa that are sufficiently rare to have been given legal protection as threatened and endangered species. Regrettably, studies of the population biology of rare woodland taxa are sparse in the published literature. In Bevill and Louda's (1999) review of 38 studies comparing rare and more common plant taxa, herbaceous taxa of wooded habitats of eastern North America were conspicuously absent. They also reported profound inconsistencies in the variables studied; most of the attributes appeared in only one or two studies. There were, however, exceptions that reported on rare woodland congeners of more common taxa, e.g., Banks (1980), Morley (1982), and Pleasants and Wendel (1989) on *Erythronium propullans* A. Gray and *E. albidum* Nutt., Fritz-Sheridan (1988) on *Erythronium grandiflorum* Benth., Mehrhoff (1983) on *Isotria*, Primack (1980) on *Plantago* spp., and Snyder et al. (1994) and Baskin et al. (1997) on *Echinacea tennesseensis* (Beadle) Small, *Iliamna corei* Sherff, and *Solidago shortii* Torr. & A. Gray.

Appendix 5.1 lists 94 endangered and threatened vascular taxa of the forested ecosystems of the eastern United States and Canada, compiled from online sources. I searched Canadian taxa using the mapping tool of Environment Canada—Species at Risk Program (http://www.speciesatrisk.gc.ca/Species/English/SearchRequest.cfm). I used this search engine for all vascular plants of Canada listed as endangered, threatened or vulnerable (special concern), extirpated, and data deficient. Taxa confined to areas west of the Mississippi River (British Columbia, Alberta, Saskatchewan, and parts of Manitoba) were eliminated. The Northwest Territories, Yukon Territory, and Nunavut had no listed taxa. The Canadian taxa in Appendix 5.1 occur in Manitoba, Ontario, Quebec, New Brunswick, Nova Scotia, Newfoundland, and Prince Edward Island. Only herbaceous vascular taxa east of the Mississippi River associated with habitats at least partially dependent on shade from a forested overstory were included.

Vascular plants of forests in the eastern United States were compiled using the U.S. Fish and Wildlife Service Threatened and Endangered Species System (TESS) (http://ecos.fws.gov/webpage/). Information on habit, distribution, and habitat was obtained from the species accounts and/or recovery plans. For the few taxa without accounts online or with data-deficient accounts, habitat information came from Gleason and Cronquist (1991) or the appropriate flora for the southern plants. Only taxa listed as endangered or threatened were used; candidate and proposed plants were not. Special concern taxa were not included, given the variation among U.S. states in the use of this definition and the differences in the number of listed taxa. For example, New Jersey lists no special concern taxa online; however, Connecticut lists 34 and Minnesota lists 133.

Some listed taxa do not occur exclusively in closed canopy forest per se but also occur within forested habitat openings, edges, or shrublands and were included (e.g., riparian woodlands, shaded stream banks, cedar glades, open woodlands, forest-river or forest-prairie ecotones, boggy areas in coniferous forests, rosemary scrub or prairielike oak savannahs). For example, I included *Geocarpon minimum* Mack. of Arkansas and Louisiana despite its restriction to "saline soil prairies" because it also occurs in savannahlike vegetation, with a low density of trees such as short-leaf pine (*Pinus echinata* Miller) and post oak (*Quercus stellata* Wang var. *mississippiensis* (Ashe) Little; U.S. Fish and Wildlife Service 1993a).

I then used two electronic search tools to find literature on any of the 94 listed vascular taxa:) Cambridge Scientific Abstracts (Biological and Medical Sciences for 1960–2000, Biological Sciences [1982–current], Biology Digest [1989-current], and Plant Science [1994–current], and the Institute for Scientific Information Science Citation Index, 1945–2000. I searched for each taxon by its binomial.

In all, 254 articles were retrieved for the 94 taxa (Appendix 5.1). I chose to avoid discussions of the pollination biology of these taxa, given a limited literature, and defer to other important reviews of the genetics of rare plants (e.g., Falk and Holsinger 1991, Glitzendanner and Soltis 2000). I eliminated articles on chromosome counts, vegetation composition and community associates, new accounts, horticulture, physiology, phytochemistry or systematics—just more than 150 articles dealing with the ecology or genetics of these 93 endangered or threatened herbs of forests of eastern North America.

One limitation of this review is the dearth of published literature on these 93 listed forest herbs. Although important information is housed in technical reports and theses, this literature is often not readily available. The problem is further exacerbated by the lack of biological knowledge critical for recovery, notably basic biology and population dynamics; results from my literature review are summarized in Appendix 5.2. While the majority of recovery plans contain excellent information on aspects of distribution and habitat of species of concern, less than half of more than 300 plans reviewed had information on fecundity, survivorship, or age structure (Tear et al. 1995). My attempt at a solution has been to review what is known about aspects of the population biology of more common herbs of forested eastern North America and present our knowledge (or lack thereof) for the listed rare taxa of Appendix 5.1 relative to these paradigms.

Reproduction and Life-History Characteristics

Modes of Reproduction

Despite the extensive use of fast-growing, short-lived weedy taxa in population studies, forest herbs also have contributed significantly to our understanding of plant reproductive strategies and resource allocation. Typically, reproduction of shade-tolerant forest herbs is by vegetative propagules as well as by seed (Bierzychudek 1982a). Recruitment by seed is generally believed to be infrequent and limited (Abrahamson 1980), a pattern also suggested for listed rare taxa of forest systems (Appendix 5.2). This generalization, however, was not supported by Bierzychudek's (1982a) survey. In at least half of the 26 taxa that she listed, the authors reported that sexual reproduction plays a major role in population persistence; successful reproduction by seed was at least sufficiently frequent to balance mortality.

The life histories of forest herbs are varied, but for rare taxa perennials dominate (Appendix 5.2). Annuals are few, accounting for only 9 taxa of the 93 taxa (9.7%); similarly, 5–6% of Wisconsin forests taxa were annuals (Struik 1965, cited by Bierzychudek 1982a). Bratton et al. 1994) noted that K-selected perennial species such as *Trillium flexipes and Dicentra canadensis* were more frequent in mature than in younger forest stands along Susquehanna River in southeast Pennsylvania and northeast Maryland. For the rare taxa of concern (Appendix 5.2), the majority are iteroparous perennials.

Phenology

The rare flora of forested eastern North America is too diverse to allow generalizations about timing of life-cycle events except within similar habits (annual, biennial, or perennial) and habitats (temperate, subtropical, deciduous, coniferous, fire-adapted, etc.). We might generalize that the phenology of shade-tolerant herbs of eastern temperate deciduous forests is typified by leaf loss or formation of flower initials the previous autumn, leaf persistence in winter, leaf emergence in spring, and flowering before or after canopy closure (Bierzychudek 1982a). Yet, even within a group of similar life forms in similar forests, such patterns are modified by extrinsic and intrinsic factors. For example, flowering asynchrony in the common woodland herb *Actaea spicata* L. may be a response to seed predation by a geometrid (Eriksson 1995). For taxa with mixed mating systems, chasmogamous flowers are typically produced in spring, followed by cleistogamous ones later in summer (Bierzychudek 1982a), as is true for federally threatened pigeon wings, *Clitoria fragrans*, of Florida scrub (Appendix 5.2). Seed set in shade-tolerant herbs of eastern forests occurs by mid- to late summer, concurrent with above-ground senescence. Seed germination occurs in autumn of the same year in some cases, but more often the following spring or autumn (e.g., for the common herb *Trientalis borealis*; R. C. Anderson and Loucks 1973; Bierzychudek 1982a).

Another intrinsic factor affecting phenology of forest herbs may be the reproductive status of the individual in the previous growing season. Although few long-term studies provide sufficient data, shade-tolerant perennials may rarely flower in consecutive years (e.g., *Allium tricoccum*; Nault and Gagnon 1993), suggesting a cost to sexual reproduction. However, protracted organ development after preformation is typical of many perennial herbs (Geber et al. 1997a) and may help limit the costs of flowering and fruiting. Reproduction in mayapple (*Podophyllum peltatum* L.) had little effect on branching but had marked effects on shoot type. Branch determination begins the year before and is completed by early spring of the subsequent year and thus does not incur costs to fruit production (Geber et al. 1997a). Geber et al. (1997b) reviewed the taxonomic and ecological distribution of organ preformation in temperate forest herbs using work of Randall (1952) for Wisconsin. These authors found a phylogenetic association of phenological traits of forest herbs, in addition to the strong phylogenetic component of their geographic ranges. They noted that the impacts of timing of development on population dynamics have been largely unexplored.

Longevity

Long-term studies of individuals are understandably rare, confounded by the difficulties of aging herbaceous angiosperms (Werner 1978). The few studies that exist, however, suggest that forest perennials can be relatively long lived. For example, individuals of *Panax quinquefolium* may live for up to 20 years, as revealed from counting bud scale scars (Anderson et al. 1993). A life span of 15–25 years has been estimated for *Arisaema triphyllum* (L.) Schott (Bierzychudek 1982b). Bierzychudek (1982a) calculated the mean life spans of shade-tolerant taxa of the temperate zone to be 14.5 ± 2.72 years (mean ± standard error, n = 18) with sexual maturity reached after several years. Rare taxa of Appendices 5.1 and 5.2 also can be long lived (e.g., Schweinitz's sunflower, *Helianthus schweinitzii* Torr. & A. Gray and some of the orchids), particularly if clonal (e.g., *Sarracenia oreophila* Wherry and *Panax quinquefolium*). Although data are limited, shorter lived rare perennials appear to be those of high light (*Dalea foliosa* (A. Gray) Barneby or fire-associated habitats (e.g., the rare endemics of Florida rosemary scrub).

Although managers may have interest in the temporal persistence of individual plants on the landscape, the effect of longevity on reproductive success is secondary to that of size. For most indeterminate growers, increases in fecundity and thus, reproductive value, are associated with increases in size. Franco and Silvertown (1996) found that for 83 species of perennial plants, there was no expected covariation between longevity (nor age at first reproduction nor life expectancy at sexual maturity) and demographic traits related to reproduction (the intrinsic rate of natural increase, the net reproductive rate, and the average rate of decrease in the intensity of natural selection on fecundity).

Breeding Biology, Mating Systems, and Pollination

It is imprudent to make generalizations about breeding biology and mating systems in herbs of forests of eastern North America. Bierzychudek (1982a) stated with caution, that, in general, most shade-tolerant temperate taxa are bisexual with hermaphroditic flowers. Herbs with partial or complete self-incompatibility are just as frequent as those that are self-compatible (in about half of the taxa listed). However, determinations of mating system are often inaccurate (see Griffin et al. 2000 for the widely distributed *Aquilegia canadensis* L.). For this reason, any blanket statement about mating systems of the 94 listed rare taxa could be spurious. Thorough information on plant breeding systems is largely lacking for many listed taxa, yet most are hermaphroditic (Appendix 5.2), with rare occurrences of monoecy (*Paronychia chartacea* Fernald and *Thalictrum cooleyi* Ahles), dioecy (Britton's beargrass,

Nolina birttoniana Nash), and mixed mating systems in *Oxypolis canbyi* (J. M. Coult. & Rose) Fernald and *Polygonella basiramia* (Small) G. L. Nesom & V. M. Bates. Of interest is the significant number of taxa for which breeding biology is unknown. Of the approximately two dozen taxa of Appendix 5.2 that have been sufficiently studied, autogamy (selfing) dominates. Relatively few of the listed taxa have cleistogamous flowers (e.g., pigeon wings, *Clitoria fragrans*, and *Polygala lewtonii* Small). The significance of these different forms of flowers on the same plant may not always be obvious (e.g., insurance of seed production through selfing of cleistogamous flowers). In *Oxalis acetosella* L., a common perennial herb of forests in Sweden, seeds from cleistogamous flowers were dispersed significantly farther than those from chasmogamous organs. This contradicts the common view that cleistogamous progeny should always be dispersed closer to the mother plant than chasmogamous progeny, although in the case of *Oxalis*, it was unclear whether chasmogamous seeds were outcrossed (Berg 2000).

Pollination of herbs of wooded landscapes is relatively sparse and is largely insect-mediated (Cain et al. 1998). Some herbs of the understory may be pollen limited (e.g., the common woodland herbs, *Hepatica acutiloba* DC., Murphy and Vasseur 1995; *Linnaea borealis*, Wilcock and Jennings 1999); others are distinguished by a high pollination rate and large numbers of seeds (e.g., federally threatened *Isotria medeoloides* Rafin., Mehroff 1983). The more widely distributed congener, *I. verticillata* Rafin., is less likely to mature successful capsules but produces more seeds per genet, given its vigorous clonal reproduction (Mehroff 1983). Our limited knowledge of the breeding biology of rare plants is more critical as we face associated precipitous declines of insect abundance (Cane and Tepedino 2001).

Vegetative and Seed Dispersal

Forest herbs demonstrate a variety of dispersal and germination strategies (Silvertown and Lovett Doust 1993; chapter 1, this volume). Long-distance dispersal can have significant consequences for demography and genetic structure (Silvertown and Lovett Doust 1993; Cain et al. 2000). Spread of forest herbs can be limited by dispersal (Peterken and Game 1984; Matlack 1994b), and many species have no obvious means for long-distance dispersal (Bierzychudek 1982a). Some dispersal is a function of ant–plant mutualisms (Handel et al. 1981), but dispersal via such modes is strictly on a local scale (Cain et al. 1998).

In general, localized dispersal of forest herbs often is < 1 m and rarely > 10 m (Bierzychudek 1982a; Matlack 1994b; Cain et al. 1998; Cheplick 1998). In an exhaustive review, Cain et al. (1998) included 14 herbaceous taxa of wooded eastern North America for which dispersal distances were reported in the literature. For these taxa, the maximum dispersal distances were 33 m for the common bird-dispersed *Phytolacca americana* of damp woods edges and interiors (Hoppes 1988) and 35 m for the ant-dispersed *Asarum canadense* L., also a common understory herb (Cain et al. 1998). Maximum dispersal distance for these select taxa averaged less than 9 m, and typical dispersal distances averaged 1.23 ± 2.814 (± standard error).

Despite limited seed dispersal, most (at least 49 of 94) of the listed perennial taxa (Appendix 5.2) demonstrate some type of vegetative spread, by rhizomes, bulbs, and fragmentation. Recent empirical evidence and modeling suggest that dispersal through clonal spread may be as great or greater than that by sexual propagules (Cain and Damman 1997). Empirical calibration of models for *Asarum canadense* yielded dispersal estimates of 10–11 km during the last 16,000 years (Cain et al. 1998). In actuality, *A. canadense* moved over an order of magnitude longer distances during this time (hundreds of kilometers). Cain et al. (1998) concluded that such occasional events were responsible for long-distance dispersal of this species during the Holocene.

Seed Banks, Dormancy, and Germination

Silvertown and Lovett Doust (1993) remind us that only two things matter to seeds: the risk of death and the chance to germinate. The chance to germinate also risks death. One way to minimize that risk until conditions are favorable for successful germination is persistence as a seed bank. Few dispute the importance of the seed bank in population and community dynamics of forests; however, the specifics of the role of the seed bank for most plant taxa may be unclear or virtually unknown (Pickett and McDonnell 1989). Buried seeds are not necessarily abundant in the soils of coniferous forests (Archibold 1989) but can be extensive in some forest soils. Seed densities are lowest in primary forests (e.g., 100–1000 seeds/m^2 in coniferous or deciduous forests; Olmstead and Curtis 1947; Bicknell 1979) or subarctic pine/birch forests of Canada where no viable seed were present in the soil (Silvertown 1987). Secondary forests of North Carolina housed 1200–3000 seeds/m^2, largely arable weeds and early successional species (Oosting and Humphreys 1940).

Baskin and Baskin (1998) reviewed the germination ecology of hundreds of vascular taxa worldwide, including the herbaceous taxa of moist, warm-temperature woodlands (temperate broad-leaved evergreen forests), temperate deciduous forests, weedy taxa of these habitats as well as nonforest areas within these forest types. Most species demonstrate some dormancy at dispersal, largely physiological dormancy (a physiological inhibiting mechanism prevents radicle emergence by the embryo) or morphophysiological dormancy (underdeveloped embryos with physiological dormancy). Requirements for breaking dormancy differ between life histories. Winter annuals typically require warm stratification (*Corydalis flavula* DC.; Baskin and Baskin 1994) compared to short periods of cold stratification for biennials and perennials. Warmer temperatures averaging 20°C typically are required for germination after dormancy is broken. The light:dark photoperiod requirements are largely unknown for most taxa, but in general, light is required or at least enhances germination. Only 2 of 23 herbaceous species of the deciduous forest with physiological dormancy germinated at higher percentages in darkness.

Individual fitness can be affected by seed dispersal and dormancy through their influence on the timing of germination. Time of dispersal may delay germination or affect the temperatures required for germination, as seen for the widely distributed *Campanula americana* L. (Baskin and Baskin 1984) and *Geum canadense* Jacq. (Baskin and Baskin 1985). Seedlings from nondormant seeds (summer cohort) of the annual *Leavenworthia stylosa* A. Gray in Tennessee cedar glades had lower survival but higher fecundity than those from dormant seeds (autumn cohort), based on differences in their timing of germination (Baskin and Baskin 1972).

Many of the listed rare floras of eastern forests have some seed dormancy; 14 of 19 taxa studied require overwintering or chilling for germination (Appendix 5.2). Seeds produced by adults in recent "good years" will dominate the seed bank (Silvertown and Lovett Doust 1993). Most seeds are buried too deeply to enter the population; thus, most recruitment occurs nearer the surface from recent additions to the seed pool. Rates and magnitude of dispersal into and recruitment from the seed pool remain unknown, although the genetic and population-level implications are profound. The presence of other genotypes in the seed bank increases effective population size, and longer periods spent as dormant seeds in the pool can increase generation time (Templeton and Levin 1979) and produce evolutionary inertia. Seeds dormant in the soil may increase the number of local patches and reduce the colonization required for stable populations. Kalisz and McPeek (1993) showed that seed dormancy can decrease risk of extinction and increase time to extinction in the common woodland annual *Collinsia verna* Nutt.

We are just beginning to understand the details of how environmental factors influence germination, particularly for rare plants. Lambert and Menges (1996) reported higher germination of federally endangered Florida golden aster (*Chrysopsis floridana* Small) in disturbed soils and when litter was removed, suggesting light requirements for germination. *Iliamni corei* Sherff, federally endangered and known from one site in Virginia, has physical dormancy due to impermeability of the seed coat to water. Germination increased from 13 to 71% if seeds were heated to fire temperatures (Baskin and Baskin 1997). Despite significant mortality of seedlings during burns, fire can promote turnover in the seed bank without significant changes in population size and helps maintain genetic variability in *Bonamia grandiflora* Hallier, a rare endemic of upland white sand scrub of central Florida (Harnett and Richardson 1989).

Biotic interactions affect reproduction of forest herbs. Mutualisms between forest herbs and ants can affect dispersal and even influence germination success (Handel 1978; Heithaus 1981; Kjellsson 1985; Casper 1987; Levey and Byrne 1993). Ants disperse 30% of spring-flowering herbs in deciduous forests of eastern North America (Lanza et al. 1992). Their preferences appear to be related to the mass and chemical composition of the elaiosome (e.g., Lanza et al. 1992 on *Trillium* spp.). There is little consistent evidence that removal of the elaiosome by ants enhances germination success. Such enhancement of germination was demonstrated for *Sanguinaria canadensis* L. but not for six other taxa (Culver and Beattie 1980, Lobstein and Rockwood 1993). It is more likely that ants move seeds to a more favorable microsite as demonstrated for *Viola hirta* L., *V. odorata* L., and *Corydalis aurea* Willd. (Culver and Beattie 1980). Germination success of *Carex pedunculata* Willd. can be greater on "nurse logs" in deciduous forests in upstate New York. Ants are attracted to *C. pendunculata* seeds by the presence of an elaiosome. However, ants can discard some seeds and transport the elaiosome to their nests in rotting logs. This ant-mediated dispersal may help seedlings of *C. pedunculata* avoid competition with parent plants and siblings on fallen logs, where this species is particularly common (Handel 1976). Ants have been implicated in dispersal of many plant species at risk in forests: *Carex juniperorum* Catling, Reznicek & Crins, *Chamaesyce deltoidea* spp. *deltoidea*, *C. garberi* Small, *Hexastylis naniflora* H. L. Blomq., *Iris lacustris* Nutt., *Polygala lewtonii* Small, *P. smallii* Smith & Ward, *Sarracenia oreophila* Wherry, *Stylophorum diphyllum* Nutt., *Trillium persistens* W. H. Duncan, and *Viola pedata* L.

Vegetative Growth

In contrast to the abundance of seeds in some forest soils, seedlings rarely are observed above ground (*Aster acuminatus*, Winn and Pitelka 1981; Hughes et al. 1988), although some taxa produce abundant seedlings in some years in some populations, including rare species such as *Arabis perstellata* E. L. Braun (Appendix 5.2). Instead, clonal growth occurs in most forest herbs, as is true of most perennials (Cain et al. 1998). The architecture of clonal growth is influenced by genotype and microsite differences (Bell and Tomlinson 1980) and was suggested to be analogous to foraging, (e.g., increased frequency of branching in areas of high resource availability; Angevine and Handel 1986 for *Clintonia borealis* Rafin.; Cain et al. 1996). The importance of vegetative growth for forest perennials cannot be underestimated. For example, asexual reproduction via ramet production maintains colonies of *Aster acuminatus* under intact canopies (Hughes et al. 1988).

Negative correlations between clonal growth and fecundity (sexual reproduction and seed set) for clonal plants suggest a trade-off between the two modes (Silvertown et al. 1993), although recent empirical and theoretical work challenges previous paradigms. Although larger clones of wild ginger (*Asarum canadense*) outperformed smaller ones (more flowers or lateral shoots per ramet), size had little impact on ramet performance (Cain and Damman 1997). Neither asexual nor sexual reproduction affected future rhizome survivorship or rhizome production. In *Asarum canadense*, experimental removals of flowers resulted in a decrease in rhizome and ramet size (Muir 1995); however, Cain and Damman (1997) found no reduction in probabilities of flowering after production of lateral shoots, thus there was little evidence for a trade-off between vegetative and sexual reproduction. They also found that survivorship of seedlings could equal that of vegetatively produced ramets (Cain and Damman 1997).

Population Dynamics

An understanding of population dynamics is pivotal for understanding plant populations, given that the same abiotic and biotic factors that affect population change also determine population size, age structure, and genetic composition (Watkinson 1986). However, generalizations from empirical and theoretical efforts are complicated by variability among the few taxa that have been studied. Significant variation in demographics (densities, size–age relationships, age-specific mortality, and probabilities of asexual and sexual reproduction) has been observed both within a taxon among sites and between closely related taxa; population sizes and their changes through time also are highly dynamic. Relatively stable population sizes through time have been reported (e.g., *Aster acuminatus*, Hughes et al. 1988); however, more often, most population sizes are quite variable temporally (*Arisaema triphyllum*, Bierzychudek 1982b, 1999; Alvarez-Buylla and Slatkin 1994).

In discussing the lack of generalizations about life histories of forest herbs, Bierzychudek (1982a) noted the need for more study of age-dependent effects (life tables) and stage-dependent effects (life cycles). She pointed to the transition matrix model as a promising approach for studying plant demography. Of the approximately 150 articles on 94 listed rare taxa of forested eastern North America that I surveyed, more dealt with some aspect of population dynamics or structure (n = 38) than any other topic. Here I briefly review recent approaches to the demography and population dynamics, with emphasis on the threatened and endangered species of eastern North America forests. Population structure and changes in size through time of these forest herbs have been studied using new empirical and theoretical approaches, although, again, few taxa have been used.

Demography

As of 1982, few models of population growth had been formulated for size-structured populations (Hartshorn 1975; Werner and Caswell 1977; Caswell and Werner 1978) or vegetative reproduction (Sarukhán and Gadgil 1974; Enright and Ogden 1979). Since that time, a significant number of studies have expanded the use of stage-based life tables and their analysis, including metapopulation dynamics, elasticity analysis, sensitivity analysis, loop analysis, and population viability analysis. A few of these studies have focused on herbs of forested communities of eastern North America.

Franco and Silvertown (1990) reviewed some 529 papers on 585 species to evaluate our knowledge of plant demography. Although species in the Nearctic were best represented, only 73 of 250 taxa (29%) were of forested systems. Semelparous perennials, annual or clonal species, most often trees of the Neotropics, were more likely to have been studied demographically than were iteroparous perennials of eastern North American forests. In their opinion, the quality of studies had improved, yet the number of studies peaked in 1985. They concluded that, numerically, plant demographic studies had not yet adequately sampled the diversity of the plant kingdom.

Size-based Projection Matrices

In 1982, Bierzychudek and others saw great promise in approaches pioneered by Tamm (1948, 1956, 1972a, 1972b)—following the fates of individually marked plants and using mathematical demographic models, such as the Leslie (1945) matrix model, to analyze such data. Matrix projection methods involve censusing mapped plants in permanent plots to construct a matrix, summarizing transition probabilities among life-history stages/classes, and using models to predict future population size, n_{t+1} from the vector $\mathbf{n}_t$ and the matrix $\mathbf{A}$. Such methods usually also require experiments with seeds to determine germination and seedling transitions. Instead of Leslie models, size-based projection matrices (Lefkovitch 1965) have been used for herbs, which are more readily classified by size than by age. These matrices can be useful for studying taxa (e.g., long-lived herbaceous perennials) whose life span and lifetime reproductive success cannot be easily determined (e.g., Morris and Doak 1998 for the alpine cushion plant, *Silene acaulis* L.).

With such models, one can compute population growth rates (λ). At the time of Bierzychudek's (1982a) writing, few studies had quantified the population dynamics of forest herbs to allow predictive inferences. The only population rates of increase computed for herbs of forested North America were for *Arisaema triphyllum* (Bierzychudek 1982b) and *Chamaelirium luteum* A. Gray (Meagher 1978). Since then, although some estimates of population growth rates have been calculated for plants (84% of 95 taxa of many habitats and growth forms in Menges 2000a), few exist for herbs of forested eastern North America, particularly rare taxa. Some estimates of λ suggest proximity to the equilibrium value of 1.0, typical of long-lived perennials and even reported for short-lived species of relatively stable habitats (Nault and Gagnon 1993). Other taxa have periods of exponential growth (e.g., *Danthonia sericea* Nutt., Moloney 1988; *Viola fimbriatula* Nutt., Solbrig et al. 1988), while taxa of variable environments, (e.g., *Collinsia verna*), can increase rapidly one year and decline the next (λ = 1.80–0.41; Kalisz and McPeek 1992).

Models based on transition matrices also can identify vulnerable stages particularly sensitive to selection and present information in a standard way, allowing comparison with other species. These matrices can be extended to forecast population sizes through time, project the fate of populations and species, and help select different management strategies. Similar analyses have evaluated the effects of sexual versus asexual reproduction (Damman and Cain 1998; Piqueras and Klimes 1998), disturbance and succession (Cipollini et al. 1994; Valverde and Silvertown 1995, 1997a, 1997b, 1998; Damman and Cain 1998), and herbivores (Bullock et al. 1994; Bastrenta et al. 1995;

Ehrlén 1995; Hori and Yokoi 1999) on population growth rates. Transition matrices have been used to predict population size through time for species at risk, including impacts from fire (Hawkes and Menges 1995; Gross et al. 1998; Menges and Dolan 1998; Menges and Hawkes 1998), habitat fragmentation (Alvarez Buylla and Slatkin 1994), trampling (Gross et al. 1998), weather (Rose et al. 1998), global climate change (Doak and Morris 1999), harvest and other management practices (Molnar and Bokros 1996; Nantel et al. 1996; Oostermeijer et al. 1996; Silvertown et al. 1996; Degreef et al. 1997, Emery et al. 1999), and other impacts on population growth relative to conservation.

Elasticity Analysis and Loop Analysis

Demographic perturbation analyses (prospective, e.g., elasticity and sensitivity analyses, and retrospective analyses, e.g., variance decomposition; Caswell 2000) are a recent addition to the ecologist's toolbox to help predict how populations change through time. These analyses aim to identify those life stages or transitions during the life cycle that most affect population dynamics (van Goenendael et al. 1994; van Tienderen 2000). These analyses, based on matrix projection models, use the vector **n** for the number of individuals in an age or stage class as well as the matrix entries a_{ij} of the transition matrix **A**, which are the transition rates between stages within a time period. These transitions depend on vital rates such as survival and fecundity (van Tienderen 2000).

Elasticities ($\partial \log \lambda / \partial \log a_{ij}$) are proportional changes in the population growth rate (λ) and thus, fitness (de Kroon et al. 1986), resulting from proportional changes in population vital rates, (e.g., parameters related to survival, growth, and reproduction) presented as matrix entries or elements (de Kroon et al. 1986; Heppell et al. 2000b; Menges 2000b). Elasticities compare the relative effects on population growth rate, with the same relative changes in the values of demographic parameters (de Kroon et al. 2000). Loop analysis extends elasticity analysis and decomposes the population growth rate into the contributions made by life-cycle pathways (van Groenendael et al. 1994; de Kroon et al. 2000).

In contrast, sensitivities ($\partial \lambda / \partial a_{ij}$) are mathematically and biologically distinct from elasticities (Caswell 1978; de Kroon et al. 2000; van Tienderen 2000). Sensitivity estimates the absolute effects of changes in demographic parameters on λ (de Kroon et al. 1986; Horvitz et al. 1997), and such estimates are "unsuitable for comparing the contributions of transitions within the matrix to population growth rates" (de Kroon et al. 2000, p. 609).

These relative contributions to population growth can be grouped in biologically meaningful ways, such as growth, reproduction, and stasis, to compare across taxa (Silvertown et al. 1993, 1996) and can provide general categorizations of populations based on their responses during different phases of the life cycle (Heppell et al. 2000a). The parameters reflect the relative contributions of the matrix elements or transitions to the population growth rate and thus the importance of certain life stages and their associated demographic rates for management and research (Heppell et al. 2000b; de Kroon et al. 2000).

These measures of changes in matrix elements, representing phases in the life cycle, and their responses to selection and demography are often applied to the management and conservation of declining populations (Silvertown et al. 1996; Heppell et al. 2000b). Recent theoretical work has emphasized that although elasticity analysis is limited, it is robust to some violations of its underlying assumptions and is a simple first step in answering important questions in evolutionary and population ecology, conservation biology, and management (Benton and Grant 1999; Grant and Benton 2000).

Elasticity and sensitivity analyses have been presented for dozens of plant taxa (see summaries in Silvertown et al. 1993; Pfister 1998), often weeds of open habitats or woody plants. A promising number of long-lived herbs associated with wooded habitats of eastern North America, however, have sufficient demographic data to allow construction of matrices and either elasticity or sensitivity analyses of the elements: *Allium tricoccum* of rich woods (Nault and Gagnon 1993), *Arisaema triphyllum* of rich, moist woods and boggy areas (Bierzychudek 1982b), *Chamaelirum luteum* of moist woods and bogs (Meagher 1982), *Danthonia sericea* of dry, sandy soil, and pine woods, chiefly on the coastal plain (Moloney 1988), *Podophyllum peltatum* of moist, open woods (Sohn and Policansky 1977; Rust and Roth 1981), *Viola fimbriatula* of dry to moist open woods, clearings and meadows, rarely along streams (Solbrig et al. 1988) and the only listed taxon from Appendices 5.1 and 5.2, *Panax quinquefolium* of rich woods (Charron and Gagnon 1991; Nantel et al. 1996), and *Pedicularis furbishiae* (Menges 1990; de Kroon et al. 2000). Silvertown et al. (1993) evaluated the relative importance of survival, growth, and fecundity to the population growth rates of woody and herbaceous perennials by comparing matrix projection models for 45 herbs and 21 woody taxa using elasticity analysis. They asked how different components of the plant life cycle contributed to population growth rate in plants with widely different life histories by summing elasticities associated with different transitions during the life cycle. Although Silvertown et al. (1993) included *Danthonia sericea* and *Pedicularis furbishiae* with herbs of open habitats, their aggregate elasticity values centered them well within the ranges for iteroparous herbs. Of 66 herbs, iteroparous taxa of forests were associated with their "survival–growth axis," reflecting a range of survivorship and growth elasticities, but limited elasticities of fecundity (Silvertown et al. 1993).

As of yet, too few elasticity analyses have been performed to allow strong inference about those phases of the life history of forest herbs that most affect population processes; although, again, the role of vegetative modes should not be underestimated. Although all seven taxa of forests listed above exhibited stable or increasing population sizes ($\lambda = 0.996 - 1.484$, mean $\pm$ standard error $= 1.1351 \pm 0.0648$), the elasticities were highly variable among taxa (Silvertown et al. 1993). The most variable transitions contributing to the life cycle were those involved in sexual reproduction (seed recruitment, seedling recruitment, and fecundity) and retrogression (decrease in size or reversion from flowering to vegetative or dormant phase). Sexual recruitment (e.g., a seed bank, germination, early seedling success) can play significant roles in long-term population growth rates, particularly for populations in temporally variable environments (Moloney 1988; Kalisz and McPeek 1992). Yet other analyses suggest that vegetative modes play major roles in population growth, although few projection matrices include both sexual and asexual reproduction. Growth and survival have been reported to affect population dynamics more than fecundity in common forest taxa (e.g., *Allium tricoccum*, Nault and Gagnon 1993; *Danthonia sericea*, Moloney 1988; *Potentilla anserina* L., Eriksson 1988) and some rare taxa of Appendices 5.1 and 5.2. Elasticities associated with growth and survival can differ among populations, in time and space, such as in more open patches of the forest overstory for the European *Primula vulgaris* Huds. (Valverde and Silvertown 1998) and between southern and northern populations of *Helianthus divaricatus* L. (Nantel and Gagnon 1999).

Despite these analyses of those phases of the life cycle most likely to affect λ, some initial evidence suggests that variable life-history stages contribute little to population growth rates. In an analysis from 17 published field studies of 30 populations of plants and animals, Pfister (1998) was unable to find evidence of that highly variable matrix terms had large effects on population growth rates. Sensitivities or elasticities for survivorship or growth always exceeded those of fecundity, suggesting that the former parameters may have larger effects on λ. Pfister (1998) concluded that natural selection might alter life histories to minimize stages with both high sensitivity and high variation.

The use of elasticities is not without controversy; their summation for interpretation of life-history theory has been debated (Silvertown et al. 1993; Franco and Silvertown 1994; Shea et al. 1994; Franco and Silvertown 1996; Oostermeijer et al. 1996). Mills et al. (1999), using both hypothetical data as well as empirical data from animals, demonstrated that population growth rates from matrices are sensitive to the extremes of the vital rates. They cautioned against use of mean values for several matrices through time and interpretation of the largest elasticities for conservation decisions. Instead, analysis of the range of variation for different demographic rates has been encouraged (Oostermeijer et al. 1996; Pfister 1998; Mills et al. 1999).

Metapopulation Dynamics

A *metapopulation* was defined by Levins (1970) as a population of populations; local dynamics and the regional processes of migration, extinction and colonization produce the ecology and genetics of populations (Husband and Barrett 1996). Such metapopulations are systems in which colonizations and extinctions occur constantly; the dynamics of such populations are a promising area for integration of genetics and population ecology (Olivieri et al. 1990). Marginal sink populations may be resurrected by propagules colonizing from high-density source populations. The relative magnitudes of extinction and colonization rates and their balance may determine the limit of a species' distribution (Silvertown and Lovett Doust 1993). Thus, to understand the population biology of rare forest herbs, we must expand our study of demography and genetics of plant populations to a landscape scale. To do so requires estimation of patch occupancy and extinction and colonization rates, both spatially and temporally (Husband and Barrett 1996), all facets of a metapopulation approach.

Metapopulation dynamics, when defined broadly as the product of local population dynamics and dispersal, may be a feature of all species (Husband and Barrett 1996). Although plants appear appropriate for metapopulation analyses (e.g., immobility, strong spatial structure, restricted dispersal), few studies adopt this approach, especially theoretical work, due to the difficulty of measuring the critical parameters of extinction, colonization, and migration (Husband and Barrett 1996). Metapopulation theory may be most applicable to those plants that meet Levins's (1970) original assumptions: (1) local populations are numerous and similar in size, (2) local extinctions are frequent, and (3) dispersal is limited and random (Gilpin and Hanski 1991; Harrison 1991; Husband and Barrett 1996).

A metapopulation approach, using measures of site occupancy, recruitment, and extinction, provides a useful framework for community-wide surveys of species in declining habitats (Husband and Barrett 1996). Metapopulation dynamics are a promising approach for some species, but one that has not yet been adopted widely. Empirical data on plants that can be used to directly address the predictions of metapopulation dynamics are relatively scarce. Of 18 studies that addressed population interactions from a regional perspective (Husband and Barrett 1996), bias was demonstrated toward annuals in ephemeral environments, and only 3 used herbs of forest or wooded ecotones (*Collinsia verna*, Kalisz and McPeek 1993; *Pedicularis furbishae*, Menges 1990; and *Plantago cordata* Lam., Meagher et al. 1978). Despite the virtues of working with annuals, demographic data sufficient for transition matrices or metapopulation analyses are available for few of the listed forest annuals in Appendices 5.1 and 5.2. A notable exception is work on the rare flora of the Lake Wales Ridge of Florida (Quintana-Ascencio and Menges 1996; M.E.K. Evans et al. 2000). Two other annuals, *Agalinis skinneriana* and *A. gattingeri* Small in Britton & A. Br., listed as endangered in Canada, may be sufficiently abundant in the United States for extensive population study.

Husband and Barrett (1996) concluded that few studies have tested theoretical models of metapopulation dynamics or addressed whether this regional perspective adds anything to our understanding of population dynamics; they noted that a gulf exists between theoretical and empirical as well as between genetic and demographic approaches to the study of plant metapopulations. In their review, few studies examined patch structure in detail; however, of those that did, local populations were clumped rather than uniform as assumed by standard metapopulation models. Many studies did evaluate the importance of migration, colonization, and extinction, but few measured it directly.

Metapopulation dynamics can offer some insights in spite of the spatial and temporal variation in species abundances. Tractable models for metapopulation analysis assume patches are identical; recent models are beginning to evaluate changes in patch size and position (Day and Possingham 1995). Studies of isolated plants or those in unevenly distributed patches, over the full range of spatial scales of their occurrence, are needed, as is the development of more spatially explicit models (Czaran and Bartha 1992; Perry and Gonzalez-Andujar 1993; Perry 1998; Winkler et al. 1999). Eriksson and Kiviniemi (1999) have developed a method for estimating extinction thresholds for metapopulations based on site occupancy (from inventories) and recruitment (and site suitability, based on sowing experiments) in Scandinavian grasslands. Eight of 18 selected taxa were determined to live below what they called the extinction threshold, that is, "the fraction of remaining suitable habitats at which a species becomes extinct" (p. 319). They also calculated a "quasi-equilibrium extinction threshold" assuming a delay in population decline and suggest that three additional taxa may experience further decline in Scandinavia in the near future.

If herbs of the forest understory do function as metapopulations, the genetic consequences of this population structure are yet undetermined. Again, the genetic theory for spatially structured populations may still be significantly ahead of our knowledge of the actual processes in the field (Silvertown 1991; Silvertown and Lovett Doust 1993). McCauley et al. (1995) suggested that metapopulation dynamics might enhance genetic structure for the weed *Silene alba* (Mill.) E.H.L. Krause. This genetic structuring, coupled with limited dispersal, may further isolate fragmented populations in the herbaceous understory.

Understanding dispersal among patches, patch quality, and patch dispersion is critical for appropriate metapopulation analyses in general and for forest herbs in particular. Eriksson (1997) offered that a metapopulation approach can help explain patterns of abundance in landscapes, specifically, that species abundance reflects colonizing ability, with seed size as a key trait for colonizing ability. A single population that acts as a source may be significant for species persistence. Although Eriksson did not work on woodland herbs per se, his work on population recruitment with three species of Asteraceae (*Antennaria dioica* Gaertn., *Hieracium pilosella* L. and *Hypochoeris maculata* L.) in a dry, seminatural grassland indicated that recruitment was limited by seed availability, dependent on seed size, and was promoted by small disturbances. Eriksson concluded that colonization influences abundance patterns in grasslands, a conclusion that undoubtedly can be extended to the forested understory.

Variation in size and spatial arrangement of patches influences population dynamics and persistence of metapopulations. Although the original models by Levins (1970) assumed that local populations within a metapopulation were identical, patches differ in age, size, and insularity (Gilpin 1991; Hanski 1991; Lande 1993; Olivieri et al. 1995). Analytically tractable models of metapopulations assume all patches are identical, but recent empirical and theoretical developments suggest local heterogeneity can impact metapopulation growth and species persistence. Site ruggedness (local spatial heterogeneity) decreased abundances of two host ant species and an endangered butterfly, based on a field-based spatial simulation model of four interacting species (including larval food plant of the caterpillars). Persistence of the butterfly depended on the arrangement of habitat quality at a local, finer-spatial scale (Clarke et al. 1997). Valverde and Silvertown (1997b, 1998) demonstrated that variation among populations influences metapopulation dynamics. In their study, changing environmental conditions during canopy closure influenced demography of the common woodland herb, *Primula vulgaris*; notably, fecundities and population growth rates were higher in patches under an open canopy with more availabile light.

Population Viability Analysis and Conservation Biology

Threats to populations, particularly small populations, include demographic stochasticity, environmental stochasticity, systematic trends (e.g., continuing habitat loss), and catastrophes (Menges 2000b). Managers must often estimate how long a rare plant species can persist in face of these threats. In its narrowest sense, population viability analysis (PVA) attempts to determine the critical population size needed given an acceptable level of risk of extinction for a determined period. PVA uses empirical data on the entire life cycle of a wild population and quantitative modeling to project future populations (e.g., the finite rate of increase [λ], extinction probability, time to extinction, or future population size or structure [Menges 2000b]).

Most PVAs for plants are based on stage- or size-class matrices (Menges 2000b), in contrast to projections based on age (Werner and Caswell 1977). Age-based parameters can be estimated from stage-based matrices (Cochrane and Ellner 1992; Menges 2000b). Some studies integrate both stage and age approaches (Morris and Doak 1998). Simulations can be based on stage-based parameters and used to estimate age-related parameters (Damman and Cain 1998).

Population viability analysis can be applied to conservation biology using population persistence and/or extinction risk estimated from empirical data and modeling (Menges 2000b). Menges (2000a, 2000b) reviewed the use of PVAs in plants, using a broad definition of PVA, specifically use of quantitative modeling to predict future population increase, probability of or time to extinction, or future population size or structure, based on empirical data of the entire plant life cycle. Most published plant PVA studies deal with one species, using about three populations on average and data collected for about 4 years (Menges 2000a, 2000b). Only 9 herbs that occur in forested communities of North America (of 99 taxa) have been used in what Menges

(2000a) called "notable PVA studies" (p. 73), using his broad definition for a PVA of inclusion of empirical data on the entire life cycle and using modeling to project future populations (Menges 2000a). These forest herbs are: *Allium tricoccum* (Nault and Gagnon 1993; Nantel et al. 1996), *Arisaema triphyllum* (Bierzychudek 1982b), *Asarum canadense* (Damman and Cain 1998), *Chamaelirium luteum* (Meagher 1982), *Cynoglossum virginianum* L. (Cipollini et al. 1993), *Collinsia verna* (Kalisz and McPeek 1992), the barely woody evergreen *Linnaea borealis* (Eriksson 1992), *Panax quinquefolium* (Nantel et al. 1996), *Pedicularis furbishiae* (Menges 1990), and *Viola fimbriatula* Small (= *V. sagittata* Ait., Solbrig et al. 1988). PVA can be used to estimate a minimum viable population, the smallest population size that will persist given some maximum level of risk of extinction. Few studies, however, actually compute minimum viable populations; a notable exception is Nantel et al. (1996) for wild leek (*Allium tricoccum*) and American ginseng (*Panax quinquefolium*).

Quantitative modeling of population sizes using metapopulation dynamics and PVA is underutilized for forest herbs of eastern North America. Such modeling can provide insight into various management practices, such as interactions with herbivores (Ehrlén 1995 for the European woodland herb, *Lathyrus vernus*), fire management (Menges and Dolan 1998 for *Silene regia* Sims of midwestern prairies; Enright et al. 1998 for the shrub, *Banksia attenuata* R. Br., of southwestern Australia), or harvest (Nantel et al. 1996 for *Allium tricoccum* and *Panax quinquefolium*).

Limitations of Models and New Approaches

The finite population growth rate, λ, calculated from a projection matrix is only validly applied to a population with a stable age or size distribution. If age or size distributions are not stable, we cannot use these population growth rates to predict future population sizes (Bierzychudek 1982a, 1999; Coulson et al. 2001). Thus, predictions of long-term population growth that use the Lefkovitch transition matrix model depend on the assumption that demographic parameters do not vary through time (Caswell 1978). Studies based on time-variant population demography (stochastic matrices) hold promise for some (Moloney 1988), but the theory is not extensive (Tuljapurkar 1984) and may be beyond the reach of most empirical work.

Some size categories may be underrepresented in populations or simply not obvious for many plant species. Easterling and colleagues (Easterling 1998; Easterling et al. 2000) have developed an integral projection model that incorporates the virtues of matrix projection models (simple structure yet can provide estimates of λ, stable distribution, reproductive value, and sensitivities of λ to changes in life-history parameters), yet does not require division of individuals into discrete size classes. Classes can be defined by a continuous variable, such as measures of length or mass. They used their model with the federally endangered northern monkshood (*Aconitum noveboracense* A. Gray), with stem diameter as the class variable. The traditional matrix model and their integral projection model produced similar estimates of population growth rate. The integral projection model produced a "sensitivity surface" that differed from the sensitivities and elasticities of the matrix model for *Aconitum*. This difference resulted, in part, in response to choice of boundaries between size or stage classes in the matrix model. Easterling et al. (2000) cautioned that size-dependent growth and survival functions must be defined. The authors stated that matrix models for discrete-state, discrete-time scenarios are easiest to fit. Again, the application of promising theory is contingent upon the strength of the supporting empirical data.

Conclusions

Bierzychudek (1982a) posed the following questions for future research: (1) what factors regulate population sizes of forest herbs, (2) how stable are population sizes of forest herbs, (3) how much site-to-site variation occurs in population behavior? Although these may no longer be the salient questions, it is useful to evaluate the progress we have made in answering these questions for forest herbs and rare taxa, in particular.

What Factors Regulate Population Sizes of Forest Herbs?

Most recovery plans present information on critical habitat for flowering adults (Schemske et al. 1994), yet fewer than half include information on population regulation (Tear et al. 1995). We need better understanding of those factors that regulate population sizes of rare plants, particularly of critical life-cycle stages, relative to anthropogenic change. Such factors include both intrinsic (internal) and extrinsic (external) as well as abiotic and biotic factors. For common forest herbs, intrinsic factors such as reproduction and pollen limitation, singly and in combination (Ågren and Willson 1992; Lundberg and Ingvarsson 1998) affect population sizes, as does demography, including that of the endangered *Panax quinquefolium* (R. C. Anderson et al. 1993). Abiotic and biotic extrinsic factors reported to regulate typical forest herbs include light (*Aster acuminatus*, Ashmun and Pitelka 1984; *Primula vulgaris*, Valverde and Silvertown 1998; *Uvularia perfoliata* L., Wijesinghe and Whigham 1997), nutrients (Brach and Raynal 1992 for *Oxalis acetosella* and *Lycopodium lucidulum* Michx.; Salomonson et al. 1994 for *Actaea spicata* and *Geranium sylvaticum* L.; Coombes and Grubb 2000 for a diversity of forest types) and disturbance (Clark 1991). Our ability to understand population regulation of rare herbs is encouraged by studies demonstrating that fire is a critical variable for rare taxa of Florida rosemary scrub (Hartnett and Richardson 1989; Quintana-Ascencio and Morales-Hernandez 1997; Menges and Hawkes 1998; Kirkman et al. 1998. Biotic factors that need study include how pollinators (Ingvarsson and Lundberg 1995; Kearns et al. 1998), herbivores (Crawley 1989), pathogens (e.g., *Danthonia spicata* Beauv. ex Roem. & Schult., Clay 1984) and endophytes (Carroll 1995; D. Smith 2000) regulate herb populations. We also need to improve our understanding of how the relationship between plants and soil affects population dynamics (Bever 1994; Bever et al. 1997, Watkinson 1998). Knowledge of population regulation is particularly needed for critical phases of the life cycle, such as the seed bank (chilling, time, pathogens, predators), germination (temperature, light, water), and those factors that stimulate episodic recruitment.

How Stable Are Population Sizes of Forest Herbs?

Determinations of population stability are made most often for annuals, short-lived herbs, or for trees. Changes in population sizes for multiple generations are more readily obtained for species with shorter life cycles. Long-lived taxa, such as trees, are easier to age, and all stages of the life cycle may be more readily obtained for larger, longer-lived taxa. As a result, estimates of population sizes through time are rare for longer-lived forest herb, particularly for most of the listed threatened and endangered taxa. The recovery plans of many taxa note that populations are highly variable among years (e.g., *Isotria medeoloides*, U.S. Fish and Wildlife Service 1993b). When available, growth rates of plant populations are highly variable among years and sites (Valverde and Silvertown 1998). Silvertown et al. (1993) presented the instantaneous (finite) rate of increase (λ) for 45 herbs, 7 of which are associated with wooded habitats of eastern North America. Although it is specious to generalize from these few studies, all but one value exceeded zero, indicating populations of all species were increasing (mean $\pm$ SE = 0.1179 $\pm$ 0.0531). Similarly, r_m = 0.12 for *Asarum canadense* (Cain et al. 1998). However, trajectories are highly variable among populations and, by definition, negative for a population at risk (e.g., $r = -0.0040$ for *Panax quinquefolium*; Charron and Gagnon 1991).

How Much Site-to-Site Variation Occurs in Population Behavior?

Although some authors have tackled the issue of variation in dynamics among populations, again, forest herbs of the eastern United States are underrepresented. Temporal and spatial variation in vital rates can be important (Valverde and Silvertown 1998), but we do not yet know how important (Coulson et al. 2001). Nantel and Gagnon (1999) used λ, confidence intervals, and elasticity values combined with log-linear analyses to evaluate how demographic variability limits the distribution of *Helianthus divaricatus* L. Their analyses suggest that variability in demography can increase extinction probabilities of local populations at their distribution limits. Presumably, such variability would have equally critical impacts on rare plants, at similar distributions limits in terms of small population size or habitat.

We need to expand our attempts to understand how environmental variation, in time and space, affects population dynamics. Megamatrix analysis holds promise and may offer new insights, despite limited use to date. In this approach, analysis of matrices from different environments can reverse predictions on the effects of different life-history stages on population growth constructed from single-environment matrix analysis (Pascarella and Horvitz 1998 for the tropical understory shrub, *Ardisia escallonioides* Cham. & Schltdl.). Such promising analyses, however, can only aid our understanding of population stability and site-to-site differences for rare plant species of forests of eastern North America if supported with sufficient empirical data.

Coda: Current Challenges and Opportunities

Since Bierzychudek's review in 1982, we have made progress in answering some of the questions she posed. Yet, we still are challenged by lack of sufficient data, the variation among populations, and our ability to model the dynamics of the herbaceous taxa of the forests of eastern North America.

The questions Bierzychudek posed in 1982 about variation in population sizes and factors affecting them can be extended for conservation of rare taxa and incorporation of new approaches in plant population biology. Schemske et al. (1994) said that the critical question for conservation biologists is, is this taxon declining, and why? Regrettably, the information needed to answer this question is lacking for most rare forest herbs, as is basic biological information on the recovery plans for the majority of rare plant and animal taxa (Tear et al. 1995). Schemske et al. (1994) evaluated 98 U.S. Fish and Wildlife Service recovery plans available for the period 1980–1992. Detailed demographic information was available for only 33 % of the species; however, autecological information was reported for 94 %. Most species did have information on number of populations, total number of individuals, geographic range, ecological requirements, and causes of endangerment. The research recommended was largely ecological in focus (e.g., factors limiting distributions, pollination biology, and autecological information). Schemske et al. (1994) concluded that rarely was the request for specific research motivated by demographic importance of one or more life-history stages.

Although one might expect there to be little extensive data on population dynamics of listed plant taxa, given their rarity, the similar paucity of large-scale demographic data on even more common herbaceous plants of forested North America in total is disappointing, yet understandable. The data sets for those few species, whether common or rare, are the result of commitments of individual researchers to long-term, extensive data sets for these taxa (e.g., *Arisaema triphyllum*, Bierzychudek 1982b, 1999); *Asarum canadense*, Damman and Cain 1998; *Panax quinquefolium*, Nantel et al. 1996; *Pedicularis furbishiae*, Menges 1990).

Plant Biology, Time, and Space: Empirical and Theoretical Challenges

Strong empirical and theoretical knowledge on demography can aid management and conservation both ecologically and logistically. Menges and Gordon (1996) suggest three different sampling and analysis methods for rare plants species, based on sound demographics and a solid understanding of limited management resources. In some cases, mapped distributions and presence/absence may be sufficient. At the other extreme are detailed quantitative assessments of abundance and analyses of population trends and hypotheses about demographic mechanisms.

Empirical challenges to working with plants include the difficulties in obtaining quantitative measures for certain phases of the life cycle, such as dormancy, the seed bank, clonal integration and growth, periodic recruitment, dispersal, local adaptation, spatial structure, and dispersion of individuals and their affects on metapopulation growth and reproduction (Husband and Barrett 1996, Menges 2000b). Such challenges need to be addressed with larger scale, long-term studies. For example, the challenge of dormancy needs to be addressed by long-term studies to avoid inflating mortality estimates or experiments (Kalisz and McPeek 1992).

Studies also need to be longer-term for accurate population projection; the typical study used for population viability analysis averaged just longer than 4 years (Menges 2000b.) Study of herbs of forested eastern North America, whether rare or abundant, needs to be longer term, given the relatively long-lived nature of the perennials and the spatial and temporal variation among populations. This time frame is well within the longevity of academics and of some conservation professionals, although it is typically beyond the time scale of a single budgetary cycle, thesis, dissertation, or grant.

A broader view of how populations vary through time is needed. Menges (2000b) encouraged modeling of disturbance cycles, environmental stochasticity and their effects on demographic parameters, projected growth rates, and extinction risk. Stochastic growth rates are slower than those from the average projection matrix and thus are more conservative (Menges 2000b). Menges (2000b) cautioned, however, that such stochastic analyses, given lack of data, assume that elements within matrices are correlated neither with each other nor through time. His suggested solution was empirical: collecting data to quantify these correlations, both among stages and among populations.

The theory as well as the empiricism present challenges. Formulation and use of models continue to be hindered by focus on short-lived colonizing species, the short-term nature of most studies, and few populations monitored. Seed dormancy, restricted dispersal, and local adaptation need to be incorporated into existing theoretical models for more accurate depiction of plant metapopulations (Husband and Barrett 1996; Cain et al. 1998). Husband and Barrett (1996) note that although most metapopulations models assume random disperal from local populations, local adaptations occur in plants, and migrants are not necessarily equally well adapted to new sink populations (Bennington and McGraw 1995; Cabin 1996; Holt and Gomulkiewicz 1997; Lonn et al. 1996). Most plant propagules dispersal locally (Harper 1977; Bierzychudek 1982a; Matlack 1994b; Cain et al. 1998; Cheplik 1998), and, as a result, distributions may be clumped rather than uniform on the landscape, complicating yet another assumption of metapopulation dynamics. Occasional events such as long-distance dispersal of woodland herbs during the Holocene (Cain et al. 1998) may have important implications for the structure of metapopulation models. Often seedling recruitment rates are modeled as constants; yet, episodic seedling recruitment is characteristic of many taxa and can be modeled (Menges 2000b; Menges and Dolan 1998 for *Silene regia* of midwestern prairies). Separate analyses are preferred for ramet and genets, which often have different demographies, as found for *Asarum canadense* (Damman and Cain 1998).

More studies need to adopt a metapopulation approach of Husband and Barrett (1996), in a broad sense, wherein populations are considered products of local dynamics, larger scale regional processes of migration, extinction and colonization, and the ecology and genetics of these populations of populations. Particularly for rare taxa, forest herb populations need to be approached in their entirety, as possibly integrated subpopulations. Husband and Barrett (1996) pointed out the need to ask basic questions such as, what fraction of available habitats are occupied, and what is the threshold number of habitats below which extinction exceeds colonization and species go extinct? Although such basic questions may be critical for effective management, they remain unanswered for virtually all rare plant species. The virtue of working with rare forests herbs is the possible greater likelihood of being able to study all populations compared to more widely distributed, abundant taxa, although the variation and linkages among populations may still be difficult to determine.

Yet another challenge to the application of metapopulation dynamics of herbs of the forest understory is the need for "a broad range of sound, well-developed and relatively error-free packages for simulating metapopulation viability" (Lindenmayer et al. 1995, p. 161). Lindenmayer et al. (1995) compared three computer packages and assessed their usefulness based on the assumptions and limitations of the program, ranging from availability and user friendliness to mathematical characteristics. All packages were structured differently and differed in their modeling of processes such as environmental and demographic variation. This resulted in disparate results from different programs, even for the same population. The widespread availability of user-friendly software for population analyses is a double-edged sword. As argued above, use of population projection matrices and their extensions (elasticity and sensitivity analysis) hold promise for management of rare populations; however, these must be interpreted with caution and supplemented with a sound knowledge of the species' ecology (Silvertown et al. 1996; Bierzychudek 1999).

We also need to incorporate explicit spatial structure of individuals within metapopulations and investigate how that dispersion affects growth and reproduction (Husband and Barrett 1996). In short, population biologists need to scale up from individual- and population-level inferences to landscape-level inferences. More accurate spatial data can aid documentation of spatial patterns. Accuracy at scales of less than a meter for location and environmental parameters is now available with geographic positioning systems and some types of remotely sensed data (e.g., LIDAR, light detection and ranging data for elevational topographies within 15 cm accuracy). Use of geographic information systems (GIS) can aid linkages among populations at the landscape scale, both empirically and theoretically, and have been applied to habitat evaluation and population models of rare species (Mann et al. 1999; Wu and Smeins 2000). For example, Sperduto and Congalton (1996) derived predictive models for the federally endangered small-whorled pogonia (*Isotria medeoloides*) using habitat variables obtained from field observation and extracted from a GIS. Clark (1991) modeled the relationship among age structure, density, and disturbance regimes at a variety of scales, from individual treefalls to the landscape level. Ultimately, we can extend our understanding of spatial patterns and temporal processes from individuals to population-, community- and ecosystem-level dynamics (Czaran and Bartha 1992; Tilman 1996, 1999).

We are called upon to use transition matrices, metapopulation dynamics, and spatially explicit modeling of rare plant populations, although judiciously. Bierzychudek (1999) revisited her earlier estimates of population growth for *Arisaema triphyllum* 15 years later. Her reworked analysis reversed earlier predictions of a population increase at one site. She cautions that any estimates based on transition matrices can be invalidated by inaccurate probabilities based on too few plants, too few years, specious assumptions of a stable age distribution, and failure to account for density-dependent vital rates. Although the theory can drive our empiricism, the tests of the models are only as good as our data.

Genetic Considerations, Demographic Information, and Conservation Biology

Demography and genetics are necessary complements for both population biology and conservation biology. Although not reviewed here, we have observed an increased contribution by genetics to our understanding of plant population dynamics in forests of eastern North America (Schemske et al. 1994; Cain et al. 2000), despite challenges with choosing a marker and its analysis (Parker et al. 1998). Useful demographic information can be extracted from genetic marker data (e.g., plant mating systems, inbreeding depression, effective population size, and metapopulation structure). Milligan et al. (1994) suggested expansion of genetic marker data beyond characterization of gene frequencies and patterns of diversity to aid conservation efforts. In their opinion, developing "genealogical-based analytical methods coupled with studies of DNA sequence variation within and among populations is likely to yield the most information on demographic processes from genetic marker data" (p. 423). Genetic diversity plays a critical role in population persistence, particularly for increasingly rare taxa (Wilcock and Jennings 1999 for *Linnaea borealis*). Not just levels of genetic variation, but how it is partitioned, can confound effects of rarity (Gitzendanner and Soltis 2000). This understanding has important implications for the success of restoration efforts (Montalvo et al. 1997; Robichaux et al. 1997) and our appropriate use of ex-situ conservation of native woodland herbs (Hamilton 1994).

Genetic knowledge must complement rather than subordinate sound demography and population biology. Genetic consequences of rarity may contribute to endangerment; however, demographic factors often are more important (Lande 1988). Schemske et al. (1994) documented that demographic research was proposed for 84% of federally endangered rare plants with recovery plans, but for only 17% of 98 was the use of projection matrices suggested. In contrast, research in genetics was suggested for only 26% of the species (e.g., information on genetic variation and development of a model to determine effective population size). A common recommendation was for knowledge of genetic variation within and between populations, without any statement of the application of this knowledge in recovery. In their words, "Demographic criteria are probably far better indicators of the biological status of rare species . . . than . . . genetic variation, whose connection to population viability has been theoretically . . . but not empirically, demonstrated" (Schemske et al. 1994, p. 594).

Call to Arms: Experimentation, Restoration, and Determination of Declines

In addition to more extensive empiricism, development of theory and knowledge of genetics, promise lies in carefully designed experimental manipulations in the field to assess influence of different factors on forest herbs, particularly ones that are rare (Handel 1978; Bierzychudek 1982a; Ashmun and Pitelka 1985; Schemske et al. 1994; Bevill and Louda 1999). Eriksson and Ehrlén (1992) demonstrated seed and microsite limitation in 9 of 14 woodland species. They later experimentally tested this hypothesis using transplants of propagules for seven taxa in Sweden and demonstrated the importance of disperal limitation in structuring forest herb communities (Ehrlén and Eriksson 2000). Such experimental approaches are even more challenging for rare plants with protected status, for which critical knowledge of basic biology, ecology, and propagation is lacking. For those with protected status, the regulatory process may make experimentation a bureaucratic hurdle few of us care to surmount. Yet many rare taxa are extremely tractable in culture (e.g., *Justicia cooleyi* Monach. & Leonard, *Panax quinquefolia, Solidago shortii*), information that is a result of or a contribution to restoration efforts. Current challenges posed by rare forest herbs today may be no greater than those posed by more common taxa during earlier days of plant population biology. "Mortality of transplants may be high, experiments may need to be large and may have high variance. . . . The student of natural populations should beware of taking refuge from such experimental uncertainty and instead learn to measure it" (Antonovics and Primack 1982, p. 73). Regrettably, we may have less time to meet those challenges for rare taxa.

Developing this knowledge for propagation and experimentation is even more critical given that preservation efforts will fail, and we will be forced to attempt to restore taxa and their habitats. The same information needed to validate models and design experiments using rare species also is needed for their preservation and recovery. Schemske et al. (1994) advocate a three-factor framework for recovery: (1) determine the biological status of a species, (2) identify the life-history stages critical to population growth, and (3) determine the underlying biological causes of variation in critical life-history stages. Similarly, Montalvo et al. (1997) presented five research areas of importance for restoration biology: (1) the role of abundance and genetic variation on colonization, establishment, growth, and evolutionary potential, (2) the influence of local adaptation and life-history traits in the success of restored populations, (3) the role of spatial arrangement of landscape elements on metapopulation dynamics and population processes such as migration, (4) the effects of genetic drift, gene flow, and selection on population persistence, and (5) the influence of interspecific interactions on population dynamics and community development. These questions and challenges posed for the conservation of rare plants are strikingly similar to the most basic ecological questions we ask of more common taxa.

Two decades later, I would still agree with Bierzychudek (1982a) that variation among populations of forest herbs precludes generalizations about their life histories, growth rates, and stability of population sizes or structures, despite Goethe's encouragement in the introductory quotation to this chapter. We still lack sufficient information, particularly for rare forest herbs, to embrace Margaret Fuller's view and simply admit we have sampled the exceptions to the rule(s). Despite lack of generalizations, evidence from forest herbs, even the rare ones, has challenged previous paradigms about trade-offs between reproductive modes and growth, including the relative primacy of sexual reproduction for population maintenance and dispersal. These taxa may not be widely dispersed or abundant but have demonstrated the dangers of underestimating the roles and significance of occasional events, vegetative spread, and temporal and spatial variation in population size and recruitment.

For those of us who might regard our progress in the past two decades since Bierzychudek's (1982a) review as limited, consider K.E. Holsinger's (pers. comm.) perspective on the difference between population biology and conservation biology.

> Population biologists choose a particular species for study, at least in part, because they think that the species they have chosen will allow them to address broad, general issues of conceptual importance in population biology. Conservation biologists have the species chosen for them by circumstances—the circumstances of endangerment. One consequence of this difference is that it may be much more difficult for conservation biologists to get complete demographic

information. Population biologists choose species that allow them to get the information they need. Conservation biologists have to figure out how to get the information they need from often-recalcitrant species. Population biologists are generally satisfied with discovering the factors that limit population size, population growth rate, or species distribution. Conservation biologists use that information to project the fate of populations/species and to decide among management strategies.

In a time of unprecedented loss of biodiversity, the dichotomy between population and conservation biologist becomes dangerously artificial and semantic. We are all called upon to be both.

Acknowledgments Several colleagues made invaluable contributions, including information on the rare species for Appendix 5.2: Paul Martin Brown and Lawrence Zettler (Orchidaceae), Kay Kirkman (*Schwalbea americana*), Eric S. Menges (Florida scrub endemics), Anton A. Reznicek (Cyperaceae), and Jeffrey L. Walck (*Echinacea tennesseensis* and *Solidago shortii*). Sincere yet insufficient thanks for patience, helpful comments, insightful discussion, and unflagging support are extended to Paulette Bierzychudek, Hal J. Daniel, Karl Faser, Frank Gilliam, Carol Goodwillie, Eric Menges, Ronald J. Newton, Steve Norton, Mark Roberts, and to other researchers whose efforts make reviews possible, as well as those students, colleagues, friends, and family who thrive on benign neglect during the process.

Appendix 5.1. Endangered (E) or threatened (T) herbaceous vascular plants of the forested communities of eastern North America.

Binomial	Common name	Family	Status	Range	Habitat	Threats
Aconitum noveboracense A. Gray	Northern wild monkshood	Ranunculaceae	T	IA, NY, OH, WI	Cold woods	Habitat loss/degradation; contamination and filling of sinkholes, grazing, trampling, logging, highway and power line maintenance, pesticides, quarrying, road building, collection
Agalinis gattingeri Small in Britton & A. Br.	Gattinger's agalinis	Scrophulariaceae	E	(Canada) ON	Dry, open woodlands; glades	Habitat loss, human encroachment, increase in soil moisture
Agalinis skinneriana (Alph. Wood) Britton	Skinner's agalinis	Scrophulariaceae	E	(Canada) ON	Open woods; rocky, open glades	Habitat loss, human disturbances, changes in water levels, invasive, changes in spp. composition from fire suppression
Aletris farinosa L.	Colicroot	Liliaceae	T	(Canada) ON	Edge of forests	Agriculture, development
Apios priceana B. L. Rob.	Price's potato-bean	Fabaceae	T	AL, IL, KY, MS, TN	Forest openings	Cattle grazing and trampling, clearcutting, roadside herbicides, lack of disturbance
Arabis perstellata E. L. Braun	Braun's rock-cress	Brassicaceae	E	KY, TN	Wooded, steep slopes; limestone outcrops	Habitat loss/modification, weedy competitors, trampling, logging, road work
Arabis serotina E. S. Steele	Shale barren rock-cress	Brassicaceae	E	VA, WV	Mid-Appalachian shale barrens	Degradation, drought, stochasticity, herbivory (deer, insects), trampling by goats, sheep
Asplenium scolopendrium var. *americanum* (Fernald) Kartesz & Gandhi (= *Phyllitis s.* var. *americana* Fernald = *P. j.* var. *a.* (Fernald) A. Löve and D. Löve)	American hart's-tongue fern	Aspleniaceae	T	AL, MI, NY, TN, Canada (ON)	Deep shade, limestone outcrops	Habitat alteration, trampling, development, logging, quarrying, infestations of defoliating insects, collecting (commercial trade)
Astragalus bibullatus Barneby & Bridges	Pyne's ground-plum	Fabaceae	E	TN	Cedar glades	Development, shade from competitors, woody plants, grazing, roadside maintenance, off-road vehicles (ORVs), dumping, drought

Astragalus robbinsii var. *jesupi* Eggl. & E. Sheld.	Jesup's milk-vetch	Fabaceae	E	NH, VT	Calcareous bed outcrops, ice scour with alders, willow, elms on stream banks	Habitat alteration (dams), collecting
Baptisia arachnifera W. H. Duncan	Hairy rattleweed	Fabaceae	E	GA	Pine woods, mixed woods	Clearcutting, reforestation
Bonamia grandiflora Hallier	Florida bonamia	Convolvulaceae	T	FL	Sand pine scrub, evergreen oaks, and sand pine (*Pinus clausa*)	Habitat destruction for development, trash dumping, invasion by exotics and weeds, ORVs, fire suppression
Cardamine micranthera Rollins	Streambank bittercress	Brassicaceae	E	NC, VA	Stream banks, moist woods near streams	Scarcity of populations, small population sizes, agriculture, development, impoundment, channelization, flooding, drought, invasives
Carex juniperorum Catling, Reznicek & Crins	Juniper sedge	Cyperaceae	E	Canada (ON)	Cedar glades (alvars)	Quarrying (alvar), grazing, fragmentation, development, utility corridors, ORVs, dumping, exotics
Carex lupuliformis Sartw. ex Dewey	False hop sedge	Cyperaceae	E	Canada (ON, QC)	Grassland/scrubland areas of maple-hickory forest; swamps; marshes	Canopy closure, climate, development, agricultural drainage, water level control
Cereus eriophorus var. *fragrans* (Small) L. D. Benson	Fragrant prickly-apple	Cactaceae	E	FL	Sand pine (*Pinus clausa*) scrub	Habitat loss; fragmentation, windthrow, development, ORVs, collection
Chamaesyce deltoidea (Engleman ex Chapman) Small ssp. *deltoidea* (= *C.d.* (Small) D.G. Burch *var. adhaerens*, = *C. adhaerens* (Small) R.C.H.M. Oudjans, = *Euphorbia d.* var. *adhaerens* (Small) R.C.H.M. Oudejans)	Deltoid spurge	Euphorbiaceae	E	FL	Pine rocklands	Development, fire suppression, invasive plants, overcollecting

(continued)

Appendix 5.1. *Continued*

Binomial	Common name	Family	Status	Range	Habitat	Threats
Chamaesyce garberi Small (= *Euphorbia g.* Engelm.)	Garber's spurge	Euphorbiaceae	T	FL	Transitions between hardwood hammocks and rock pinelands	Habitat loss, development, storms, fire supression
Chimaphila maculata Pursh	Spotted wintergreen	Pyrolaceae	E	(Canada) ON	Dry-mesic oak-pine woodland	Habitat loss, ORVs
Chrysopsis floridana Small (= *Heterotheca floridana* (Small) R. W. Long)	Florida golden aster	Asteraceae	E	FL	Open areas of sand pine-evergreen oak scrub, beach dunes	Habitat loss, development, mowing, dumping, grazing, ORVs
Clitoria fragrans Small	Pigeon wings	Fabaceae	T	FL	Scrub, turkey oak barrens, edges of high pine (sandhills)	Habitat loss to citrus groves and development, fire suppression and canopy closure
Crotalaria avonensis Delaney & Wunderlin	Avon Park harebells	Fabaceae	E	FL	Scrub, possibly high pine, turkey oak barrens	Habitat loss to development, road-side stabilization, dumping, ORVs, herbivory
Dalea foliosa (A. Gray) Barneby (= *Petalostemum foliosa* A. Gray)	Leafy prairie-clover	Fabaceae	E	AL, IL, TN	Prairie cedar glades, limestone barrens, full sun	Habitat loss, drought
Drosera filiformis Rafin.	Thread-leaved sundew	Droseraceae	E	(Canada) NS	Swamps	Habitat loss through drainage of bogs; human activities
Echinacea laevigata S. F. Blake	Smooth coneflower	Asteraceae	E	GA, NC, SC, VA	Open woods, cedar barrens	Collecting, development, encroachment by woody vegetation, road construction and maintenance
Echinacea tennesseensis (Beadle) Small	Tennessee purple coneflower	Asteraceae	E	TN	Cedar glades	Development
Eriogonum longifolium var. *gnaphalifolium* Gand.	Scrub buckwheat	Polygonaceae	T	FL	Scrub and high pine (sandhills)	Development; inadequate or wrongly timed fire

Eryngium cuneifolium Small	Snakeroot	Apiaceae	E	FL	Sand pine-evergreen oak scrub, FL rosemary scrub	Fire suppression, development, ORVs, trampling
Erythronium propullans A. Gray	Minnesota dwarf trout lily	Liliaceae	E	MN	Rich deciduous/mixed woods	Development, logging, expanded agriculture, conversion of floodplains to cropland, disturbance of uplands
Euphorbia telephioides Chapm.	Telephus spurge	Euphorbiaceae	T	FL	Savannah	Habitat loss, development, silvicultural practices, fire timing, roadside maintenance
Eurybia divaricata (L.) G. L. Nesom (= *Aster divaricatus* L.)	White wood aster	Asteraceae	T	Canada (ON, QC)	Deciduous forests	Habitat loss, insect grazing, garlic-mustard
Galactia smallii A. Herndon	Small's milkpea	Fabaceae	E	FL	Rock pineland	Habitat loss, fire suppression, invasion by exotic plants, overcollecting
Geocarpon minimum Mack.	(No common name)	Caryophyllaceae	T	AR, LA, MO	Sandstone glades	Succession, fire suppression?, forestry practices, ORVs, grazing (but can clear competitors)
Helianthus eggertii Small	Eggert's sunflower	Asteraceae	T	AL, KY, TN	Rolling uplands	Habitat alternation and loss; development, succession, conversion to pasture/croplands, herbicide use along roadsides
Helianthus schweinitzii Torr. & A. Gray	Schweinitz's sunflower	Asteraceae	E	NC, SC	Clearings and edges of moist woods	Roadside maintenance, mowing, lack of disturbance (fire, native grazers), development, invasion by exotics
Helonias bullata L.	Swamp pink	Orchidaceae	T	NJ, DE, MD, VA, NC, SC, GA	Swampy forested wetland	Habitat loss to development, off-site water withdrawal, discharged, siltation, nutrient addition; succession, collecting, limited genetic variability; limited dispersal, survival, growth rate, flowering
Hexastylis naniflora H. L. Blomq.	Dwarf-flowered heartleaf	Aristolochiaceae	T	NC, SC	Bluffs	Timber harvesting, development, conversion of woodland to pasture, reservoir/pond construction, dumping, insecticide use
Hydrastis canadensis L.	Goldenseal	Ranunculaceae	T	Canada (ON)	Moist areas of deciduous woods and floodplain forests	Habitat loss (timber production, development, expansion of agriculture), harvesting, lack of disturbance (fire, flooding, wildlife)
Hypericum cumulicola (Small) W. P. Adams	Highlands scrub hypericum	Hypericaceae	E	FL	FL rosemary scrub, scrubby flatwoods	Fire suppression, development, ORVs, trampling

(*continued*)

Appendix 5.1. *Continued*

Binomial	Common name	Family	Status	Range	Habitat	Threats
Iliamna corei Sherff	Peter's mountain mallow	Malvaceae	E	VA	Mixed forest	Fire suppression, collecting, vandalism, competition with weedy species, and predation by insects and feral goats
Iris lacustris Nutt.	Dwarf lake iris	Iridaceae	T	MI, WI, Canada (ON)	Transition between boreal forest and lakeside	Development, road widening, chemical spraying and salting, ORVs; collection
Isotria medeoloides Rafin.	Small-whorled pogonia	Orchidaceae	T (E, Canada)	CT, DE, GA, IL, MA, ME, MI, NC	Open deciduous woods, damp mixed woods	Collection, trampling, loss of habitat, ATVs
Isotria verticillata Rafin.	Large-whorled pogonia	Orchidaceae	E	Canada (ON)	Rich, moist deciduous woods	Habitat loss, beaver activity (flooding), logging, trampling, collection
Jacquemontia reclinata House ex Small	Beach jacquemontia	Convolvulaceae	E	FL	Disturbed or sunny tropical maritime hammocks	Habitat loss, urbanization, invasive plants, mowing, herbicides, park maintenance
Justicia cooleyi Monach. & Leonard	Cooley's water-willow	Acanthaceae	E	FL	Hardwood forests, wet hammocks, swamps	Habitat loss, urbanization, conversion of forested habitats to pasture, silviculture, limestone mining, exotic plant species, roadside maintenance, alterations in hydrology
Lesquerella lyrata Rollins	Lyrate bladderpod	Brassicaceae	T	AL	Cedar glades	Small population sizes, herbicide use, road improvement, quarrying, development
Liatris ohlingerae B. L. Rob.	Scrub blazingstar	Asteraceae	E	FL	FL scrub, scrubby flatwoods	Habitat loss to development, conversion to citrus groves and agriculture, dumping
Liparis liliifolia A. Rich. ex Lindl.	Purple twayblade	Orchidaceae	E	(Canada) ON	Open oak savannah, secondary successional deciduous/mixed forest	Habitat loss due to control of natural disturbances and conversion to agriculture, pesticides, collection

Lupinus aridorum McFarlin ex Beckner	Scrub lupine	Fabaceae	E	FL	Sand pine scrub	Habitat loss, development, road construction/maintenance, conversion to pastureland, traffic, collection
Lysimachia asperulaefolia Poir.	Rough-leaved loosestrife	Primulaceae	E	NC, SC	Long-leaf pine-pocosin ecotone	Fire supression, drainage, development
Macbridea alba Chapm.	White birds-in-a-nest	Lamiaceae	T	FL	Savannah with long-leaf pine and/or runner oaks (mesic flatwoods)	Habitat loss, development, silvicultural practices, timing of fire
Marshallia mohrii Beadle & F. E. Boynton	Mohr's Barbara button	Asteraceae	T	AL	Moist, prairielike openings in woodlands	Agricultural development, road maintenance/expansion, conversion to pasture/agriculture, drainage, seeding with forage grasses, woody succession
Mimulus glabratus var. *michiganensis* (Pennell) Fassett	Michigan monkey-flower	Scrophulariaceae	E	MI	Eastern white cedar (*Thuja occidentalis*) openings	Habitat loss, alteration of water regimes and temperatures
Nolina brittoniana Nash	Britton's beargrass	Agavaceae	E	FL	FL scrub, high pine, hammocks	Habitat loss
Oxypolis canbyi (J. M. Coult. & Rose) Fernald	Canby's dropwort	Apiaceae	E	GA, MD, NC, SC	Wet pineland savannahs	Loss/aleration of wetland habitat, lowering water tables, ditching, draining, road construction, swallowtail herbivory?, unnecessary collecting
Panax quinquefolium L.	American ginseng	Araliaceae	E	Canada (ON, QC)	Rich, moist, mature deciduous woods	Small population size, habitat loss/degradtion, logging, overharvesting
Paronychia chartacea Fernald (= *Nyachia pulvinata* Small)	Papery whitlow-wort	Caryophyllaceae	T	FL	FL scrub, scrubby flatwoods	Fire suppression, development, ORVs, trampling
Pedicularis furbishiae S. Wats.	Furbish's lousewort	Scrophulariaceae	E	ME, Canada (NB)	Where mixed forest and river meet; shaded	Habitat loss, ample habitat is not being colonized naturally, dumping, litter, gravel pit operations, forestry practices

(continued)

Appendix 5.1. *Continued*

Binomial	Common name	Family	Status	Range	Habitat	Threats
Pilosocereus robinii (Lem.) Byles & G. D. Rowley	Key tree cactus	Cactaceae	E	FL, Cuba	Treelike cactus, tropical hammock	Habitat destruction from development, unique habitat specificity, small population sizes
Pinguicula ionantha R. K. Godfrey	Godfrey's butterwort	Droseraceae	T	FL	Depressions in pine flatwoods	Habitat degradation due to lack of prescribed burning, shading by planted pines
Plantago cordata Lam.	Heart-leaved plantain	Plantaginaceae	E	Canada (ON)	Streams running through heavily wooded areas, silver maple (*Acer saccharinum*) swamps	Habitat destruction/alteration, water quality (eutrophication, siltation), needs shade from wooded buffer, conversion to agriculture, alteration of natural streams, cattle grazing/trampling
Polemonium van-bruntiae Britton	van Brunt's Jacob's ladder	Polemoniaceae	T	Canada (QC)	Riparian meadows, swamps, partially shaded with seasonal flooding	Habitat loss to development, road/residential construction (development), agriculture, plant succession, ORVs, alterations that produce permanent flooding
Polygala lewtonii Small	Lewton's polygala	Polygalaceae	E	FL	FL scrub, high pine (sandhills)	Habitat loss to development, citrus groves, pastures, fire suppression, ORVs
Polygala smallii Smith & Ward	Tiny polygala	Polygalaceae	E	FL	Open areas in pine rocklands	Habitat loss, roadside maintenance, development, fire suppression, invasion by exotics
Polygonella basiramia (Small) G. L. Nesom & V. M. Bates (= *P. ciliata* var. *b.* (Small) Horton)	Wireweed	Polygonaceae	E	FL	FL scrub, scrubby flatwoods	Development, ORVs, trampling, fire suppression
Pycnanthemum incanum Michx.	Hoary mountain mint	Lamiaceae	E	Canada (ON)	Open, dry, sandy-clay habitats in open deciduous woods	Habitat loss, invasion by shrubs, shoreline erosion and slumping
Sarracenia oreophila Wherry	Green pitcher-plant	Sarraceniaceae	E	AL, GA, NC	Mixed oak or pine flatwoods	Collection for commerical sale, development, fire suppression

Sarracenia rubra ssp. *alabamensis* (Case & Case) D. E. Schnell (= *S. a.* ssp. *wherryi* Case & Case)	Alabama canebrake pitcher-plant	Sarraceniaceae	E	AL	Swamps	Agricultural development, fire suppression and growth of woody competitors, roadside herbicide use, water table alteration, invasive species
Schwalbea americana L.	American chaffseed	Scrophulariaceae	E	FL, GA, LA, MS, NC, NJ, SC	Open, moist pine flatwoods	Fire suppression, development, agricultural and forestry practices
Scirpus ancistrochaetus Schuyler	Northeastern bulrush	Cyperaceae	E	MA, MD, NH, PA, VA, VT, WV	Forested wetlands	Requires fluctuating water levels
Scutellaria floridana Chapm.	Florida skullcap	Scrophulariaceae	T	FL	Scrubby oak vegetation	Development, silvicultural practices, timing of burns (wildfires), woody plant encroachment, grazing, small population size
Scutellaria montana Chapm.	Large-flowered skullcap	Scrophulariaceae	E	GA, TN	Canopy of mature hardwoods	Development, silvicultural practices, timing of burns (wildfires), woody plant encroachment, grazing, small population size
Silene polypetala (Walter) Fernald & Schubert	Fringed campion	Caryophyllaceae	E	FL, GA	Deciduous hardwoods, ravines	Residential development, logging, invasives (*Lonicera japonica, Lygodium japonicum*)
Sisyrinchium dichotomum Bickn.	White irisette	Iridaceae	E	NC, SC	Clearings and edges of upland woods	Roadside/road maintenance, residential development, fire suppression, elimination of large native grazers, aggressive exotics (*Pueria lobata, Lonicera japonica, Microstegium vimineum*)
Solidago shortii Torr. & A. Gray	Short's goldenrod	Asteraceae	E	KY	Cedar glades, oak-hickory openings	Restricted distribution, limited numbers, development, inadvertent trampling, habitat alteration, overcollecting, destructive fires, fire suppression
Solidago speciosa var. *rigidiuscula* Torr. & A. Gray	Showy goldenrod	Asteraceae	E	Canada (ON)	Prairielike areas under oak canopy	Habitat destruction through agricultural and residential development
Spigelia gentianoides Chapm. ex A. DC.	Gentian pinkroot	Loganiaceae	E	AL, FL	Mixed pine-hardwood	Habitat destruction and alteration by forestry practices, proximity to recreational activities, overcollecting

(continued)

Appendix 5.1. *Continued*

Binomial	Common name	Family	Status	Range	Habitat	Threats
Stylophorum diphyllum Nutt.	Wood-poppy	Papaveraceae	E	Canada (ON)	Rich, damp, wooded calcareous soils, ravines, bluffs	Habitat destruction, forest clearing
Symphyotrichum anticostense (Fernald) G. L. Nesom	Anticosti aster	Asteraceae	T	(Canada) NS, QC	Banks of fast rivers through boreal forest	Habitat destruction, ORVs, development, regulation of water levels
Symphyotrichum sericeum (Vent., G. L. Nesom)	Western silver-leaved aster	Asteraceae	T	Canada (MN, ON)	Open oak savannahs	Habitat loss to development, recreation, quarrying, fire suppression, pasture enhancement, haying, invasion of grasslands by aliens and woody taxa; 30% seed loss to weevils
Tephrosia virginiana Pers.	Virgina goat's-rue	Fabaceae	E	Canada (ON)	Open oak and pine woods on ridges	Fire suppression and resultant natural canopy growth, roadside maintenance (sand removal, herbicides), weevils
Thalictrum cooleyi H. E. Ahles	Cooley's meadowrue	Ranunculaceae	E	FL, NC	Savannahs, woodland clearings	Fire suppression, agricultural development, small populations, roadside maintenance, logging, low genetic variability?
Thelypteris pilosa var. *alabamensis* Crawford (= *Leptogramma p.* var. *a* (Crawford) Wherry)	Alabama streak-sorus fern	Aspleniaceae	T	AL	Rock surfaces, crevices along streams; cove-type hemlock-hardwoods	Bridge construction, dam construction and resultant flooding, logging interrupts natural seepage, loss of canopy, vandalism, limited range, small population sizes
Trichophorum planifolium (Spreng.) Palla.	Bashful bulrush	Cyperaceae	E	Canada (ON)	Open canopied deciduous/mixed forests	Habitat loss, urbanization, erosion control, trampling, shading by woody taxa
Trifolium stoloniferum Muhl.	Running buffalo clover	Fabaceae	E	AR, IN, KY, MO, OH, WV	Open forest-prairie ecotones	Disappearance of large herbivores in woodlands
Trillium flexipes Rafin.	Drooping trillium	Liliaceae	E	Canada (ON)	Mature deciduous woodlands	Habitat destruction, collecting, disease, deer grazing

Trillium persistens W. H. Duncan	Persistent trillium	Liliaceae	E	GA, SC	Deciduous or mixed woods of ravines or gorges	Limited range and distribution, overcollecting, logging
Trillium reliquum J. D. Freeman	Relict trillium	Liliaceae	E	AL, GA, SC	Undisturbed hardwoods	Logging, road construction, agricultural conservation, development, urbanization, stone quarrying, invasives (*Lonicera japonica*, *Pueraria lobata*)
Triphora trianthophora Rydb.	Nodding pogonia	Orchidaceae	E	Canada (ON)	Rich, humid, deciduous woodlands	Trampling, soil compaction, collection, storm damage to humus layer, grazing, logging
Viola pedata L.	Bird's-foot violet	Violaceae	T	Canada (ON)	Savannahs with deciduous forests	Habitat loss due to agriculture, competition from woody taxa, fire suppression, mowing, herbicides
Warea amplexifolia Small	Wide-leaf warea	Brassicaceae	E	FL	Dry long-leaf pine (*Pinus palustris*) and scrub	Habitat loss, development (citrus, residential), collecting, vulnerable to disturbance given low numbers and annual life cycle, limited genetic variability
Warea carteri Small	Carter's mustard	Brassicaceae	E	FL	High pine (sandhills), scrubby flatwoods	Fire suppression, development, ORVs, trampling, fire suppression
Woodsia obtusa (Spreng.) Torr.	Blunt-lobed woodsia	Aspleniaceae	E	Canada (ON, QC)	Sugar maple (*Acer saccharum*) forests, extensive rock outcroppings	At northern limit of its range in Canada, habitat alteration from human recreation and trash dumping
Xyris tennesseensis Kral	Tennessee yellow-eyed grass	Xyridaceae	E	TN, AL, GA	Thinly wooded habitat, seep-slopes, springy meadows or gravelly shallows of small streams	Habitat loss, agricultural development, logging, quarrying, road construction, woody plant encroachment (natural succession), diversion of seeps or ground water

Appendix 5.2. Summary of biological characteristics of 94 threatened or endangered herbs of forest habitats of eastern North America from recovery plans, species accounts and the published literature

Binomial	Habit	Phenology	Longevity	Vegetative reproduction	Mating system	Pollination	Dispersal	Seed	Germination	Seed dormancy
Aconitum noveboracense[a]	P	Ju–Se		Tubers		Bumblebees				
Agalinis gattingeri[a]	A	Au–Se; Se–Oc		None		Insects	Wind?			
Agalinis skinneriana[a]	A	Au–Se; Se–Oc		None?						
Aletris farinosa[a]	P	Ju–Jl		Rhizome buds		Bumblebees, beeflies	Wind		Many small	
Apios priceana[a]	P?									
Arabis perstellata	P	Mr–My; My–Ju	Veg. for 2 yr; 5 yr		Unknown	Probably insects; unknown	Wind, gravity		Seedlings can be abundant	Seed bank?
Arabis serotina	facultative B	Jl–frost, then dies		Rhizomes?		Unknown; one Syrphidae		12–730 (14,000)		Seed bank?
Phyllitis scolopendrium	P			None					Spores on cool, calcareous soils	
Astragalus bibullatus[a]	P	Ap–My; My–Ju								
Astragalus robbinsii var. jesupi	P	My–Ju; Ju–Jl				*Bombus* sp.			Attempts low; mortality high	
Baptisia arachnifera[a]	P	Ju–Jl								

Bonamia grandiflora	P vine	Fire dependent		Taproot; connections below ground						Seed banks
Cardamine micranthera[a]	A	Ap–My	< 1 yr	None		Ants?				
Carex juniperorum[a]	P	Ap–My; My–Ju	> 10 yr	Densely caespitose	Mostly selfing?	Wind	Ants			
Carex lupuliformis[a]	P?	Ju–Oc; Jl–No	Indefinite?	Rhizomes?		Wind	Flooding, at least in large part			
Cereus eriophorus var. *fragrans*	P	Ap–Se (2 peaks); My and Se		fragmentation		Night blooming	Birds, rodents, tortoises?	15,000/fruit; persistent fruit (8 mo)		
Chamaesyce deltoidea ssp. *deltoidea*	P	Mr–De (var. *adhaerens*); My–No (var. *deltoidea*)	long lived from root system		Monoecious; highly variable	Bees, flies, ants, wasps?	Explosively dehiscent? Ants?			
Chamaesyce garberi	P		Short lived			bees, flies, ants, wasps?	Explosively dehiscent? Ants?			
Chimaphila maculata[a]	P	Jl–Au		Shoots off larger stems		Insects	Wind			
Chrysopsis floridana	P	Late No–De; De onward	Short lived	Rosettes at end of rhizomes?					Bare sand for germination	
Clitoria fragrans	P	CH: My–Ju; CL: late Se; fire stimulated	Long lived	fire stimulated or clonal	CH, CL; inverted anthers and stigma	CH: insects				

(continued)

Appendix 5.2. *Continued*

Binomial	Habit	Phenology	Longevity	Vegetative reproduction	Mating system	Pollination	Dispersal	Seed	Germination	Seed dormancy
Crotalaria avonensis	P	Mid-Mr–Ju		Substantial taproot; resprouts post-fire				18/legume		
Dalea foliosa	P	Jl–Au; Oc	Short lived	None	Strongly exserted anthers					
Drosera filiformis[a]	P	mid-Jl				Insects		8 capsules/plant; 70 seeds/capsule		
Echinacea laevigata	P	My–Jl; late Ju–Se		Rhizomes	Poor reproductive success	Butterflies, moths, beetles, bees	Birds, small mammals, wind dispersed < 3 m			
Echinacea tennesseensis[a]	P	mid-My–Oc; Se–De (spring)	At least 6 yr	None	Obligately outcrossed	*Bombus*, *Apis*, *Junonia*, *Colias*, *Pieris* spp.	Ineffective; limits colonization	Unknown	Nearly equal in light and dark	Cold stratification required
Eriogonum longifolium var. *gnaphalifolium*	P	My–mid-Oc	Unknown; high post-fire survival	Woody taproot		Hymenoptera spp.	Seedlings observed close to parent		In summer in open sand	
Eryngium cuneifolium	P	Au–Oc	Short lived	Very weak resprouter after fire	Mixed (self-compatible)	Generalist insects	Limited, gravity	20–30 per fruit	Winter and spring if rainy	Persistent seed bank
Erythronium propullans[a]	P			Bulb offshoots				Rare		

Euphorbia telephioides	P	Ap–Jl		Stout storage root	Cyathium					
Eurybia divaricata[a]	P	Fall; fall–winter		Rhizomes						
Galactia smallii	P	All year; fire stimulates and synchronizes				Bees, wasps, *Leptotes* sp., *Cassius* sp.				
Geocarpon minimum	A	Mr–Ap; germinates No	(3–4) 4–6 wk				Surface water flow			
Helianthus eggertii[a]	P	Au–Se; Se–Oc		Rhizomes				5–25/head	<25%	Exposure to cold
Helianthus schweinitzii	P	Au–frost	Long lived; decades	Rhizomes; tubers			Limited or unknown	Readily germinable in greenhouse		
Helonias bullata	P	Mr–My		Clonal root growth	6–12% of plants flower		Limited; passive, animals			
Hexastylis naniflora[a]	P	Mid-Mr–Ju; mid-My–?; next yr				Flies, thrips	Ants		In spring	Cold
Hydrastis canadensis[a]	P	Ap–My; Jl–Au		Rhizomes						
Hypericum cumulicola	P	Ma–De (fl, fr); peak Ap–Oc	1–several yr	No post-fire resprouting	Self-compatible; 10% autogamy	Small solitary bees, native bumblebees	Limited	4000 fruit/plant	Fall-winter-spring	persistent seed bank
Iliamna corei[a]	P									

(continued)

Appendix 5.2. *Continued*

Binomial	Habit	Phenology	Longevity	Vegetative reproduction	Mating system	Pollination	Dispersal	Seed	Germination	Seed dormancy
Iris lacustris[a]	P	My–Ju (Oc); mid-Jl–Au		Rhizomes, most repro.	Self-compatible; 80–90% set self, 70–71%	Bees, flies but no pollen	Ants; water dispersal of rhizomes		Low (5/192 plots had seedlings), late My–Jl	Overwinter; no cold requirement in genus
Isotria medeoloides	P	My–Ju	Dormancy 4–(10–20) yr	Infrequent rootstock	Self-compatible	Autogamous; mechanical	Wind	Thousands per 1–2 capsules per plant	Likelihood low	Overwinters; short lived
Isotria verticillata[a]	P	(end Fe in FL) end My–Ju; Se–Oc	Clones ≥ 40 yr	Extensive clones	Geitonogamous/xenogamous	Solitary bees	Wind	Poor, < 1% of population	Require a specific fungus	Overwinters; short lived
Jacquemontia reclinata	P; vine, woody at base	No–My, Au–De?		Rhizomatous			"Prolific"		Low in field; readily in greenhouse	
Justicia cooleyi	P	Au–De (Mr)		Rhizomes			Insects?		Readily in greenhouse	
Lesquerella lyrata	A	Mid-Mr–Ap; mid-My	< 1 yr						Disturbance	
Liatris ohlingerae	P	Jl–Oc	Probably decades	Moderate post-fire resprouter from corm	All disks; self-incompatible	Lepidoptera	Wind and secondary along sand	Easily grown from seed		
Liparis liliifolia[a]	P	Ap–Jl	≥ 10 yr	Pseudobulb		Insect visitors	Wind and water			

Lupinus aridorum	B or short-lived P	Mr–Ap	1–several yr; declines after 1st flowering	Susceptible to root rot					Fire induced?	Little to 50–yr seed bank?
Lysimachia asperulaefolia	P	Mid-My–Ju; Ju–Oc		Highly clonal	Xenogamous	Solitary bees; pollen limited				
Macbridea alba	P	My–Jl		Fleshy rhizomes						
Marshallia mohrii	P	Mid-My–Ju; Jl–Au			Protandrous; obligate outcrossing?					
Mimulus glabratus var. *michiganensis*	P; (semi-) aquatic	Mid-Ju–Au (Oc)		Stolons	Low pollen viability; self-compatible? autogamous?		Greatly limited; fragmentation			
Nolina brittoniana	P	Ap	Leaves live several yr; flowering annually	Clonal spread, bulb-like rootstock	Subdioecious; not apomictic	Generalist insects	Wind or gravity, limited	Abundant	Easily	
Oxypolis canbyi	P	Mid-Au–Oc		Stoloniferous rhizomes	Bisexual and/or unisexual	Black swallow-tail butterflies?	Strong-winged schizocarp			
Panax quinquefolium[a]	P	Aged 3–8 yr; Ju–Au; Au–Se	Long lived	Rhizome fragmentation; largely seed	Autogamy, xenogamy	Syrphidae, Halictidae	Close to parent; farther by birds	0.55% chance of reaching maturity	< 25 cm for *P. trifolium*	18–24 mo
Paronychia chartacea	var. *minima* = A; var. *chartacea*= P	Late summer or fall; Se–Oc	Short lived, 1–several yr	None	Dioecious				Post fire	Probably persistent seed bank

(continued)

Appendix 5.2. *Continued*

Binomial	Habit	Phenology	Longevity	Vegetative reproduction	Mating system	Pollination	Dispersal	Seed	Germination	Seed dormancy
Pedicularis furbishiae	P	Flower @ 3 yr; mid-Jl–mid-Au; Se		none; root hemiparasite	self-compatible?	obligate insect pollination; bumblebees	wind or water; beneath or close to parent	25 inflorescences each with 7–17 capsules of 25 seeds	stratification	1 yr viability
Pilosocereus robinii	P	yr-round, Jl–Oc; Au		Mostly veg.; wind-thrown branches	Fruit set in absence of large pollinators	Sphingid moths?	Birds? (*Cardinalis cardinalis*)	Encased in a soft, white pulp		
Pinguicula ionantha	P	Mr–Ap								
Plantago cordata[a]	P	Mid-Ap			Can self	Wind	Wind, water	Mean = 86 capsule per plant	Seedlings appear a few weeks after	None
Polemonium van-bruntiae	P	Mid-Ju–Jl		Horizontal rhizome; importance varies	Cross	Honeybees, bumblebees	Winter winds, spring flooding			
Polygala lewtonii	P	Fe–My	5–10 yr	None	CH and 2 types of CL; autogamous?	Lepidoptera, Diptera, Hymenoptera	Ants			
Polygala smallii	B	Yr-round, summer peak; yr-round	Short lived; < 1 yr	Taproot		Not observed	Ants?, water	Aril-like growths on capsule		

Polygonella basiramia	P	Fall; fall-winter	Short lived, 3 yr from seed to seed set	No post-fire resprouting	Gynodioecious; 1:1 F: H	Small Halictidae specific to genus, Eumenidae wasps, Bombyliidae	Limited	Post-fire, F>H, 218 vs. 32 seeds per stem	On plant or shortly after dispersal; some persist if buried; high germination rates	Innate and conditional
Pycnanthemum incanum	P	Mid-Jl–mid-Se		Rhizomes; largely by seed in Ontario						
Sarracenia oreophila	P	Mid-Ap–early Ju; Au	Rhizomes live for decades	Rhizomes; root stocks; principal mode		Ants, beetles, bees, *Sarracenia* flies; queen *Bombus*, esp. *B. pennsylvanicus*		Valvate capsule responds negatively to humidity		Germinable in greenhouse
Sarracenia rubra ssp. *alabamensis*	P	Mid-Ap–early Ju		Rhizomes						
Schwalbea americana	P	Ap–Ju (South), Ju–mid-Jl (North); early summer (South), Oc (North)		Hemiparasite	Autogamous	Bees, esp. *Bombus*	Wind; ants unlikely	Numerous, enclosed in a sac-like structure	90% inculture; seedlings limited to 2 cm	Stratification unnecessary
Scirpus ancistrochaetus[a]	P	Mid-Ju–Jl; Jl–Se		From nodes and culms of recumbent stems	Probably selfing	Wind	Water, animal fur?		Dry storage and stratification	Seed bank

(continued)

Appendix 5.2. *Continued*

Binomial	Habit	Phenology	Longevity	Vegetative reproduction	Mating system	Pollination	Dispersal	Seed	Germination	Seed dormancy
Scutellaria floridana	P	My–Ju		Swollen storage roots or rhizomes		Nectar guides				
Scutellaria montana	P	Mid-My–Ju; Ju–Jl		Perennation as rootstocks		Apoidae Hymenoptera	< 2 mi?	10–40% fruit set		over-winters
Silene polypetala	P	Late Mr–My		Stolon-like rhizomes; leafy						
Sisyrinchium dichotomum	P	Late My–Jl						4–6 flowers per stem; 3–5 seeds/capsule		
Solidago shortii[a]	P	Mid-Au–early No; late Se–No	Individuals required 3–9 yr to flower	Rhizomes	Xenogamous	Goldenrod soldier beetle?, Halicitidae?	Wind and gravity	250–1700 seeds per flowering ramet	Up to 66% in nature; in darkness, not limiting	Cold stratification required
Solidago speciosa var. *rigidiuscula*[a]	P	Late Se–early Oc		Short rhizomes		Bees, wasps, flies, lepidopteran	Wind?	>50% of stems bear flowers		
Spigelia gentianoides[a]	P	My–Ju				Stamens stay inserted		Single stemmed; few-flowered		
Stylophorum diphyllum[a]	P	Ap–early Ju		Rhizomes		Insects; seed set in isolated plants	Ants?		In subsequent spring	
Symphyotrichum anticostense[a]	P							Can flower in yr 1	In spring	

(*continued*)

Symphyotrichum sericeum[a]	P	Early Au–mid Se; 3–4 wk after		Rhizomes		Bees, insects		Main repro. mode; low set esp. in dry years		
Tephrosia virginiana[a]	P	My–Ju		Rhizomes					Collected seed have 100% germination	
Thalictrum cooleyi	P	Mid-June; Au–O; partial shade delays flowering	Several yr in culture	Rhizome	Dioecious; 3: 1 male: female	Wind and/ or insects?	Lacks effective mechanism; e.g., no male plants at one site in 8 yr	Low; fewer, larger seeds than congeners	20% if stratified	Stratifycation; short-lived seed
Thelypteris pilosa var. *alabamensis*	P; evergreen	Spores year-round?	5 yr in culture	Rhizomes						
Trichophorum planifolium[a]	P	Spring; late Jl–Au	Indefinite?	Short rhizomes		Wind				
Trifolium stoloniferum	P	Mid-Ap–Ju;		Long stolons; rhizobia				"Good" on favorable sites	100% if scarified	
Trillium flexipes	P	Mid-My–early Ju	Flower at 10 yr	Yes	Selfs	Flying insects				
Trillium persistens	P	Mid-Mr–mid-Ap; Jl	> 6 yr			Bees, wasps, flies, ants, butterflies, beetles?	Ants (eliasome)		First spring produce root and no stem?	

(continued)

Appendix 5.2. *Continued*

Binomial	Habit	Phenology	Longevity	Vegetative reproduction	Mating system	Pollination	Dispersal	Seed	Germination	Seed dormancy
Trillium reliquum	P	Early spring; early summer		Tuberous rhizome						
Triphora trianthophora[a]	P	Late Jl–Dec (FL), Au–Se	Colony > 100 yr	Tuberous rooted	Dry conditions limit flowering	Flower lasts 4–6 h; insects	Wind	1–7 flowers, open 1–2 at a time (8 per stem); thousands of seeds per capsule	Contact with a specific fungus	Plant dormant except 4 wk/yr
Viola pedata	P	Mid-My–mid Ju, end Se–mid-Oc		Rhizome		Butterflies, bumblebees	Projected 25–510 cm, ants			
Warea amplexifolia	A	Mid-Au–early Oc; late Se-mid No	< 1 yr	None		Butterflies, bumblebees	Near parent, released by wind			≥ 2–4 yr seed bank
Warea carteri	A	Se–Oc; Oc–No	Annual, < 1 yr	None	Protandrous; autogamous; largely selfing	A great diversity of insects; can limit seed set	Passive release	Numerous per silique; passive dispersal	Not fire stimulated; light requiring; fall-winter-spring	Persistent seed bank?, ≥ 2 yr

Woodsia obtusa[a]	P	Spores mature by late July	Rhizomes	Bisexual gametophyte	water, wind	All gametophytes produced a sporophyte in culture	84% germination of spores in culture
Xyris tennesseensis	P	Jl–Se; flowers wither by afternoon	Bulbous bases; lateral buds in crown leaf axils			High; light requiring	Open wet areas needed

A = annual, B = biennial, P = perennial; Ja = January, Fe = February, Mr = March, Ap = April, My = May, Ju = June, Jl = July, Au = August, Se = September, Oc = October, No = November, De = December; CH = chasmogamous, CL = cleistogamous. Phenology is presented as months of flower; months of fruiting. Terminology on mating system reflects that of the original reference (autogamy = transfer of pollen within the same flower, geitonogamy = transfer between flowers of the same genet, xenogamy = transfer of pollen between different genets [outcrossing]).

[a]Recovery plan not reviewed or not available.

III

Community Dynamics of the Herbaceous Layer across Spatial and Temporal Scales

6

The Herbaceous Layer of Eastern Old-Growth Deciduous Forests

Brian C. McCarthy

The study of old-growth forests has long been a fascination of plant ecologists. In all likelihood this was an outgrowth of the Clementsian view of succession (Clements 1936; Weaver and Clements 1938) in which communities were perceived to change in an orderly and predictable way and culminate in a terminal climax community (chapter 10, this volume). Throughout much of eastern North America, old-growth forests were the epitome of the climax community (Braun 1950; Whitney 1987). Old-growth forests have been of continued interest in the study of plant ecology in large part because they are generally free of the myriad of disturbances caused by humans, thereby allowing the study of natural processes. This unique quality has generated an enormous research database on many aspects of eastern old-growth forests (Nowacki and Trianosky 1993). Due to the massive disturbances associated with European settlement, the landscape in eastern North America underwent enormous change (M. Williams 1989; Whitney 1994; Irland 1999). Lumbering, agriculture, grazing, and anthropogenic wildfires all altered the landscape such that now most deciduous old-growth forest exists only as small relicts. More recently, some have argued that no forest in eastern North America is free of anthropogenic disturbance (direct or indirect). There is certainly a need to better understand the ecology of human-dominated ecosystems (Vitousek et al. 1997; Grimm et al. 2000); however, natural systems will always be needed to serve as the controls or benchmarks in such studies.

The Clementsian concept of succession and the notion of a terminal old-growth climax have now been largely disputed, few ecologists currently accept this as a viable concept (McIntosh 1985), although the notion continued to be propagated in the biological literature throughout the early 1980s and remains in some introductory ecology textbooks. Likewise, few, if any, communities exhibit an orderly and predictable pattern of development, and communities rarely if ever achieve a stable equilibrium due to natural disturbances (White 1979). Concomitant with changing ideas about community succession, ecosystem ecology began to flourish and generated a renewed interest in old-growth forest ecosystems (Franklin et al. 1972). These ecosystems provided considerable information about ecosystem development. Moreover, they provided permanent benchmarks in the landscape that could be used to evaluate management and conservation activities.

Faced with the recognition of an emerging biodiversity crisis in the 1980s, there was a large, albeit uncoordinated, effort instituted throughout the eastern United States to develop criteria for the identification of old-growth forests, to find these relicts in the landscape, and establish some form of preservation for their continued survival and study (e.g., T. L. Smith 1989). Forest managers were becoming increasingly aware of the need to manage forests for multiple reasons, including biological diversity. This period led to a focus on the characteristics of old-growth forest and what made them different from second-growth forests. Oddly, it was not until the early 1990s, with the controversial work of Duffy and Meier (1992), that we began to gain a heightened concern regarding the effects of forest management activities on the forest herb layer. In spite of its shortcomings (chapter 14, this volume), this landmark contribution spurred an interest in community-level understanding of the herb layer, which had before then been largely population based. The relative lack of attention to the herb layer relative to the overstory is curious given that it is the layer of greatest diversity in most hardwood forests throughout eastern North America (Braun 1950).

My focus in this chapter is fourfold: (1) to clarify the factors that constitute old-growth forests in most mesic eastern deciduous forests, (2) to describe what we know about composition, structure, and diversity of the herb layer in old-growth forests, (3) to evaluate the differences between old-growth and second-growth forests with respect to the herb layer, and (4) to assess the linkages among stability, diversity, and habitat invasibility. Although the herb layer contains both woody and herbaceous species (chapter 1, this volume), I restrict my focus to the herbaceous component only.

The Nature of Eastern Mesic Old-Growth Forests

Before European settlement, forest vegetation of eastern North America could hardly be considered free from anthropogenic disturbance (Denevan 1992). Native Americans have been clearing hardwood forests since at least the last ice age (at least 10,000 years; Dickens 1976). However, Native American settlement appears to have been largely restricted to floodplain areas, and the long-term effects are questionable (M. Williams 1989). Evidence of the influence of Native Americans on the upland vegetation of the eastern deciduous forest, particularly with respect to fire, is conflicting (Day 1953; Russell 1983; Patterson and Sassaman 1988). Observations suggest that Native Americans did clear some portion of the uplands for various purposes using a slash (girdle) and burn technique that the early European settlers subsequently adopted. Regardless, the influence of Native Americans in the uplands pales in comparison to the subsequent land clearing and land conversion associated with European settlement. Early descriptions of presettlement eastern hardwood forest from the late 1700s and early 1800s suggest a composition and structure very different from today's forests (Walker 1983). Many forests were cleared by Europeans for grazing pasture or crop agriculture. Catastrophic fire frequently followed logging activities. Today, large parts of the land base are being converted for development. As a result, there is a paucity of primary hardwood forest left in eastern North America. Those stands that continue to exist today were preserved by a family lineage, were not easily accessible to loggers, or were the result of surveying errors (Auten 1941; McCarthy 1995). Recent concerns regarding the biodiversity crisis have pointed to old-growth forests as potential reservoirs of genetic material and possible rare species reserves. Moreover, old-growth forests as an ecosystem are generally considered endangered in much of the eastern landscape.

Whitney (1987), Hunter (1989), Parker (1989), Martin (1992), and McCarthy (1995) all provide technical definitions and descriptions of old-growth hardwood forest in the eastern United States. A simple definition is difficult, if not impossible, to employ. Different authors have adopted different criteria for designation. Some use a functional definition associated with stand dynamics (e.g., Oliver and Larson 1996). Others use a historical criterion such as lack of evidence of direct human impact (i.e., logging; Duffy and Meier 1992). Martin (1992) and others have designated a suite of structural criteria for mesic forests. These criteria relate to species identity, structure, coarse woody debris, and forest floor characteristics, among other characteristics. Keddy and Drummond (1996) espouse this criteria-based approach and provide a summary of features that should be monitored, and potentially managed for, in eastern deciduous forests.

Although all approaches have their merits, I strongly endorse the use of specific, quantifiable criteria in old-growth designations to reduce unproductive semantic debates. For example, Martin (1992) argues that mesic forests should contain most of the following attributes in order to be considered old-growth: a moderate to high richness and diversity (S and H, respectively) of tree species ($S > 20$; $H' > 3.0$); uneven ages with a wide distribution of size and age classes (reverse *J* distribution); some large (> 75cm dbh) canopy tree species; some large, high quality merchantable trees (i.e., of high economic importance suggesting no logging or high-grading); some trees older than 200 years; overstory density > 250 trees/ha (for stems > 10 cm dbh); overstory basal area > 25.0 m^2/ha; strong presence of coarse woody debris (logs and snags) in multiple age and decay classes; tree-fall gaps from windthrow along with pit-and-mound topography; plants and animals that prefer or reach their optimum in old-growth; undisturbed soils with a thick organic layer and presence of soil macropores; and little or no evidence of human disturbance (e.g., stumps, skid trails, wire fencing, cultivated plants, foundations).

Although an abundant and diverse spring herbaceous flora is one of the hallmarks of temperate deciduous forests, this feature has not been widely used in the old-growth designation debate. Keddy and Drummond (1996) make specific recommendations regarding spring ephemerals (sensu Givnish 1987) of mature deciduous forest (table 6.1). They note that, given the variability in herb communities associated with latitude and environment, it would be inappropriate to use the presence of specific indicator species for the designation of high-quality old-growth forests. But a number (richness) of indicator species from the local spring ephemeral guild may be useful in assessing stand quality (table 6.1): ≥ 6 for high quality (old-growth), 2–5 for intermediate, < 2 for poor (Keddy and Drummond 1996). Although this approach seems useful, additional studies are necessary to include a broader subset of vegetation types (beyond *Fagus-Acer*) and guilds (functional groups) in the herbaceous flora (i.e., summer species). Moreover, simple observation in the central Appalachians indicates that six of these ephemeral species are present in even the most degraded second-growth deciduous forests of the region (McCathy pers. obs.).

Table 6.1. Spring ephemerals from seven old-growth forests, predominantly *Fagus-Acer,* located throughout the Kentucky, Ohio, Pennsylvania, and Michigan region

	Locale					
Species	KY	PA	OH	OH	OH	MI
Allium tricoccum			*	*	*	
Caulophyllum thalictroides	*	*		*		*
Claytonia virginica	*		*	*	*	*
Dentaria diphylla	*	*		*		
Dentaria laciniata	*		*		*	*
Dicentra canadensis			*	*	*	*
Dicentra cucullaria			*	*	*	
Erythronium americanum	*			*		*
Maianthemum canadense						*
Podophyllum peltatum	*		*	*	*	
Polygonatum biflorum	*	*	*	*	*	
Sanguinaria canadensis	*		*		*	
Tiarella cordifolia	*	*		*		
Trillium grandiflorum	*	*	*	*		*
Total number of species	10	5	9	11	8	7

Six species have been proposed as a minimum target number in this functional group to indicate the old-growth condition or quality. Presence or absence of an individual species is probably unimportant. From Keddy and Drummond (1996). Nomenclature follows Gleason and Cronquist (1991).

Composition, Structure, and Dynamics of the Herb Layer

Beyond simple descriptions of vegetation (e.g., Oosting 1942; Wistendahl 1958), the composition, structure, and dynamics of species in the herbaceous layer of mesic old-growth forests only began to receive significant attention in the early 1980s. Most foresters, land managers, and ecologists have historically ignored the understory layer in favor of the more economically important overstory. Indeed, most definitions and characteristics of old growth (as the ones presented here) are rooted primarily in tree species composition and forest structure or disturbance characteristics. There is an important need to understand the factors affecting herb layer dynamics in old-growth forests. Historic views of succession (i.e., climax concept) would suggest that forest understories should be largely stable over long periods. Modern views of succession would suggest that the understory is quite dynamic and not in equilibrium. However, long-term temporal studies of the herb layer are few (Brewer 1980; Davison and Forman 1982), as are geographically broad studies of spatial pattern (Rogers 1981, 1982). The relationships between temporal and spatial pattern (scale) in forest herb communities is only recently attracting attention (McCarthy et al. 2001; Small and McCarthy 2002), even though these notions have been around in ecology for some time. In fact, Small and McCarthy (2002) suggest that previous studies of the herb layer in species-rich forests may have been undersampled by at least an order of magnitude, thus making inference about pattern weak.

Brewer (1980) provided one of the first long-term (50 year) studies of herb community dynamics in an old-growth *Fagus-Acer* stand (Warren Woods) in southern Michigan. Warren Woods was found to have a luxuriant (i.e., dense cover) but a relatively species-poor herb layer. Some species decreased and some increased over the study period. Brewer concluded that changes were not related to exogenous factors (e.g., climate) but rather related to changes in overstory dynamics stemming from a major disturbance more than 150 years before (believed to be a catastrophic fire). Indeed, what was being observed were long-term successional changes associated with old-growth development. Species with the greatest increase (> 200%) during this time included *Asarum canadense* L., *Boehmeria cylindrical* L., *Epifagus virginiana* L., *Laportea Canadensis* L., *Osmorhiza claytoni* (Michx.) C. B. Clarke, *Polygonatum pubescens* (Willd.) Pursh., and *Viola* species, many of which are slow-growing perennials and tolerant of deep shade. He did not find windthrow to be a major influence with respect to diversity.

Davison and Forman (1982) also provided one of the few long-term studies of forest herb community dynamics. They conducted a 30-year study of herb and shrub dynamics in an old-growth oak forest (Hutcheson Memorial Forest; HMF) in central New Jersey. Their goal was to evaluate the stability of the understory layer over the 30 years and assess patterns of community change. They hypothesized that a stable climax forest should have a stable understory. They found that the herb layer declined from 33 species in 1950 to 24 in 1969 and 26 in 1979. Moreover, cover increased dramatically from 8% to

60%. *Podophyllum peltatum* L. increased in cover by 700%; *Circaea quadrisculcata* L. by 500%. The sharp increase in the former species was a partial explanation for the decrease in diversity. *Podophyllum* has a strong clonal nature and can displace other species. In addition, *Lonicera japonica* Thunb., an introduced vine, increased substantially and was believed to negatively influence the native species diversity. Davison and Forman also noted that a long period of fire suppression might have influenced the observed species composition. Increasing canopy gaps were believed to influence the understory by increasing light and moisture conditions.

McCarthy et al. (2001) established a long-term study of forest herb dynamics in an old-growth mesophytic forest dominated by white oak in southeastern Ohio (Dysart Woods; DW). Dysart Woods is comparable to HMF (Davison and Forman 1982) in many ways, including history, composition, structure, diversity, and natural disturbance regimes. Perhaps most important, DW is also experiencing a decline of white oak in the overstory. Many of the DW oaks are in excess of 85 cm dbh and 400 years of age (Rubino and McCarthy 2000) and are in a disease decline spiral (Manion 1991) associated with multiple predisposing and inciting agents (e.g., acidic deposition, drought stress, *Armillaria* root rot). This will likely result in dramatic long-term changes in the understory, but these patterns will only be revealed in the context of long-term permanent plot-type ecosystem data where both biological and environmental components are examined.

Within-year variability in the herb community is familiar to everyone who has worked in the eastern deciduous forest. McCarthy et al. (2001) provided a quantitative example of these short-term temporal (phenological) differences and local spatial differences. The vegetation varied dramatically among early spring, early summer, and late summer samples. Certainly, this would be expected between the vernal herbs and the summer species, but the differences between the two summer samplings were somewhat unexpected. Curiously, the phenological differences within seasons have only been rarely quantified in old-growth stands (Goebel et al. 1999). Most studies (e.g., McCarthy and Bailey 1996; Olivero and Hix 1998) use only a single sample period in midsummer and may not be representative. Further, McCarthy et al. (2001) also discovered a dramatic topographical aspect effect. Highly dissected topographies, like those found in much of the central Appalachians, may have dramatically different floras on north-and south-facing slopes (fig. 6.1). The interplay between spatial and temporal dynamics at various scales remains largely unexplored.

Rogers (1981, 1982) evaluated the understory layer of old-growth forests throughout Ohio, Indiana, Michigan, Wisconsin, and Minnesota. He noted that there had been few studies of the herb layer that were both quantitative and geographically extensive. He found that soil fertility, not climate, was the most important difference in explaining regional patterns of spring ephemerals. Within stands, soil drainage and microtopography had the most important influence on vernal herb diversity and abundance (higher on soil mounds than pits). Species with large perennial organs tended to be uncom-

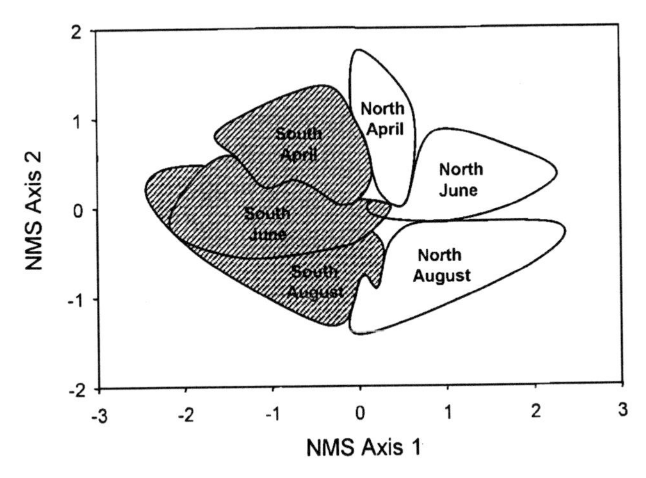

Figure 6.1. A nonmetric multidimensional scaling (NMS) ordination of 2-m² quadrats used to sample the herbaceous vegetation of Dysart Woods, an old-growth mesophytic forest in southeastern Ohio (n = 210). The first two NMS ordination axes account for 33.0% and 24.1% of the variability, respectively. Axis I separates quadrats on the north-facing slope from those on the south-facing slope. Axis II is a temporal gradient separating spring, early summer, and late summer floras. Individual points are omitted for visual clarity; isopleths are drawn to encompass all sample points.

mon in forests that were relatively recently disturbed (Rogers 1982). Among the summer herbs, Rogers (1981) found 23% to be widespread and common (present in $\geq$ 30% of stands), another 23% to be common but distributed unevenly, 15% uncommon but widely distributed, and the remaining 40% to be both uncommon and highly restricted. This suggests that species-specific generalizations will be difficult or impossible to make and that current and future studies should focus on life-history types or guilds in analyzing community data.

Interest in spatial and temporal variation has returned the concept of diversity to center stage in understory ecological research (Wilson et al. 1998). However, few studies have explored the relationship of scale to patterns of understory diversity in eastern deciduous forests. In fact, the method of assessing, interpreting, and reporting understory diversity is less than clear. An objective evaluation of impending parameters is critical because many estimators of diversity are very sensitive to sample size, area measured, species rarity, grain of the environment, and so on. We recently explored the relationship between micro- (2 m²) and mesoscale (70 m²) herb diversity in DW (Small and McCarthy, 2002). We examined standard estimators of richness, evenness, and diversity, as well as species area curves, and more recent techniques such as SHE (richness [S], diversity [H], and evenness [E]) analysis (Hayek and Buzas 1997, 1998) and species richness estimators (E. P. Smith and van Belle 1984; Palmer 1990; Colwell and Coddington 1994; Chazdon

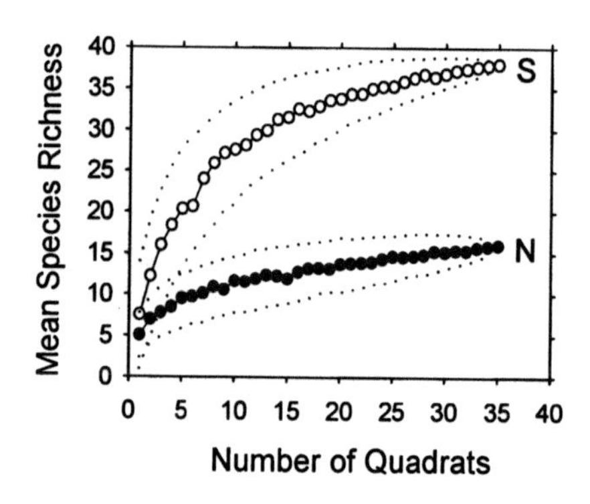

Figure 6.2. Bootstrap-derived species area curves (±95% CI, dotted lines) for the thirty-five 2-m² understory sampling quadrats established on north-facing (N) and south-facing (S) slopes at Dysart Woods for the June (early summer) sampling period. These data suggest that 70-m² (35 plots) is a minimum sampling area; however, sample size estimation equations indicated > 200-m² (100 plots) would be necessary under most conditions to capture 95% of the species.

et al. 1998). Although scale clearly had an effect on the reporting and interpretation of diversity, we found the species area curves and species richness estimators to be the most enlightening (fig. 6.2). We sampled thirty-five 2-m² plots on one northeast- and one southwest-facing slope—a relatively large sample compared to most studies. Each plot was sampled three times during the growing season to assess phenological variability (fig. 6.1). Based on the diversity patterns in this oak-dominated mesophytic old-growth forest and bootstrapped 95% confidence intervals around our estimators, we determined our sampling to be inadequate in this species rich stratum. In fact, sample size estimation equations suggest that 100–300 2-m² quadrats are necessary to describe and estimate species richness within any one stand. This sampling intensity is dramatically higher than anything published in the literature. This suggests that previous studies of the herb layer in temperate hardwood forests need to be reconsidered. The likelihood that representative samples were obtained is low, and the strength of inference is potentially quite weak.

Old-Growth Versus Second-Growth Herb Layer Dynamics

In the absence of anthropogenic disturbance, the factors that affect herb distribution and abundance are numerous and include topographic and soil features such as elevation, aspect, soil quality (Bratton 1976; Elliott et al. 1997), overstory composition (Hicks 1980), stand structure attributes such as basal area and cover (Ford et al. 2000), herbivory (Alverson et al. 1988; van Deelan et al. 1996; Brown and Parker 1997; Rooney and Dress 1997a), and timing or size of gap disturbance (Collins and Pickett 1987; Clebsch and Busing 1989; Reader and Bricker 1992a). The relationship between disturbance, particularly anthropogenic disturbance in the form of logging, and herb community structure is still not well understood and has only recently begun to be explored in detail (Duffy and Meier 1992; Gilliam and Turrill 1993; Bratton et al. 1994, Gilliam et al. 1995; Meier et al. 1995, 1996; Goebel et al. 1999; Ford et al. 2000; Gilliam 2002; Roberts and Zhu 2002; chapters 11 and 14, this volume; others reviewed in Battles et al. 2001).

Duffy and Meier (1992) argued that second-growth stands approaching 90 years since disturbance still did not exhibit a similar flora to paired old-growth stands. Old-growth stands had significantly greater cover and species density in all nine pairs of stands. They suggested that the explanations could include (1) recovery was so slow that 87 years was an insufficient time to observe change, (2) logged forests may never recover to match primary forest because of different climate conditions now relative to the past, and (3) many herbaceous species require the microtopography generated by pit-and-mound dynamics so recovery must await the death and decay of the overstory.

The influence of landscape factors, such as slope aspect, is known to have an important effect on vegetation (Wolfe et al. 1949), yet the few studies that examine this component are not in agreement. Olivero and Hix (1998) examined the ground flora in second-growth and old-growth stands in southeastern Ohio. Half the plots were located on mesic, northeast-facing slopes and the other half on dry, southwest-facing slopes. First, they found significant differences in both of the main effects (i.e., between old-growth and second-growth stands and between mesic and dry stands). However, they also discovered an interactive effect in that species richness and Hill's diversity differed between old-growth and second-growth on northeast-facing plots but did not differ on southwest-facing plots. McCarthy et al. (2001) found the exact opposite pattern within DW. In this old-growth forest, all measures of diversity were greater on the southwest-facing slope compared to the northeast-facing slope (but abundance was greater on the latter). Goebel et al. (1999) conducted another study comparing an old-growth forest and second-growth forest in southeastern Ohio; however, they restricted the study to two stands in each study area on southwest-facing slopes only, but sampled both spring and summer vegetation. In contrast to the study of Olivero and Hix (1998), Goebel et al. did find differences in southwest-facing slope floras in old-growth versus second-growth forests.

From an experimental design standpoint, all of the previous studies (Olivero and Hix 1998; Goebel et al. 1999; McCarthy et al. 2001) are problematic in one way or another, particularly in the context of pseudoreplication (Hurlbert 1984). Olivero and Hix (1998) sampled only once in midsummer and then only eight 2-m^2 plots were sampled in each of 32 study areas. Goebel et al. (1999) sampled both the spring and summer herb vegetation but pseudoreplicated at the forest level so they had only one replicate treatment (primary vs. secondary) and then only a relatively small sampling effort within any one time period. Based on data from Small and McCarthy (in press), considerably more sampling is required to describe within-stand herb communities in a species-rich forests. McCarthy et al. (2001), while maintaining a moderate within-stand sample size to explore phenological patterns, pseudoreplicated at the stand level and thus had only one of each of the representative slope aspects in the one forest. In all likelihood, the limitations of time, money, and energy impinge on the scale of any one study. Investigators need to draw boundaries within the limits of their questions and available resources. The necessity to sample large numbers of plots multiple times within one season is very resource intensive, particularly when the flora is so phenologically dynamic that complete species turnover can occur within weeks. Clearly, many more studies are needed over a larger geographical area for longer periods of time before we can begin to draw conclusions or conduct informative meta-analyses.

Several studies have now used successional chronosequences examining understory diversity recovery after major disturbances such as logging (Gilliam et al. 1995; Ford et al. 2000). Using this approach, neither Gilliam et al. (1995) nor Ford et al. (2000) was able to demonstrate any clear pattern in herb diversity with respect to stand age. These results could suggest that clearcutting mature second-growth stands does not significantly impact understory herb diversity. However, as Duffy and Meier (1992) and Meier et al. (1995) point out, several centuries may be required to develop the characteristics associated with old-growth outlined previously in this chapter. Thus, there may be a long time between stand maturity after clearcutting (75–100 years) and the old-growth condition (300–400 years) whereby herb community structure returns to its primary condition (chapters 11 and 14, this volume).

After disturbance, community composition and structure will be initially directed largely by propagule dispersal. Spatial properties such as patch isolation (landscape connectivity) and microenvironment (chapter 14, this volume) along with temporal factors such as seed banks, will be important determinants in the rates of recolonization. Unfortunately, most of the old-growth forest patches in the eastern deciduous forest exist in a nonforest matrix. For example, both DW in Ohio and HMF in New Jersey are surrounded by agriculture and/or suburban development. Without connectivity, disturbed forests will be slow to regenerate the previous community structure. Unfortunately, studies of temperate old-growth forest seed banks are uncommon. In contrast to general successional patterns found in the seedbank literature (e.g., Roberts and Vankat 1991), Leckie et al. (2000) found that weedy aliens of adjacent disturbed landscapes did not dominate the seedbank in an old-growth hardwood forest in Quebec. In fact, the seedbank was found to be fairly diverse and contained many shade-tolerant species. Leckie et al. also found that many of the species in the above-ground vegetation were also present in the seedbank. McCarthy et al. (unpublished data) found a similar pattern at DW. Many shade-tolerant forest species were found in the seed bank, and this could serve as an important propagule source after disturbance. However, unlike Leckie et al. (2000), McCarthy et al. (unpublished data) found little similarity (< 20% in most sample periods) between above-ground and below-ground species constituencies (fig. 6.3).

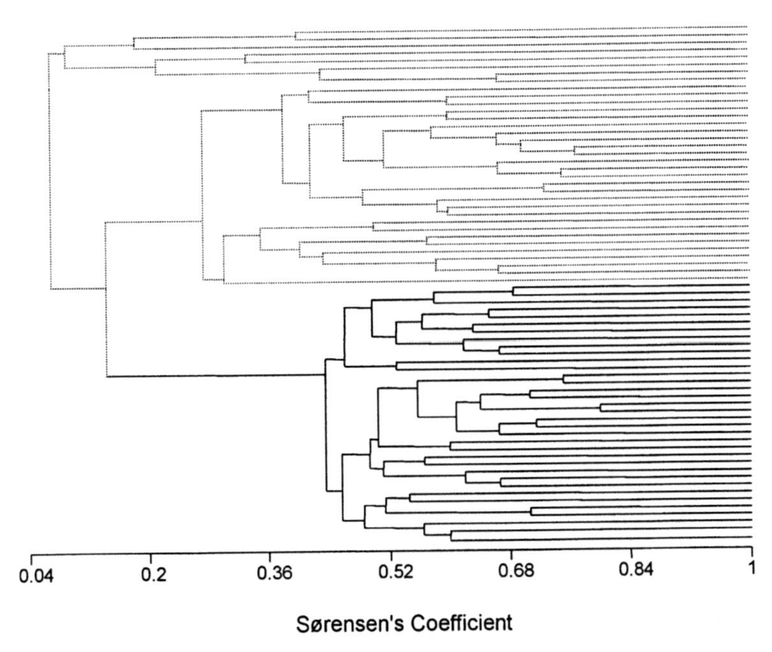

Figure 6.3. Unweighted pair group mean analysis, using Sørenson's coefficient of similarity, to compare the seedbank (solid lines) with the above-ground vegetation (dotted lines) in the 35 quadrats on the north-facing slope (June vegetation data shown) at Dysart Woods, Ohio.

Stability, Diversity, and Habitat Invasibility

Concern over biodiversity loss with respect to ecosystem function has regenerated interest in the links among stability, diversity, and habitat invasibility (Levine and D'Antonio 1999). The notion of stability can be traced to early ideas about the development of the community as espoused by Clements and his followers. According to early thinking, at some point, all communities achieved a homeostatic equilibrium (the climax community). This idea continues today in many respects under the rubric "balance of nature." A community or ecosystem is often seen as stable when no change can be detected in species over time. However, as Connell and Sousa (1983) point out, it is important to distinguish between the degree of constancy of the numbers of organisms versus the constancy of species (presence or absence). The former is quantitative, the latter qualitative. Quantitative issues include both the *resistance* of a system to disturbance (the ability to remain at equilibrium when faced with potentially disturbing forces), as well as the *resilience* of a system (the ability to return to equilibrium after being disturbed). Qualitative issues focus on *persistence* (the ability of a species to not become locally extinct). Thus, both qualitative and quantitative features of a system must be examined.

Old-growth forests have been, and will continue to be, a critical component to our development of ecological theory, particularly with respect to issues of system stability. My observations in the central and southern Appalachians suggest that old-growth herb layers exhibit all three forms of stability: resistance, resilience, and persistence. This is confirmed by at least one empirical field study (Brothers and Spingarn 1992). But long-term data in forests that have been perturbed are needed before this notion can be better assessed. In this regard, old-growth forests will remain invaluable benchmarks to compare against a predominantly managed (perturbed) forest landscape now existing in eastern North America. The predominant disturbance in most eastern old-growth forests, with the advent of fire suppression in the twentieth century, is gap disturbance. However, empirical tests of the effects of gap disturbance on the understory community in second-growth forests have not been particularly revealing. Collins and Pickett (1987) found that small (single-tree) and large (multi-tree) gap openings in a northern hardwood forest had little influence on species richness or cover. This suggests that the herb community is resistant to the disturbance regime most prevalent in eastern old-growth forests and therefore relatively stable. In a similar study, Reader and Bricker (1992a) were unable to show a positive relationship between gap size and perennial herb abundance over a 2-year period in hardwood forests of southern Ontario. They concluded that herbs might be under too much competition with advanced regeneration in larger gaps to allow increases in herb abundance.

For years, there has been the tacit assumption that there is a link between stability and diversity, probably first proposed by Elton (1958), who stated that "the balance of relatively simple communities of plants and animals is more easily upset than that of richer ones" (p. 145). However, the empirical evidence for this notion is less than apparent (McNaughton 1988), particularly in communities of vascular plants. Tilman (1996) recently examined this relationship using long-term plot data in grasslands and concluded that diversity did stabilize community and ecosystem processes (e.g., biomass), but not population processes (e.g., individual species abundance). However, the ultimate interpretation of these results is difficult due to confounding effects found in many ecological experiments (Huston 1997). Huston (1994) suggests that in a dynamic equilibrium context (assuming low levels of disturbance), high-diversity communities are likely to occur under conditions of low population growth rates, areas low-diversity communities are likely to be found under conditions with high population growth rates. But the relationship between stability and complexity may simply be a consequence of the type of environment in which high-diversity communities exist, rather than an inherent property of diversity and complexity (Huston 1994). Many old-growth relicts are found on relatively high-quality sites. Empirical tests of the diversity–stability hypothesis are lacking for old-growth forests.

More recently, biologists have become concerned about the prevalence of biological invasions and their impact on natural systems (Mooney and Drake 1986; McKnight 1993; Williamson 1996; Luken and Thieret 1997; chapter 13, this volume). *Invasibility* is defined as the ease with which new members become established in a community. It is widely believed that high native diversity decreases the invasibility of communities (Lodge 1993). The origins and rationale for this notion have been summarized well by Levine and D'Antonio (1999). Constructed community studies where diversity has been directly manipulated have produced both positive and negative relationships with invasibility, both in field (Palmer and Maurer 1997) and microcosm (McGrady-Steed et al. 1997) experiments. As far as I am aware, there has been no empirical test of this notion in forest understory communities. Simple observations of old-growth forests are conflicting. HMF and DW described earlier are two interesting case examples. Both are old-growth forests dominated by white oak, have similar disturbance histories, and are of similar composition and structure. Yet, HMF has been plagued with invading species in the understory (Davison and Forman 1982; McCarthy, pers. obs.) and DW has not. In a recent vegetation study of DW, McCarthy et al. (2001) found no nonindigenous species in the herbaceous layer! But clearly the woods are surrounded by many nonindigenous species, some quite invasive (e.g., *Lonicera japonica, Alliaria petiolata*). A key difference may have been the profound effect of gypsy moth (*Lymantria dispar* L.) attacks at HMF in the early 1980s; whereas DW has not yet been reached by the gypsy moth. A biological disturbance by one invasive organism may govern the invasibility of the entire community, depending on the strength of the disturbance. Under normal disturbance regimes and climatic fluctuations, old-growth forests may contain stable understories. Considerably more observations and experimental work are needed in this area.

Summary

I have reviewed and explored several features of the herbaceous layer in eastern old-growth deciduous forests. After many years of debate, we are now coming to consensus on how to define old-growth forest. Technical definitions with specific quantifiable criteria are now available, but the herb layer has not commonly been included in these definitions. It is important to incorporate the herb layer in these definitions because this is usually the most diverse vegetative layer in these ecosystems. More research is needed to understand the herbaceous layer diversity patterns along chronosequences that include old growth. The use of functional groups (in place of species) may be helpful in setting targets for preservation or conservation. This will permit better standards to emerge and ultimately assist with land management decisions.

Because of the temporal variability of forest understory communities, single samplings of the vegetation are inappropriate in most cases; yet many examples remain in the literature. There also appears to be a large emphasis on the spring vernal herbs (perhaps because of their lower diversity), apparently at the expense of the rest of the summer flora. In fact, we know little of the complete herbaceous flora and its phenology in most forests. As I have described, given the high diversity in this layer, we need to better study the sampling methods and sample sizes required to capture at least 95% of the species. Furthermore, if we are to ever understand spatial and temporal patterns in the understory, we must expand our observations beyond the level of the stand and single-year study. As far as forest herb communities are concerned, I cannot overemphasize the need for long-term, permanent plot studies arranged in a stratified fashion throughout the landscape.

Finally, in recent years, biologists have become acutely aware of the threat of invasive, nonindigenous species to many natural ecosystems. However, our understanding of the relationships among diversity, stability, and invasibility are still weak and confined to specific types of microcosms or ecosystems. Forest systems have not been carefully examined in this light, especially old-growth forests. Despite a shifting emphasis toward a better understanding of the managed forest landscape, old-growth forests will remain an important component in our understanding of forest pattern and process. These systems serve as the benchmarks in our heavily disturbed eastern forest landscape.

7

Habitat Heterogeneity and Maintenance of Species in Understory Communities

Susan W. Beatty

What determines species distributions and species richness in forested communities? This question has been debated for decades. Rather than ruling out possible explanations, we have continued to propose new hypotheses. Four explanatory factors often appear in the literature: disturbance, patterns in physical factors (soils, microclimate), biological processes (e.g., competition, dispersal, colonization, and extinction), and history (land use, anthropogenic impacts, and successional age). In this chapter I address the effects of all four factors in determining the species composition and richness of northeastern deciduous forest communities in the context of spatially and temporally heterogeneous environments.

Disturbance in plant communities has been connected with the maintenance of species richness, community stability, and a variety of life history strategies (Watt 1947a; Whittaker 1969; F. E. Smith 1972; Drury and Nisbet 1973; Grubb 1977; Platt and Weis 1977; Whittaker and Levin 1977; Connell 1978; Grime 1979; Huston 1979; Denslow 1980; 1985; Pickett 1980; Collins and Pickett 1987; Foster 1988a; Foster and Boose 1992; Cooper-Ellis et al. 1999). Current disturbance theory includes the two following hypotheses: (1) richness is maintained in communities where the disturbance is more frequent than the time required for competitive exclusion to operate, and (2) communities with the highest richness should be those affected by moderate disturbance intensity and/or frequency (intermediate disturbance hypothesis) (Huston 1979; Pickett and White 1985). Numerous studies have indicated that there is a complex interaction between controls exerted by fine-scale, local processes and coarse-scale, regional processes (Glitzenstein et al. 1986; Pastor and Broschart 1990; Beatty 1991; Lertzman 1992; Frelich and Reich

1995; Lertzman and Fall 1998). We need to establish the interaction between potential controls on richness and the scales at which they operate within the spatial heterogeneity inherent to many communities (Wilson et al. 1999).

Variation in the physical environment has long had the attention of plant ecologists in explaining species distributions and, more recently, species diversity (Kupfer and Malanson 1993; Allen and Walsh 1996; Parker and Bendix 1996; Bendix 1997; Clark et al. 1999; Pitman et al. 1999). In forested ecosystems, individual treefalls may be responsible for maintaining spatial habitat heterogeneity and for supporting gap-phase species within a mature community (Stephens 1956; Lyford and MacLean 1966; Veblen et al. 1979; White 1979; Runkle 1981, 1982). However, Hubbell et al. (1999) argued that a recruitment limitation (lack of widespread dispersal to gaps) in tropical forests "appears to decouple the gap disturbance regime from control of tree diversity" (p. 554) and said that we need to reexamine the gap-dynamics models elsewhere. In glaciated regions of forested North America, treefall disturbance creates a remnant microrelief on the forest floor that can affect spatial heterogeneity for hundreds of years after the direct disturbance gap effect has disappeared (Lyford and MacLean 1966; Beatty and Stone 1986; Schaetzl et al. 1989; Schaetzl and Follmer 1990). This mound-pit microtopography is responsible, in part, for determining understory species distributions at the micro-scale (Beatty 1984, 1991; Canham 1984; Beatty and Sholes 1988; Peterson et al. 1990) and is discussed here as a "disturbance legacy" rather than as a direct disturbance effect.

A crucial question in plant-community ecology continues to be whether species composition results from assembly rules (e.g., dominance–diversity relations) that limit the number of species in an area through interactions among species and with the environment, or from dispersal/colonization dynamics that limit whether a species reaches a site (Wilson 1999a). Many studies have shown the importance of biological interactions in determining species composition and richness of communities (Grace and Tilman 1990; Tilman 1994; Hubbell 1997; Malanson 1997a, 1997b; Dalling et al. 1998; Duncan et al. 1998). Explanations of species distributions have frequently invoked competition as a driving force (Lack 1947; Hutchinson 1959; Daubenmire 1966; Werner and Platt 1976; Grace and Wetzel 1981; Tilman 1982; Schoener 1983), but some have argued that stable communities have evolved so that competition is minimal or absent (Connell 1980; Givnish 1982; Parrish and Bazzaz 1982). With respect to understory layer interactions, one of the few experimental studies with forest herbs that demonstrated competitive segregation of congeneric forest herbs used a transplantation technique (W. G. Smith 1975). Using nonexperimental studies in forests, Rogers (1983) found neither evidence of spatial interference competition between functionally similar spring-ephemeral taxa, nor of temporal interference competition between spring-ephemeral and summer-species groups (Rogers 1985). In moist forest understory communities, distributional patterns may result from differences in species' colonizing abilities, longevity (as in clonal integration), and/or space preemption on available microsites (Harper et al.

1961; Drury and Nisbet 1973; Grubb 1977; Rogers 1982). Treefall gaps in closer proximity to old field/forest edges were found to have higher species richness, yet also contained more exotic species (Goldblum and Beatty 1999). In the deciduous forests of the northeastern United States, Beatty (1991) found that active seed dispersal corresponded more closely to microsite species composition than did the buried seed component.

History of a site can also be important in determining the trajectory of community development (Hibbs 1983; Foster 1988b; Whitney 1990; Peterson and Carson 1996; Nekola 1999). Understory species composition and richness may vary depending on the age of the stand, prior land-use treatment (plowed or not), and types of surrounding communities supplying a colonization source during the successional development. The clearing of forests since the 1800s for exploitation of wood resources and creation of agriculture/livestock grazing land has been widespread (Delcourt and Delcourt 1996). As a result, nearly all old-growth forests in some areas have been logged sometime since the time of European settlement. The patterns of forest regeneration are patchy in the landscape, giving rise to a mosaic of stands abandoned at different times (Odum 1943; Russell 1958). Adjacent land cover has been identified as a factor in determining species richness for birds (Cubbedge and Nilon 1993; Green et al. 1994; Parish et al. 1994), as well as for floodplain tree species (Everson and Boucher 1998). Matlack (1994b) investigated the relationship between species composition and distance from nearby older stands that were sources of dispersal propagules. He found that the migration rates of plant species depended on distance from source populations as well as on dispersal mode.

The importance of scale in ecological systems has been known for decades (Greig-Smith 1952; Stommell 1963; Schumm and Lichty 1965; Pielou 1969; Meentemeyer and Box 1987; Meentemeyer 1989), but we are still seeking a synthesis (O'Neill and King 1998). Fine-scale habitat heterogeneity is commonly overlooked in the study of coarser scale vegetation processes (Beatty 1984). In this chapter I consider the importance of microscale dynamics of herbaceous and woody species (dispersal, colonization, and persistence) in determining forest community pattern and process.

Study Site

Over the last 20 years, I have been conducting a combination of observational and experimental studies in eastern deciduous forest of east-central New York State, at the Edmund Niles Huyck Preserve and Biological Field Station (42° 31' latitude, 74° 09' longitude). The preserve is 2000 ha in area today, and consists of parcels of land acquired intermittently since 1931 when it was established. Before settlement in 1781, some land was cleared for low-intensity agriculture, sheep grazing, haying, and logging for sawmill production. There are also areas thought to be old-growth hemlock forest (Odum 1943). The oldest second-growth deciduous forests are now about 150 years old. Mature deciduous forest in which much of my research has been conducted is second-growth forest located on east-facing, upland slopes around Lake Myosotis (elevation 450–500 m). This area was part of the original preserve land acquisition and has been undisturbed by anthropogenic manipulations since that time.

The forest communities at the preserve are representative of deciduous and mixed forests throughout much of the northeastern United States. The dominant tree species are sugar maple (*Acer saccharum* Marsh.) and American beech (*Fagus grandifolia* Ehrh.), but white ash (*Fraxinus americana* L.), red oak (*Quercus rubra* L.), hop-hornbeam (*Ostrya virginiana* (Mill.) K. Kock), striped maple (*Acer pensylvanicum* L.), and occasionally eastern hemlock (*Tsuga canadensis* (L.) Carr.) are also found in the forests (see also Beatty 1984). The understory consists of a few shrub species, seedlings of tree species, and about 85 species of herbs, forbs, and bryophytes. The acid mull soils are moderately well-drained silt loams (Lordstown series) with glacial till parent material overlaying shale bedrock (1–5 m depth). Other vegetation types on and around the preserve include old fields of varying ages, young successional forest (ca. 25–50 years old), old-growth hemlock stands, spruce/pine plantations (Goldblum 1998), floodplain forest, lake-edge marsh, fen, and residential gardens.

Methodology

In 1978 I established 144 microsite plots on mounds, pits, and undisturbed forest floor and have monitored them on a yearly basis. Each plot is circular (to fit the shape of the microsite features) and 0.65 m^2 (average size of features). Since that time I initiated several additional studies and experiments, a summary of which is given here. In all cases, I preserved the initial sampling design, using the same-sized plots and usually stratifying the sampling by microsite. Plant species composition (density and coverage) was measured for all sample sites; I segregated data into finer categories on occasion, such as differentiating between first-year seedlings (identifying cotyledons) and established plants for both herbaceous and woody species. I also included bryophytes in the vegetation composition data, using only coverage as a measure of species importance in a plot. As of 2000, I had a total of 447 microsite plots being monitored in these forest communities. Generally, I also monitored soil and microenvironmental parameters for plots (e.g., soil temperature, moisture, pH, extractable cations and nitrate, summer rainfall amounts, air temperature, photosynthetically active radiation, canopy coverage, recent treefall activity, and standing tree mortality within 30 m of a plot), although not every year in every plot.

Experimental studies have included: (1) a removal experiment to test for competitive interactions (n = 80 experimental and control plots), (2) a leaf litter removal experiment to determine the impacts on establishment and survivorship (n = 40 experimental and control plots), (3) a seed trap study that quantified dispersal pathways by which microsites are colonized (n = 100 sample points), (4) an environmental variability experiment to test hypotheses about species shifting under extreme environmental conditions (n = 80 experimental and control plots), and (5) an experiment creating mound and pit microsites to determine micro-successional changes in new microsites (n = 40 experimental and control plots). In the competition experiment I removed *Aster divaricatus* L. (dominant mound species) from replicate microsites using two techniques: only clipping above-ground parts and applying herbicide to cut stems after clipping. Clipping was repeated biweekly throughout the growing season for the 4 years of the study; herbicide killed plants in the first year, after which cotyledons were removed each year. In the leaf litter removal experiment (Beatty and Sholes 1988) we removed litter by hand and kept litter out of sites using wire-mesh cages for 8 years. In the seed trap study, I used a nondrying spray-on adhesive on filter paper in petri plates located on the soil surface or buried down to the surface, and then either covered or not covered with a fine-mesh cage that allowed light and moisture input but no seed rain. This partitioned seed inputs into aerial versus overland flow. Buried seed was also germinated in the greenhouse (Beatty 1991). The environmental variability experiment was designed to either increase or decrease the available moisture to a microsite by a factor of two. Based on measured rainfall in the forest communities, collected rainfall was added to some microsites to mimic a wetter year. Other microsites had canopy covers, which allowed PFD penetration but prevented direct input of ambient rainfall to mimic a drier year. Soil moisture was monitored to assess how effective the treatments were over the 3 years of the experiment. Finally, in 1979 I created 10 pairs of mounds and pits by digging up a 0.65 m^2 area of soil and depositing it to mimic a treefall (some burial of A horizon, some mixing with B horizon; sensu Beatty and Stone 1986). These artificial microsites have been monitored since 1979, along with nearby control mound-pits pairs.

Disturbance Legacy Effects on Understory

Treefall is a common disturbance in northeastern deciduous forest. As a complement to the literature on the canopy regeneration dynamics as a result of gap formation, my work has followed the longer-term effects of the soil disturbance created by the uprooting of the trees. This process creates persistent mound and pit microsites, which I estimated to be on the order of 200–300 years old (created before old-growth clearing). Thus, microtopography becomes a significant feature of the forest floor in these communities, representing a legacy of disturbance that has little to do with the original treefall effect (Nakashizuka 1989). In my study sites, mounds and pits together compose 60% of the forest floor area. Previous work (Beatty 1984) established that many understory species populations are concentrated in one main microsite, with some species being distributed among all microsites.

This creates a pattern of segregation such that the mounds have almost twice the species richness as pits, with little spatial overlap in many of the species distributions. Spatial heterogeneity has been suggested to be important in maintaining species richness in communities (Whittaker and Levin 1977) by allowing for species segregation among microsites and providing some buffering against extreme environmental conditions. One of the major questions my research has addressed is whether the presence of this microsite heterogeneity is a benefit or a liability to the community, in terms of species richness. As a benefit, it could provide a range of environmental conditions available to species so that some portion of a species population is likely to survive in even extremely stressful years. It may decrease competitive interactions via spatial segregation. It may enhance species richness by providing a variety of microsites to support species of widely varying environmental tolerances. As a liability, it may limit space available to species, possibly increasing competition in the microsites to which some species are restricted, and even limit species richness through exclusion or lack of space to support a viable population. To gain some understanding of what kind of role heterogeneity plays in deciduous forests, I experimentally tested several hypotheses about dynamics within and between mound and pit microsites.

Heterogeneity as a Liability?

Competitive Effects

Since mound microsites have higher richness than pit microsites, a valid question is whether the species packing is subject to competitive interactions that may exclude species from these sites. In the competition experiment, *A. divaricatus* was removed (above-ground and below-ground removal gave the same results), and responses of species were monitored in subsequent years. Before any manipulation, microsites were monitored for 2 years in all plots, establishing that there were no significant differences between pretreatment experimental and control plots (fig. 7.1). In the first year after removal, no significant changes in species cover were seen in 75% of species in control plots. However, in experimental plots, 83% of the species changed significantly, and the majority of these declined. Experimental sites also had new colonization in the first year, which increased richness significantly over controls. However, by the third year, many of those declining species became locally extinct, and richness was significantly lower than in control sites (fig. 7.1).

My conclusion is that although mounds may support twice as many species as pits, the continued coexistence of these species is mediated by the presence of the dominant *A. divaricatus*. I believe this dominant species may be acting as a keystone competitor. This effectively counters the liability factor of competitive exclusion by a dominant in spatially limited microsites because its

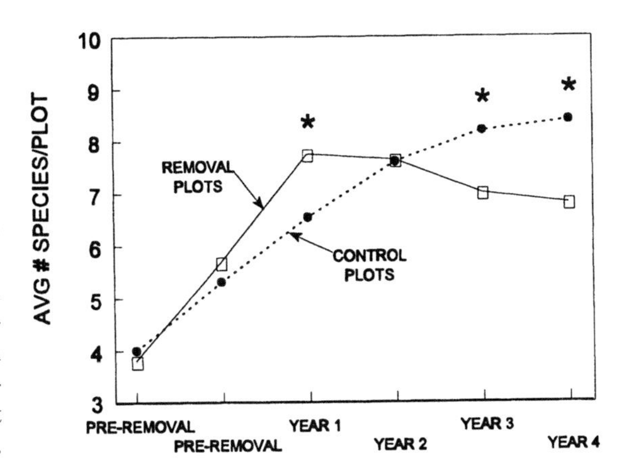

Figure 7.1. Species richness changes in experimental plots (solid line) where the dominant herb was removed versus changes in control plots (dashed line). Preremoval years had no manipulations; years 1–4 had continuous removal in experimental plots. *Significant difference between richness in control and experimental plots in that year (paired t test, $p < .05$).

presence instead promotes coexistence of a number of species, although with lower population sizes.

Environmental Limitations

Pit microsites have many fewer species than mounds, yet seem to have more favorable environmental conditions, such as higher available moisture and nutrients, higher pH, more moderate temperature fluctuations, and a better-developed A horizon (Beatty 1984; Beatty and Stone 1986). Pits also have much greater accumulation of leaf litter, which may act as physical and chemical limiting factors to establishment, particularly for small seeded species (Koroleff 1954; McPherson and Thompson 1972; Al-Mufti et al. 1977; Sydes and Grime 1981a, 1981b). We tested the hypothesis that leaf litter limits pit species composition by removing leaf litter from experimental pits after leaf drop in 1983 (mounds do not have appreciable litter, so were not manipulated) (Beatty and Sholes 1988). Plots were sampled for species composition during the growing season before removing litter. The litter was not allowed to accumulate in these sites for 8 years, after which I monitored the effects of reaccumulation of litter.

Before litter removal (1983), plots to be used for experimental and control pits were not significantly different in species richness (paired t tests). However, in the first year after removal of leaf litter, pits significantly increased in species richness to the point of being the same as adjacent mounds (fig. 7.2). During the 8 years of the experiment, species richness in litterless pits fluctuated in the same manner as in mounds, remaining not significantly different from mounds. In 1991, leaf litter was allowed to reaccumulate in experimental pits at the end of the growing season. During the 1992 season, species richness in experimental pits declined to levels seen in the control pits and remained at that level.

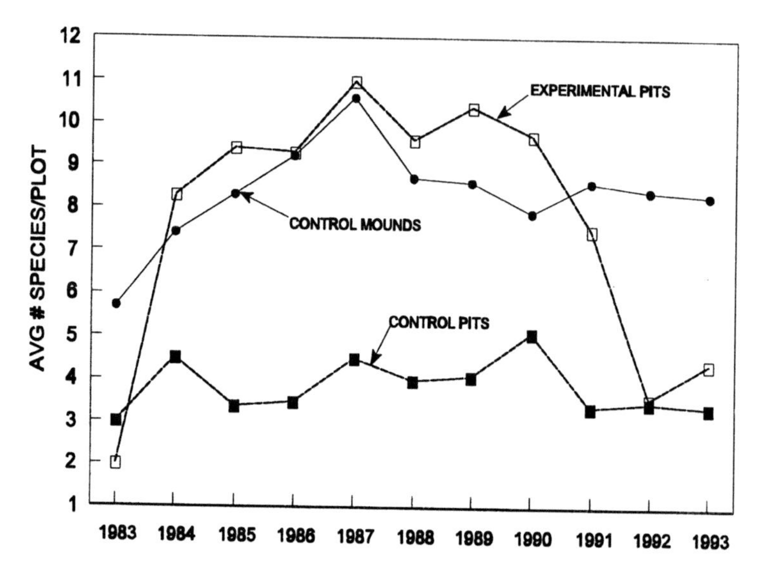

Figure 7.2. Species richness in mounds and pits in an experimental litter-removal study. The 1983 data are prior to any manipulations; litter was eliminated in experimental pits (open squares) for years 1984–91; litter was allowed to reaccumulate after 1991. Control pits (filled squares) and control mounds (filled circles) are shown for comparison. Control mounds are significantly different from control pits for all years (paired *t* test, $p < .05$). Experimental pits are significantly different from control pits ($p < .05$) for years 1984–91.

The mechanism behind such dramatic changes in species richness (and composition) was the successful germination and establishment in litterless pits. Even before litter removal, seed rain added many species to the pit seedpools that were not found in the flora of pits (Beatty 1991). However, these species either did not germinate or did not survive long enough to emerge above the litter layer. In the absence of litter, these species successfully established and survived the course of the experiment. The species that established in litterless pits were largely species that typically occurred on mounds (Beatty and Sholes 1988). Upon allowing litter to reaccumulate, almost all of these species were lost. Therefore, the leaf litter not only prevents establishment but also survivorship of adults.

Pit microsites do appear to be liabilities in the community, potentially limiting species richness by restricting space available to many species not able to tolerate the extreme conditions created by deep leaf-litter accumulation. However, although the richness of pits increased due to non–pit-species colonizing the experimental sites, the existing pit species declined (Beatty and Sholes 1988). By the end of the experiment, about 25% of the sites had lost one or two pit species, while maintaining the invaders (and thus a high richness). Thus, instead of limiting richness in the forest understory, unmanipulated pits allow the persistence of species that otherwise would not compete favorably in the understory community where many other species are present.

Succession in New Microsites

Both of the previous experiments examined dynamics on old, well-established mound and pit microsites. What are the dynamics on newly formed soil disturbances, and how do they differ from those in the mature microsites? Are newly created microsites rapidly colonized by forest species, or do exotic invaders preempt the space and delay establishment of mature forest species? On a more philosophical level, what is the quality of richness when enriched by species exotic to the forest community?

In 1978 I created pairs of artificial mounds and pits by digging up soil and redepositing it adjacent to the pit, as would happen in an uprooting event. Nearby natural mound–pit pairs were marked as controls. By monitoring species composition over the years, I found that there was an initial rapid colonization of new microsites by nonforest herbaceous species (e.g., *Eupatorium rugosum* Houtt., *Geum canadense* Jacq., *Solidago juncea* Ait., *Rubus allegheniensis* Porter), followed by a gradual increase in species more typical of the older forest microsites (controls). By the seventh year after creation (1985), the species richness and composition of new mounds was not significantly different from that of control mounds (fig. 7.3). The micro-

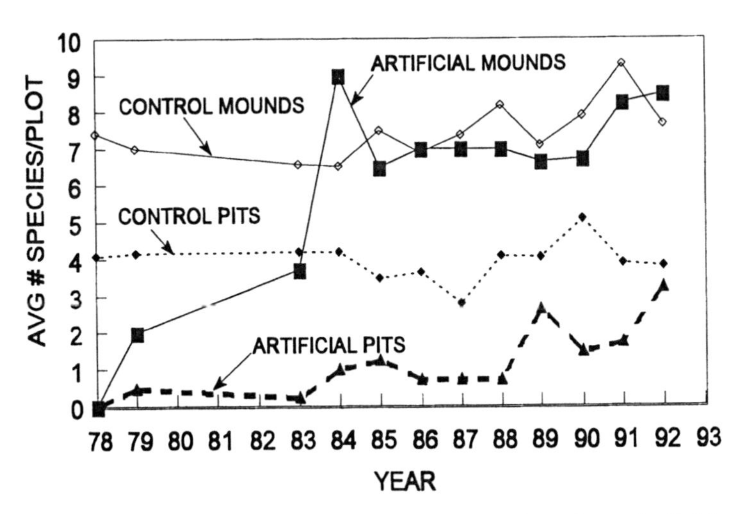

Figure 7.3. Species richness in artificial and control mounds and pits. Artificial mounds (filled squares) and pits (filled triangles) were created in 1979 and are compared with control mounds (open diamonds) and control pits (filled diamonds). Artificial mounds are significantly different from control mounds (paired t test; $p < .05$) in 1978–84. Artificial pits are significantly different from control pits ($p < .05$) for all years except 1992.

successional process took longer in pits, which only reached richness and compositional similarity to controls in year 14 (1992).

Although, initially, new microsites are largely occupied by species exotic to the mature forest community, there is a fairly rapid replacement of these species by forest understory species. The initial pulse of exotic species is accentuated in treefall gaps, where newly forming mounds and pits are also in a higher-light environment (Goldblum 1997; Goldblum and Beatty 1999) and are unstable substrates for several years as the root mass disintegrates (Beatty and Stone 1986; Nakashizuka 1989). As with treefall gap dynamics in general, new soil microsites provide a temporary space for exotic species colonization, but these species are lost to the understory community as the site recovers. Thus, overall richness is increased in space (fugitive persistence among new sites), but not in time, and there appears to be no limitation to mature forest species as a result.

Heterogeneity as a Benefit? Buffering Against Environmental Variability

It has been established that species composition and richness of the forest understory is affected by the heterogeneity of microsites. Species on mounds coexist in a seemingly high-competition environment tempered by dominance, whereas species in pits persist by tolerating limiting conditions deleterious to other species. The liability of spatial constriction into microsites is offset by the benefit of providing conditions where more species can persist and coexist. However, what happens when environmental conditions fluctuate? How do the microsites change, and what effect does this have on the dynamics of species maintenance in the forest community?

Using long-term data collected on control microsite plots over a 20-year period, I looked at the variability in species population sizes in the forest community (fig. 7.4). The variance in species density was used as a measure of change in population size from year to year. Species were first grouped into two main categories, based on their spatial distribution among microsites: ubiquitous species occurred in more than one microsite (although usually concentrating in a particular one); restricted species were found only in one microsite (i.e., only on mounds or only in pits). Species were then placed in subcategories, based on the specific microsite (mounds or pits) in which they mainly or exclusively occurred. The question was whether species with more limited spatial distributions show a greater variation in population size over time. The answer is yes: the average variance among species in the ubiquitous category was considerably less than that for the restricted species (fig. 7.4). In the 20-year period there were several drought years (e.g., 1980, 1985, 1987–88) and several very wet years (e.g., 1979, 1984, 1986), during which species restricted to one microsite may have experienced greater losses if conditions were stressful enough in the microsite. In contrast, I hypothesized that species with some individuals in more than one microsite may have

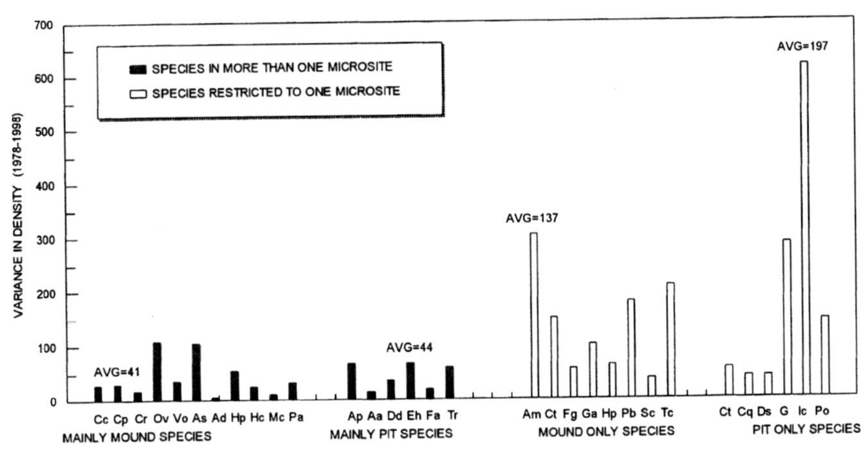

Figure 7.4. Variance in species density, 1978–98; higher values indicate greater variation in population density over the 20-year period. Species abbreviations are (1) mainly mound species: Cc = *Carex convoluta* Mackenz., Cp = *Carex pedunculata* Muhl., Cr = *Carex retrorsa* Schwein., Ov = *Ostrya virginiana*, Vo = *Veronica officinalis* L., As = *Acer saccharum*, Ad = *Aster divaricatus*, Hp = *Hieracium pratense* Tausch, Hc = *Hypnum curvifolium* Hedw., Mc = *Maianthemum canadense* Desf., Pa = *Prenenthes altissima* L.; (2) mainly pit species: Ap = *Acer pensylvanicum* L., Aa = *Arisaema atrorubens* (Ait.) Blume, Dd = *Dentaria diphylla* Michx., Eh = *Epipactus helleborine* (L.) Crantz, Fa = *Fraxinus americana*, Tr = *Trillium erectum* L.; (3) mound-only species: Am = *Aster macrophyllus* L., Ct = *Carya tomentosa* Nutt., Fg = *Fagus grandifolia*, Ga = *Galium aparine* L., Hp = *Hypericum punctatum* Lam., Pb = *Polygonatum biflorum* (Walt.) Ell., Sc = *Solidago caesia* L., *Tc* = *Tsuga canadensis*; (3) pit-only species: Ct = *Caulophyllum thalictroides* (L.) Michx., Cq = *Circaea quadrisulcata* (Maxim.) Franch. & Sav., Ds = *Dryopteris spinulosa* (O. F. Muell.) Watt, G = *Geranium maculatum* L., Ic = *Impatiens capensis* Meerb., Po = *Polystichum acrostichoides* (Michx.) Schott.

differential survivorship in different microsites, thus ensuring that a greater number of individuals survive the stressful years. I consider these species to be "shifters" because their survivorship shifts between microsites depending on environmental conditions, similar to the shifting microsite mosaic hypothesis of Whittaker and Levin (1977). This suggests that the presence of different microsites provides many species the opportunity for greater survivorship in the face of environmental variation, thus serving as a benefit to the maintenance of richness in the forest community.

Because my conclusion was based on observational data, I decided to experimentally test aspects of this microsite shifting hypothesis in the field by experimentally manipulating moisture regimes for replicate sets of microsites (refer to Methods section for details). I hypothesized that extreme conditions should have opposite effects on richness in mound versus pit microsites (a wet year might be good for mounds and bad for pits), which should affect survivorship. In experimental mound sites the species richness in wet and dry treatments fluctuated almost inversely (fig. 7.5a). The wet treatment increased in richness during the experiment and the dry treatment had no change, while controls declined in richness. The year after treatments were stopped, wet mounds declined in richness, while dry mounds increased, as did controls. Similarly, experimental pit sites had slightly inverse responses to wet versus dry treatments (fig. 7.5b). Wet pits did not change in richness, whereas dry pits increased in richness during the experiment, and controls declined. The effect of simulating drought conditions in a pit gave similar results to that of the leaf litter removal experiment, with new colonization by nonpit species. Although I expected exaggerated wet conditions in pits and exaggerated dry conditions in mounds to both lead to species decline, I found instead that extreme environmental conditions (of the experiment) promoted colonization (wet mounds and dry pits increased in richness) but had little liability for survivorship of existing species. This result helps explain the success of ubiquitous species that may colonize additional microsites in favorable years but not lose individuals in their main microsites in bad years. Restricted species are not likely to colonize other microsites, so they do not have this population buffering capacity. It is certainly possible that the mois-

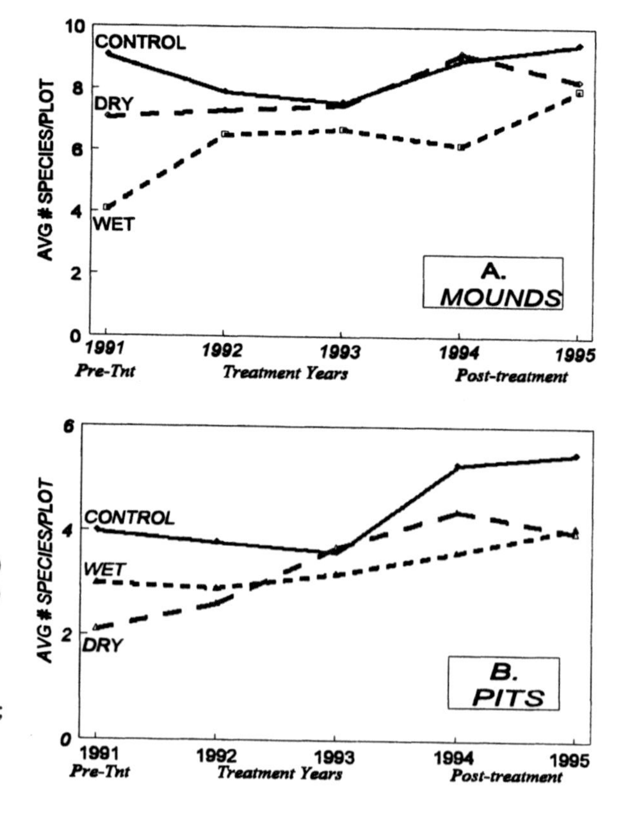

Figure 7.5. Species richness in experimental plots, with wet (twice ambient precipitation) and dry (half ambient precipitation) treatments. Results are shown for (A) mound microsites and (B) pit microsites. Data from 1991 are pretreatment; data from 1992 to 1993 are treatment applications; data from 1994 to 1995 are post-treatment.

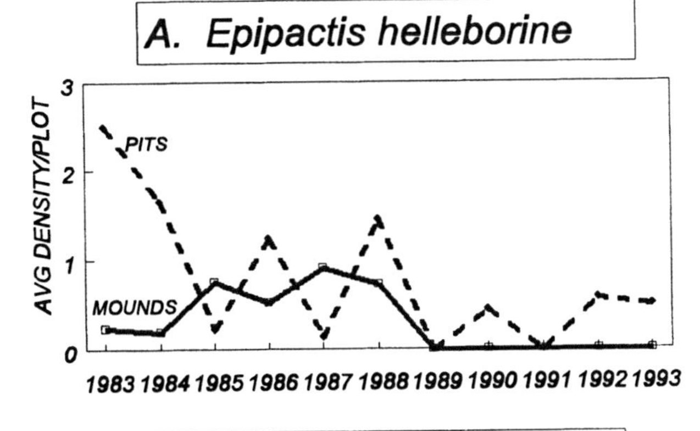

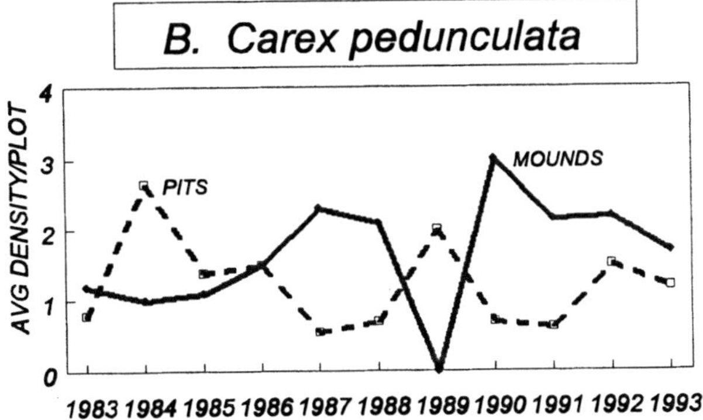

Figure 7.6. Individual species fluctuations in average density per plot over a 10-year period. Species density is given for mound (solid lines) and pit (dashed lines) microsites. (A) *Epipactis helleborine* is mainly found in pit microsites; (B) *Carex pedunculata* is mainly found in mound microsites. Variance in density values over a 20-year period is shown for these species in figure 7.4.

ture conditions of my experiment did not reach levels sufficiently extreme to adequately test the hypothesis that survivorship would be affected in stressful years. However, the result that ubiquitous species can expand their distributions in moderate wet–dry years was a complementary finding. Thus, microsite heterogeneity does offer some buffering against environmental fluctuation, whether extreme or moderate.

In looking at fluctuations of some of the ubiquitous species over a 10-year period, it is apparent that there is often an inverse response in species density in mounds versus pits in a given year (fig. 7.6). For the two species shown in figure 7.6, an increase in density in mounds is often accompanied by a decrease in density in pits in the same year, and vice versa. The observational data support the experimental data in suggesting that ubiquitous species can withstand environmental variation and maintain more stable forest-level population sizes by taking advantage of the microsite heterogeneity.

Because local microsite colonization and extinction are mechanisms by which species richness and population sizes fluctuate from year to year, an examination of these events over a similar time period is useful. The data in figure 7.7 are from control microsites (for the litter removal experiment) that were monitored from 1984 to 1993. Species colonization and extinction rates

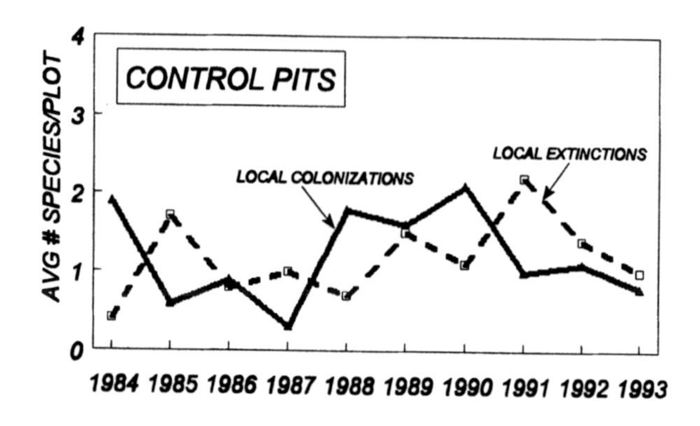

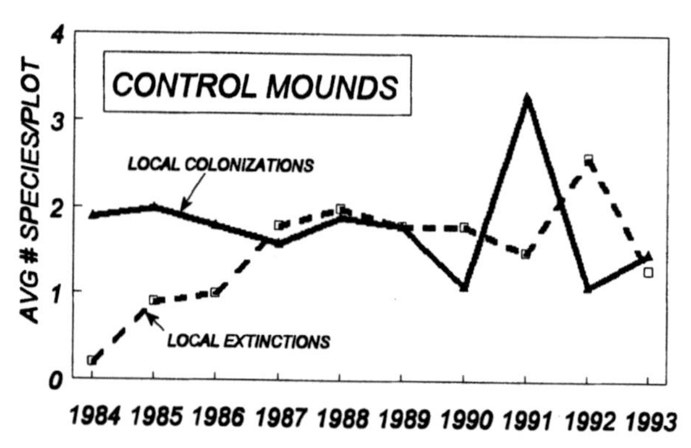

Figure 7.7. Average number of species disappearing from a plot (local extinctions; dashed lines) or appearing newly in a plot (local colonizations; solid lines) for control microsites over a 10-year period.

(species/plot/year) often vary inversely over this 10-year period, in a pattern similar to that of fluctuations in individual species density in mounds and pits in figure 7.6. If colonization rates are high in a given microsite (mound or pit), extinction rates are often low (favorable conditions). In stressful years for a given microsite, extinction rates go up, while colonization rates go down (more so in pits than mounds). This supports the idea that the mechanism of the individual species fluctuations in density in a given microsite are partly a result of new colonization or extinction events as environmental conditions vary from favorable to unfavorable over the years. Finally, to further reinforce this pattern, the fluctuation in average richness for all control mound and pit microsites shows the same inverse trend (fig. 7.3, control curves). Often an increase in richness in one microsite is accompanied by a decrease in the other microsite. This, then, also buffers richness fluctuations in the forest as a whole, given that a local extinction of a species on a mound is likely to be balanced by a local colonization by that same species in a pit. Although not all species in the forest community are shifters, the proportion (ca. 50%) is high enough to have an impact on forest richness and population sizes.

In conclusion, heterogeneity appears to be more of a benefit to the forest community than a liability. It serves to segregate species that might otherwise out-compete one another. It provides a range of environmental conditions that may be exploited as climatic conditions vary from year to year, therefore serving as a spatial buffer to extreme environmental change. Those species able to grow in more than one microsite are less likely to become locally extinct, thus maintaining a higher species richness in the forest community.

Models for the Future

The various studies in this chapter differ from most studies in disturbance ecology in terms of the scale at which the data are collected and in the stratification by known microsites in the community. Rather than averaging across space, this work seeks to identify sources of variability at the fine scale. It is my hope that this approach can be applied in numerous other community types, seeking some common ground for testing hypotheses about species richness dynamics. To that end, I have developed two hypotheses relating to the function of heterogeneity in space and time. To determine the generality of these hypotheses, they need to be tested in as many communities and geographic regions as possible. The "intermediate heterogeneity hypothesis" presents the idea that heterogeneity can be both a benefit and a liability to a community, depending on scale and intensity of that heterogeneity. The "heterogeneity cycle hypothesis" proposes a model of heterogeneity in forest communities that relates to land use and successional age of a stand.

Intermediate Heterogeneity Hypothesis

For the forest communities I have been studying, as summarized here, it is clear that spatial heterogeneity in the form of mound and pit microtopography does not pose a limitation to species richness but enhances richness on several levels. Other work, however, has found that extreme microsites create conditions limiting to many species (Falinski 1978). Therefore, it seems that the degree or intensity of heterogeneity may be important in gauging its impact in a community. Based on limited sampling in areas at the Huyck Preserve that had been cleared and plowed, I found that this activity completely eliminated the mound–pit microtopography, leaving a homogeneous forest floor compared to the mature forest communities. Without mound and pit microsites, the forest had fewer total numbers of species (ca. 15% fewer than average richness in adjacent forest of similar age), a higher proportion of species with large population sizes, and fewer rare species. A current research project is addressing whether this decline in richness might be due to lack of propagules to colonize the site, lack of specific microsites for some species (such as pit obligates that are poor competitors), or greater dominance by several species (similar to the results of my *A. divaricatus* removal experiment on mounds).

In 1977, I undertook some preliminary sampling before initiating the stratified microtopography design used in the work reported in this chapter.

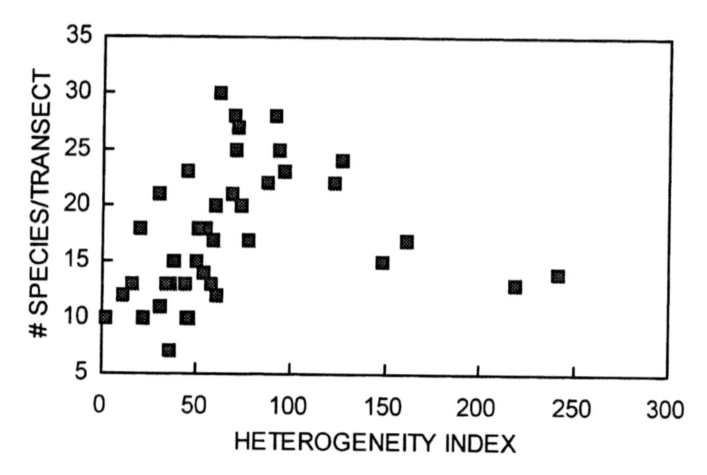

Figure 7.8. Species richness in 10-m² belt transects sampled in mature forest communities. Heterogeneity index is the variance of deviation in microelevation above and below the average slope of the transect. A high value indicates greater height and depth to mound and pit microsites, respectively.

I wanted to test the hypothesis that greater heterogeneity correlated positively with higher richness. I sampled 40 belt transects (10 m × 1 m), measuring species composition and micro-elevational variability (deviation above or below the average slope for the transect). All transects were in mature forest with a range of mound heights and pit depths in microtopography. No flat sites were sampled at the time. The results did not support the hypothesis (fig. 7.8). Instead, there was a peak in richness at an intermediate level of heterogeneity and a decline for areas having more extreme differences in mound–pit relative height (which would give a greater variance in microelevational departures from the average slope). At extreme levels of heterogeneity, physical limitations may exclude many species (i.e., heterogeneity as a liability). For instance, pits may accumulate litter to a much greater depth, retain moisture to the point of saturation for longer into the summer, and maintain much cooler soil temperature, all of which would affect germination and survivorship even for pit obligates. Coupled with the finding that forests with a lack of microtopography may tend toward lower richness, it appears that an intermediate spatial heterogeneity presents the condition where sufficient microsite differentiation exists to both support and segregate species, allowing coexistence and higher richness. In addition to continued work in the forests I have been studying for 23 years, there is a need for data sets from other communities to support or refute this hypothesis.

Heterogeneity Cycle Hypothesis

Land-use history and stand successional age may also explain species-richness dynamics. In all the mature stands that I have sampled, there is a well-developed microtopography, the stand ages are between 85 and 125 years, and each was grazed (not plowed) before abandonment with subsequent successional development. However, in the mosaic of forests and old fields of the Huyck Preserve, there are stands that vary in age from very young stands (ca. 25 years) to much older stands (150 years or more) and that vary in land-use history. Plowing versus grazing creates different scales and intensities of heterogeneity in these forests. Plowing obliterates microtopography, leaving a much more uniform soil surface with lower heterogeneity. Grazing preserves much of the previous microtopography, so that a young successional forest will have some degree of spatial heterogeneity. In both cases, as the forest regrows, trees will reach sufficient age and size to uproot and begin creating new mound–pit microsites. This process accumulates with stand age, adding to and often reworking the microtopography and degree of heterogeneity. However, the stand regenerating on a plowed surface begins with very low heterogeneity, whereas the one on a grazed surface initially has a moderate to high heterogeneity (a legacy from previous forests before clearing). It may take several generations of treefall activity for the plowed site to redevelop the heterogeneity of the grazed sites. In contrast, a truly old-growth stand would have possibly the greatest development of microtopography and degree of heterogeneity.

For the northeastern U.S. landscape, I propose three pathways that are important in the development of spatial heterogeneity in forests (fig. 7.9). The old-growth model (center box in fig. 7.9) continuously generates and accumulates microtopography through the process of treefall and gap regeneration. These mounds and pits likely last centuries (Lyford and MacLean 1966). A land-use disturbance cycle involving clearing and/or grazing followed by abandonment (lower portion of fig. 7.9) could lead to a period of decreased heterogeneity with redevelopment through the successional recovery process. This should produce an intermediate level of heterogeneity compared to the old-growth stand. Finally, a land-use disturbance cycle that involved plowing and complete loss of microtopography (upper portion of fig. 7.9) would result in a successional stand with little or no heterogeneity initially, and a long process of reaccumulation of microrelief to follow. This stand would have the lowest level of spatial heterogeneity. If there were repeated forest-clearing activities, a cycle of declining heterogeneity could result, depending on the nature of the land use. In addition, when assessing the effect of stand age on composition and richness, this effect of land-use on heterogeneity may play a role that is independent of succession. Chapter 14 (this volume) discusses other land-use practices and their effects on substrates.

Based on studies in deciduous forest communities, I propose a general heterogeneity cycle hypothesis which states that many communities may have persistent forms of spatial heterogeneity that come about through cycles of creation and destruction of heterogeneity and that the position of a community in this cycle may impact how or whether heterogeneity has any effect on richness. In work I have done in chaparral ecosystems in the Mediter-

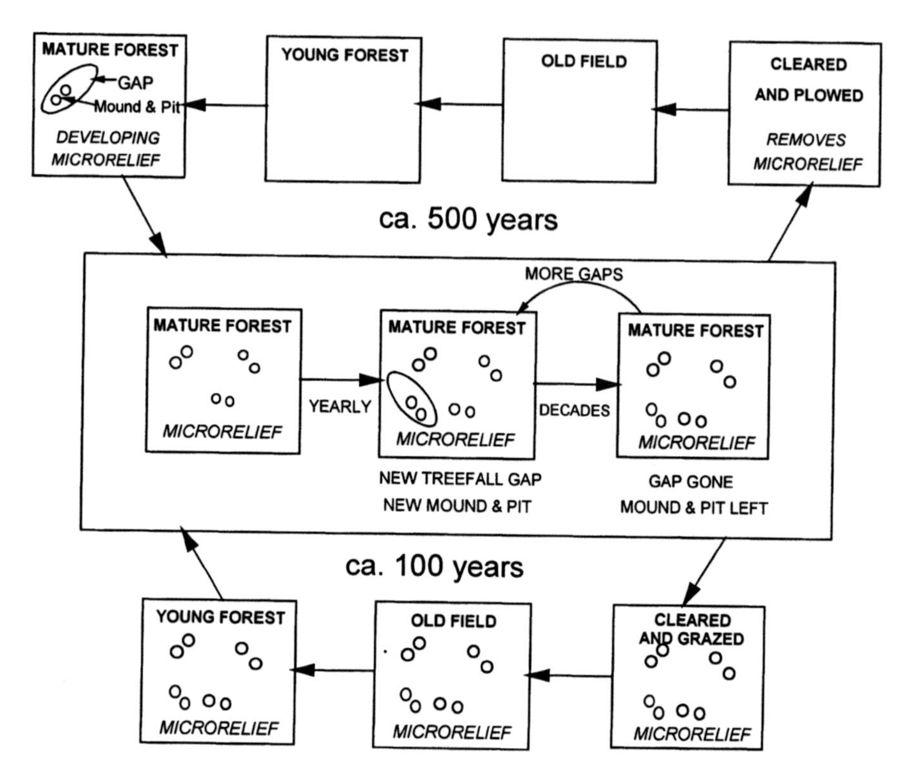

Figure 7.9. Cycles of change in northeastern deciduous forest landscapes. Microsite heterogeneity is generated internally within mature forests via the treefall uprooting process. Exogenous anthropogenic disturbances initiate longer term cycles of regrowth and can alter the fine-scale nature of the soil substrate.

ranean climate of southern California (Beatty 1987a), I discovered a cycle of heterogeneity related to shrub effects on soil and the fire cycle. Different chaparral shrub species affect soils differently, creating a strong pattern of heterogeneity in physical and chemical properties, which intensifies as a stand becomes older (Beatty 1987b). However, fire acts to reset the heterogeneity level to almost zero (very homogeneous surface) (Beatty 1989). Therefore, the fire return interval is important not only to species adaptations to fire recovery, but to the degree of generation of surface soil heterogeneity. In addition, those chaparral communities on the channel islands, which have much longer fire intervals (well over 150 years for Santa Cruz Island), may have a much better developed soil heterogeneity, which may play a role in understory species dynamics (this needs to be tested). These older island stands are up to 30 m in height, have a more open canopy, and have a substantial understory (Beatty and Licari 1992). I suggest that investigation of fine-scale heterogeneity in soils of coniferous forests, which also have fire cycles, might be another community in which to test the heterogeneity cycle hypothesis. As for deciduous forests in the eastern United States, just looking at stand age and species-richness relationships may give an unrealistic view of a successional process, if the land use and heterogeneity are confounding factors not taken into account. Because heterogeneity may indeed be important in maintaining species richness of communities, knowing how the land use history interacts with the development of heterogeneity can be a useful management, conservation, and restoration tool.

Conclusions

The species richness of forest understory communities can be affected by numerous spatial and temporal factors. Depending on the geographic location of the forested community, factors operating at a coarse scale, such as disturbance (treefall, massive blowdown, fire, insect outbreak, land use change), can substantially alter the successional development of a community (chapter 14, this volume). On a finer-scale and usually a shorter time frame, an underlying spatial heterogeneity within a community can affect local distributions, richness, and yearly variation in population sizes. At an even finer scale, dynamics within microsites themselves (competition, colonization and extinction events) can determine the presence or absence of species or affect species reproductive success and productivity. A landscape mosaic exists because of the stand-destroying or perturbing effects at the coarser scale, and this mosaic also can affect the finer-scale, within-community dynamics by providing varying sources of species for colonization of new and existing microsites. The heterogeneity in a landscape and within a community is an important feature affecting species richness in understory communities.

The results and predictions made in this chapter have great potential for management, conservation, and restoration as well as preservation. Because communities chosen (obtained) for preservation are often patches within mosaic landscapes (lots of edge effect, little interior), the exogenous factors (physical and biological) are likely to be very important. Knowing whether target forest species have suitable microsites may be crucial to their maintenance in a spatially limited community. In addition, how do exogenous factors affect these microsites (e.g., colonization by exotics, overland flow deposition or erosion, wind redistributing leaf litter, browsing or herbivory effects that might shift competitive interactions in a microsite)? In a restoration plan, one could consider whether producing additional microsites might promote species maintenance. An important consideration is whether an increase in species richness is always a positive development. Many factors can increase richness in the understory of a forest community, but often this increase is due to the addition of species exotic to the forest community or the geographic region (chapters 6, 13, 14, this volume). In many cases in the arid western United States, exotic species exploit conditions or microsites not occupied by natives and gain a foothold in a community, with negative impacts for many natives. However, in the moist deciduous forest of the northeast, the few exotic species are usually considered naturalized and an established part of the forest community. None of the naturalized species in my study sites are aggressive colonizers, nor do they appear to affect other forest species distributions. The nonforest exotics that colonize new mound–pit formations do not persist beyond 15 years (in my experimental study), so disturbance does not facilitate permanent invasion of exotics. Because much attention has been paid to the importance of maintaining a natural disturbance regime in conservation efforts, I suggest we also pay attention to whether we need to maintain a certain heterogeneity regime (intensity and cycle), and what impacts this might have on native and non-native species composition.

Summary

This chapter has explored some of the mechanisms by which heterogeneity may act as a liability and as a benefit to maintaining species in the community. In most cases, for the deciduous forest communities in the northeastern United States that I have studied, spatial heterogeneity (mound–pit microrelief) appears to provide a means by which species are maintained in the community. This maintenance may be a result of the presence of a specialized microsite in which a species finds a tolerable environment with minimal competition (pits), or the presence of a keystone competitor in a rich microsite (mounds) that allows coexistence of numerous species with lower population sizes by preventing other potential dominants from exerting greater influence. Those species whose spatial distributions include multiple microsites also have the added benefit of being buffered against moderate to severe environmental fluctuations. Colonization increases in years where conditions are favorable in a given site. Even though localized extinction may also result in unfavorable years, the loss is not uniform among microsites, thus ensuring that some individuals survive. Those species restricted to one microsite, however, suffer a greater variability in population size with environmental fluctuation and have a higher risk of local extinction. For these species, the heterogeneity may limit available space and serve as a liability to those species.

Given the role of heterogeneity in these forest communities and the prevalence of land use changes over the past 200 years in the northeastern United States, the interaction of these factors should be taken into account in any study of richness of a community. The proposed intermediate heterogeneity hypothesis suggests that an intermediate level of heterogeneity will promote greater species richness in a community. Extreme microsites may shift the balance toward a liability effect, restricting species composition. Lack of microsites may increase overall competition with greater species overlap and may not provide necessary safe sites for some species. The proposed heterogeneity cycle hypothesis suggests that the current level of heterogeneity in a community is likely the result of past events that either create/enhance the number of microsites or destroy/obliterate microsites. Plowing and intense compaction from overgrazing obliterate mounds and pits so that a second-growth forest on such land will have a much lower spatial heterogeneity for the understory. An old second growth stand on land not plowed will retain microtopography from the original stand and continue to create new treefall mounds and pits with age. This stand will have a higher spatial heterogeneity. Therefore, stand age alone should not necessarily be used to predict richness patterns through succession, as there are other confounding factors affecting that richness.

8

Interactions Between the Herbaceous Layer and Overstory Canopy of Eastern Forests

A Mechanism for Linkage

Frank S. Gilliam
Mark R. Roberts

A clear understanding of the nature of interactions between vegetation strata of forest ecosystems has implications for both basic and applied studies in forest ecology. For example, community mapping studies may rely on the dominant overstory strata as interpreted from aerial photographs or satellite imagery to represent the entire community. Conversely, site classification systems may use herbaceous layer vegetation to represent communities and their relationships to environmental factors (Cajander 1926; Rowe 1956; Pregitzer and Barnes 1992). In such studies, an understanding of relationships between vegetation strata is necessary if interpretations are to be made concerning the entire community.

This information is generally lacking in disturbed stands because site classification studies are usually carried out in mature stands. As increasing proportions of forested areas become disturbed, it is a critical to know whether the same overstory–understory relationships that occur in mature stands are also found in disturbed stands. Also, the need to manage for biodiversity, a need now recognized by national and global forestry organizations (Burton et al. 1992; Roberts and Gilliam 1995a), requires a shift in focus away from the small number of commercially important tree species toward all plant species, including noncommercial tree and herbaceous species. Thus, the nature of interactions among forest strata and their response to forestry practices need to be understood to maintain biodiversity (Gilliam and Roberts 1995).

In this chapter we have three primary objectives. First, we determine via literature review what is known of interaction among forest strata, with a specific focus on overstory–herbaceous layer interactions in eastern deciduous forests. We present contrasting views of the nature of these interactions, from one that sees a quantifiable linkage among strata to one that sees little true interaction occurring. Second, we develop a mechanistic explanation for patterns of linkage in forest ecosystems, with emphasis on eastern deciduous forests. Finally, we examine data from two different forest types, a central Appalachian hardwood forest and a successional aspen forest of northern lower Michigan, for evidence supporting or refuting this explanation.

The Nature of Linkage Among Forest Strata

Studies of secondary succession often have emphasized changes in only dominant species in a single stratum of vegetation. For example, studies of old-field succession in the North Carolina Piedmont (e.g., Oosting 1942; Keever 1950; 1983; Christensen and Peet 1984; De Steven 1991b) have focused on the shift from herb-dominated communities early in succession to pine-dominated and then hardwood-dominated communities later and thus have not examined the herbaceous layer that develops beneath the woody overstory canopy during these later stages of old-field succession (but see chapter 9, this volume). Indeed, despite the increasing number of studies of the herbaceous layer in forest communities (see table 1.1), few consider interactions among strata in any forest type.

Influence of the Overstory on the Herbaceous Layer

There is little argument that the forest overstory and herbaceous layer exert reciprocating influences on each other. The forest overstory has a direct effect on the availability of resources to herb layer species, the most obvious being to decrease the quantity and alter the quality of light reaching the forest floor (chapter 3, this volume). Other effects would include decreasing nutrient and moisture availability by competitive uptake by fine roots of trees. Although he did not speculate about mechanisms, Rogers (1981) found nearly 20% higher species richness and around 70% higher cover for the herb layer of mature mesophytic stands (Minnesota to Michigan) with little or no *Fagus grandifolia* Ehrh. compared to that of stands with *F. grandifolia* codominance in the overstory. Crozier and Boerner (1984) reported tree species-specific differences in microhabitat conditions (e.g., different levels of soil nutrients caused by differences in stemflow chemistry) that resulted in spatial variation in cover of dominant herb species. Hill and Silander (2001) found that spatial dynamics of dominant ferns of mixed hardwoods-hemlock stands of Connecticut varied significantly among dominant tree canopy species (i.e., there was a species-specific effect of individual trees on distribution of ferns). They ascribed these differences in fern species dynamics to different light regimes among contrasting stand types.

Other studies have emphasized the importance of detrital inputs to the forest floor by overstory species. Whitney and Foster (1988) found substantial differences in percent occurrence of numerous herb layer species in conifer and hardwood stands in New England, suggesting that, in addition to variation in soil moisture and light regimes, such differences resulted from contrasts in physical and chemical characteristics of litter of conifer versus hardwood species. Nemati and Goetz (1995) made similar conclusions for herb layer differences in stands of *Pinus ponderosa* Dougl. ex Laws. versus *Quercus gambelii* Nutt. Saetre et al. (1997) found that abundance of herb layer species was lower under pure *Picea abies* (L.) Karst. stands than under mixed *P. abies*/*Betula* stands. They proposed that foliar litter from *Betula* (and associated higher fertility) was the most important cause of such pronounced stand-related differences in the herb layer. McGee (2001) determined that decaying logs of dead overstory species provided sites for early establishment critical for some, but not all, herb layer species, suggesting an additional influence of the overstory on species composition of the herbaceous layer.

Influence of the Herbaceous Layer on the Overstory

Although the influence of the herbaceous layer on the overstory of forests may not be as obvious as that of the overstory on the herb layer, it is potentially as profound (Maguire and Forman 1983). In chapter 1 we discussed the temporal and spatial dynamics of resident and transient species of the herb layer, with resident species being the truly herbaceous (nonwoody) annual, biennial, and perennial component of the herb layer, and transient species being predominantly seedlings and sprouts of tree species. Although resident and transient species exhibit contrasting life histories, they share common resources when co-occurring in the herb layer. Maguire and Forman (1983) demonstrated that cover and composition of herbaceous species (residents) determined, in part, the density and distribution of seedlings of dominant tree species (transients) in an old-growth hemlock-hardwood forest.

Because of their typical growth characteristics that often include a dense spreading of fronds, fern species can have a particularly profound effect on survivorship and growth of juveniles of forest overstory species. Horsley (1993a) demonstrated a species-specific interference of *Dennstaedtia punctilobula* (Michx.) limiting growth of seedlings of *Prunus serotina* Ehrh. in hardwood forests of the Allegheny region of Pennsylvania. George and Bazzaz (1999a, 1999b) showed that high fern cover in a New England hardwood forest can function as a species-specific filter, affecting emergence and survival of tree species. This has particular ecological significance in the context of response of eastern hardwood forests to disturbance (chapter 11, this volume).

Linkage as Reported in the Literature

Several studies have reported significant relationships between species' patterns in overstory and herbaceous strata. Using canonical correlations between principal components axes, Gagnon and Bradfield (1986) concluded that tree and herb strata of coastal Vancouver Island forests were linked via their response to predominant site gradients. Roberts and Christensen (1988) combined canonical correlation analysis with detrended correspondence analysis to examine vegetation strata of successional aspen stands of northern lower Michigan; they attributed significant correlations between strata to soil factors and disturbance regime. Hermy (1988) demonstrated correlations between strata in deciduous forests of Belgium, concluding that the degree of correspondence between compositional patterns was directly proportional to β diversity. Working in upland hardwood forests of northwestern lower Michigan, Host and Pregitzer (1992) determined that significant tree–herb linkages resulted from similar responses of strata to environmental and historical factors, citing moisture availability as especially predominant. Nemati and Goetz (1995) described a linkage between *Pinus ponderosa*/*Quercus gambelii* overstory and herbaceous understory by correlating canonical variable scores of the overstory to those for the herb layer. They concluded that linkage resulted from a variety of factors, but emphasized the importance of canopy-mediated changes in environmental conditions for herb layer species, such as light availability and soil acidity. Gilliam et al. (1995) studied interactions between the overstory and herb layer in second-growth hardwood forests of West Virginia. They found evidence of linkage for mature stands, but not for young (20-year old, even-aged) stands, and concluded that linkage changes through secondary succession, becoming tighter through time after disturbance.

Other studies, however, have concluded that forest strata do not form significant linkages. Looking at β diversity in undisturbed *Fagus grandifolia* communities of Great Smoky Mountains National Park, North Carolina, Bratton (1975) also found significant responses of species diversity to a moisture gradient. In contrast to Host and Pregitzer (1992), however, Bratton determined that overstory and herbaceous understory strata responded to this gradient in a manner that was neither linear nor parallel between strata.

Sagers and Lyon (1997) found that species associations in riparian forests of the Buffalo National River, Arkansas, were strongly influenced by gradients of pH and elevation. They concluded that forest strata largely responded to these gradients independently. They referred to this independent response as "incongruence" and suggested several possible reasons for its occurrence, including (1) environmental gradients appearing continuous at the landscape scale may be discontinuous at the local scale, and (2) each forest stratum may respond to a disturbance (in this case, largely flooding of riparian zones) in ways distinct from other strata (Lyon and Sagers 1998).

One of the more compelling arguments against the existence of linkage among forest strata was presented by McCune and Antos (1981). They reviewed earlier work in Europe (e.g., Lippmaa 1939) and North America (e.g., Cain 1936) that rejected the notion of linkage and instead urged a "synusial approach" to studying forest communities—taking a view that each stratum of a forest comprises a community (synusia) to be considered a distinct unit of vegetation (Oosting 1956). More important, McCune and Antos (1981) tested for linkage among five strata in forest stands of Swan Valley, Montana, using correlation of dissimilarity matrices, Bray-Curtis polar ordination, and cluster analysis. The different strata changed in composition across environmental gradients neither at the same rate nor in the same pattern (McCune and Antos 1981). They further concluded that apparent linkages, when found, can be artifacts of diversity across large sample areas. By sampling over large areas, it might be possible to encounter, for example, a conifer stand with its associated herbaceous component and a hardwood stand with its contrasting associated herbaceous component. Ordination of such data would produce two discontinuous clusters of plots and might lead to a spurious conclusion that the herbaceous and overstory components are closely linked.

Thus, part of the debate over the existence of linkage among forest strata appears to have arisen from studies that have addressed the question at different spatial scales. Those working at the landscape scale (e.g., McCune and Antos 1981) have not found linkage to occur. We suggest that linkage is a phenomenon that does indeed occur in forest ecosystems but that linkage operates at spatial scales smaller than the landscape.

Linkage may arise in two ways: (1) from similarities among strata in response to the same environmental factors (e.g., Gagnon and Bradfield 1986; Hermy 1988; Roberts and Christensen 1988, Host and Pregitzer 1992) or (2) from direct and reciprocating influences of the overstory and the understory on each other. The second mechanism has been documented for many forest types at fine spatial scales (e.g., 1-m^2 plots) as the influence of individual canopy trees on herbaceous layer species Everett et al. 1983; Turner and Franz 1986; Joyce and Baker 1987; Tyler 1989; Nemati and Goetz 1995; Berger and Puettmann 2000; Hill and Silander 2001) via various mechanisms including changes in soil acidity and fertility, light availability, or physical effects of litter under canopy trees.

A Mechanism for Linkage Between Forest Strata

It is notable that the reciprocating effects between overstory and herbaceous layer suggested in studies such as Maguire and Forman (1983) and Gilliam et al. (1995) occurred in mature forests. However, up through the thinning phase of secondary forest succession (chapter 9, this volume), the overstory and herb strata respond to different sets of environmental factors. Overstory composition at this time in succession is governed largely by competition for light (described by Bormann and Likens [1979] as the exploitive strategy), whereas herb layer growth and composition is determined largely by availability of moisture and nutrients (Gilliam and Turrill 1993, Morris et al. 1993; Wilson and Shure 1993). As the stand approaches steady state, the overstory becomes dominated by shade-tolerant species that were able to survive beneath the initial canopy of intolerant species. The new canopy is more closed and stratified and alters light conditions for herb layer species (chapter 3, this volume; Brown and Parker 1994). Seedlings and sprouts of woody species often increase (relative to herbaceous species) in the herb layer because of more light-limited conditions (Wilson and Shure 1993; Gilliam et al. 1995; Walters and Reich 1997). Woody species also exhibit greater relative abundance at this time because of an increase in the number of juveniles from late successional overstory species that are typically prolific seed producers. The result of successional change, then, is that the two strata start responding to similar environmental gradients, establishing and intensifying the linkage between overstory and herbaceous layer (Gilliam et al. 1995). This leads us to pose the following hypothesis as a mechanism of linkage between forest strata: Linkage among forest strata arises from parallel responses of strata to similar environmental gradients.

One of the challenges of testing such a mechanism is establishing the appropriate environmental gradients to which species may be responding. Direct gradient techniques are possible by stratifying sampling along a known gradient, such as elevation in a forested watershed (Barbour et al. 1999). This, however, has the limitation of presuming that the chosen gradient (e.g., elevation) is indeed an overriding factor influencing both species composition and gradients of other important environmental factors (e.g., moisture and soil nutrients). Indirect gradient techniques are also possible, wherein a multivariate method, such as detrended correspondence analysis (DCA), is used to generate ordination axis scores for plots. Axis scores then are correlated to environmental variables measured at each plot (McCarthy et al. 1987; Roberts and Christensen 1988; Gilliam et al. 1993; Sagers and Lyon 1997). This has the obvious limitation of presuming that ordination axis scores and environmental variables are related in an inherently linear fashion.

Palmer (1993) assessed the advantages of yet another multivariate analytical approach, canonical correspondence analysis (CCA) (ter Braak 1986). In addition to pointing out the numerous improvements of CCA over DCA, Palmer (1993) showed that the output from CCA contains an important feature that is germane to testing our gradient-based hypothesis of linkage. Because CCA performs a least-squares regression of plot scores (species' weighted averages) as dependent variables onto environmental variables as independent variables, CCA is a form of direct gradient analysis (Palmer 1993). In addition to generating ordination diagrams with plot and species locations, CCA also generates environmental vectors originating from the center of the ordination space. The lengths of these vectors represent the gradient lengths of each measured environmental variable, such that vector length is proportional to the importance of an environmental gradient in explaining species' patterns. Thus, shorter lines represent gradients of lesser importance, longer lines represent gradients of more importance. Accordingly, whether the herbaceous layer and overstory are responding to environmental gradients in a similar fashion may be assessed by performing CCA on each stratum separately for a given stand age and then comparing vector lengths of herb layer versus overstory on a gradient-by-gradient basis.

Using such an approach, we examined two data sets for evidence of whether herbaceous and overstory layers respond to similar gradients. The data sets are each from a different site and study: the Fernow Experimental Forest in Tucker County, West Virginia, and the University of Michigan Biological Station in northern lower Michigan. Because these were carried out as unrelated studies, they were not done using identical sampling methods. Nonetheless, the important environmental gradients thought to control species distributions were quantified as appropriate in each study, including stand structural variables, as well as soil nutrient and moisture variables.

Study Sites

Fernow Experimental Forest

The Fernow Experimental Forest (FEF), an approximately 1900 ha area of largely montane hardwood forests in the Allegheny Mountain section of the unglaciated Allegheny Plateau, is located in Tucker County, north-central West Virginia. Mean annual precipitation is approximately 1430 mm, with most precipitation occurring during the growing season (Gilliam and Adams 1996a). Four contiguous watersheds were selected for this study: WS7 and WS3 were about 20 year-old, even-aged stands that developed following clearcutting (hereafter "young" stands); WS13 and WS4 were uneven-aged stands (>80 years old, hereafter "mature" stands).

Study watersheds at FEF support primarily mixed hardwood stands, with dominant trees varying with stand age. Early-successional species, such as *Betula lenta* L., *Prunus serotina*, and *Liriodendron tulipifera* L. are dominant in young stands, whereas late-successional species, such as *Acer saccharum* Marshall and *Quercus rubra* L., are dominant in mature stands (table 8.1). Dominant herbaceous layer species vary less with stand age and include *Laportea canadensis* (L.) Wedd., *Viola* spp., and several ferns, including *Dryopteris marginalis* L. Gray and *Polystichum acrostichoides* Michx. Schott. (table 8.2).

Soils are similar among study watersheds. These are relatively thin (<1 m in depth), acidic, sandy-loam Inceptisols of two series: Berks (loamy-skeletal, mixed, mesic Typic Dystrochrept) and Calvin (loamy-skeletal, mixed, mesic Typic Dystrochrept) (Gilliam et al. 1994). Soils of the study watersheds are generally acidic, but are high in organic matter, resulting in high cation exchange capacity (table 8.3).

University of Michigan Biological Station

The study was conducted within a five-county (Cheboygan, Emmet, Charlevois, Otsego, Montmorency) region of northern lower Michigan. Climatic conditions are relatively uniform throughout the area, with an average an-

Table 8.1. Important overstory (live woody stems ≥ 2.5 cm diameter at 1.3 m height) species of young versus mature stands of Fernow Experimental Forest, West Virginia

Species	Stand age class[a]	
	Young	Mature
Acer pensylvanicum L.	5	5
A. saccharum	33	90
Betula lenta	19	—
Fagus grandifolia	5	14
Fraxinus americana	10	2
Liriodendron tulipifera	16	5
Prunus serotina	76	20
Quercus prinus L.	—	12
Q. rubra	7	30
Robinia pseudoacacia L.	6	—
Sassafras albidum (Nutt.) Nees.	6	—

Data are mean importance values (sum of relative basal area and relative density) for two watersheds per age class. Nomenclature follows Gleason and Cronquist (1991).

[a]Young = 20 years; mature > 80 years.

Table 8.2. Important herbaceous layer (vascular plants ≤ 1 m tall) species of young versus mature stands of Fernow Experimental Forest, West Virginia

Species	Stand age class[a]	
	Young	Mature
Acer pensylvanicum L.	7	17
Dryopteris marginalis	30	6
Laportea canadensis	17	27
Polygonatum biflorum (Walter) Elliot	—	10
Polystichum acrostichoides	9	15
Prunus serotina	9	5
Rubus spp.	12	4
Sassafras albidum (Nutt.) Nees	8	—
Smilax rotundifolia L.	14	9
Viola spp.	20	20

Data are mean importance values (sum of relative cover and relative frequency) for two watersheds per age class. Nomenclature follows Gleason and Cronquist (1991).

[a]Young = 20 years; mature > 80 years.

nual precipitation of 770 mm and an average annual temperature of 6.2–6.7°C. Precipitation is distributed relatively evenly throughout the year (Albert et al. 1986).

Soils of the study area are Spodosols derived from parent materials of contrasting glacial origin. Study sites encompass a broader range of soil conditions than FEF, from dry-mesic sites with soils of the Rubicon series (sandy, mixed, frigid Entic Haplorthods), derived from glacial outwash deposits, to mesic soils of the Montcalm series (sandy, mixed, frigid Alfic Haplorthods), derived from glacial till (Roberts and Richardson 1985; Roberts and Christensen 1988). In general, these soils are acidic and low in organic matter, with extractable nutrients supplied largely from organic constituents (table 8.3).

The presettlement forests within the region were predominantly northern hardwoods such as *Fagus grandifolia* and *A. saccharum* on mesic sites and

Table 8.3. Means of environmental variables used in canonical correspondence analysis of young versus mature stands at Fernow Experimental Forest (FEF), West Virginia, and University of Michigan Biological Station (UMBS)

	FEF[a] Stand age class		UMBS[a] Stand age class	
Variable	Young	Mature	Young	Mature
Elevation (range in m)	725–860	735–870		
Tree basal area (m^2/ha)	22.5	42.8*	10.8	22.8*
Tree density (stems/ha)	2099	854*		
Texture (%)	67.3	65.7		
Sand				
Clay	10.8	9.0		
Silt	22.0	25.4		
Bulk density (g/cm^3)[b]			1.06	1.13*
WAI (%)[c]			2.62	2.93
Organic matter (%)	13.8	12.6	5.1	3.8*
Cation exchange capacity (μmol_c/g)	45.5	40.1		
pH	4.39	4.32	5.04	4.58*
Nutrients (μg/g)				
NO_3	1.3	0.9		
NH_4	2.1	1.9		
PO_4	0.8	0.4	6.0	4.3
Ca	12.6	6.4	826	496
Mg	2.1	2.4	60.8	41.0
K	2.3	2.2	54.3	33.4*

Means of all measured variables, except for elevation, are compared between stand age classes with a *t*-test. Data summarized from previous studies at FEF (Gilliam 2002) and UMBS (Roberts and Gilliam 1995b).

[a]Young = 20 years (FEF), 1–20 years (UMBS); Mature >80 years (FEF), 55–167 years (UMBS).

[b]Disturbed.

[c]Water availability index (water content by weight between −0.033 MPa and −1.5 MPa moisture potential).

*Significant difference between age classes for a given variable at $p < .05$; others are nonsignificant at $p < .05$.

Table 8.4. Important overstory (live woody stems ≥ 2.5 cm diameter at 1.3 m height) species of young versus mature stands of University of Michigan Biological Station

	Stand age class[a]	
Species	Young	Mature
Acer pensylvanicum	—	4
A. rubrum	26	46
A. saccharum	10	33
Amelanchier spp.	6	—
Fagus grandifolia	—	10
Fraxinus americana	—	6
Ostrya virginiana	—	4
Pinus resinosa	—	6
P. strobus	—	5
Populus grandidentata	60	40
P. tremuloides	—	21
Quercus rubra	6	15

Data are mean importance values (sum of relative basal area and relative density) for 25 (young age class) and 36 (mature age class) stands. Nomenclature follows Gleason and Cronquist (1991).

[a]Young = 1–20 years; mature = 55–167 years.

coniferous species including *Pinus resinosa* Aiton, *P. strobus* L., and *Tsuga canadensis* (L.) Carriere on the dry-mesic sites (Kilburn 1957). From 1850 to 1920, extensive logging of the pine and hardwood forests occurred, followed by wildfires (Gates 1930; Kilburn 1957). Thus, the mature stands (55–167 years-old) in the present study are second-growth stands that originated from cutting and burning. The young stands (≤ 20 years old) originated from clearcutting (without burning) of these mature, second-growth stands.

In our sample, *Populus grandidentata* Michx. had the highest importance value for trees in young stands, in contrast to mature stands where *Acer rubrum* L. and *A. saccharum* shared dominance with *P. grandidentata*. Mature stands contained a greater variety of secondary species than young stands, including *Fagus grandifolia, Quercus rubra, Fraxinus americana* L., *Pinus strobus*, and *P. resinosa* (table 8.4).

Herbaceous layer dominants in young stands included several early successional species, such as *Pteridium aquilinum* (L.) Kuhn, *Rubus ideaus* L., *R. allegheniensis* T.C. Porter, and *Fragaria virginiana* Duchesne (table 8.5). Dominant species in the herb layer in mature stands included seedlings of shade-tolerant to mid-tolerant tree species, such as *A. saccharum, Fagus grandifolia, Ostrya virginiana* (Miller) K. Koch, *Fraxinus americana*, and shade-tolerant herbaceous species, such as *Maianthemum canadense* Desf.

Table 8.5. Important herbaceous layer (vascular plants ≤ 1 m tall) species of young versus mature stands of University of Michigan Biological Station

	Stand age class[a]	
Species	Young	Mature
Pteridium aquilinum	43	9
Acer saccharum	13	45
Rubus ideaus	10	—
Fragaria virginiana	9	—
Rubus allegheniensis	8	—
Fraxinus americana	7	12
A. rubrum	6	14
Ostrya virginiana	5	15
Maianthemum canadense	3	20
Fagus grandifolia	1	15

Data are mean importance values (sum of relative cover and relative frequency). Nomenclature follows Gleason and Cronquist (1991).

[a]Young = 1–20 years; mature = 55–167 years.

Field Sampling

Fernow Experimental Forest

Fifteen circular 0.04-ha sample plots were established in each watershed. In each plot, all woody stems ≥ 2.5 cm diameter at about 1.3 m in height (dbh) were tallied, identified, and measured for diameter at breast height to the nearest 0.1 cm. The herbaceous layer was sampled by identifying and estimating cover of all vascular plants ≤ 1 m in height within ten 1-m^2 circular subplots in each sample plot, using a visual estimation method as described in Gilliam and Turrill (1993). Subplots were located within sample plots using a stratified-random polar coordinates method (Gaiser 1951).

Methods for sampling mineral soil also have been described previously (Gilliam and Turrill 1993, Gilliam et al. 1994). Briefly, two 10-cm deep samples were taken from each plot and placed into separate bags; thus, values for each plot represent the average of two soil samples. Each sample was sieved (2-mm screen), air-dried, and analyzed for pH (1:1 weight:volume, soil:water), 1 N KCl-extractable calcium, potassium, magnesium, and phosphorous (with plasma emission), 1 N KCl-extractable NO_3 and NH_4 (with flow-injection colorimetry), and organic matter (loss-on-ignition method). Particle size (texture) was determined for each soil sample using the hydrometer method.

University of Michigan Biological Station

Data were taken from 0.1-ha (20 × 50 m) plots, one plot located in each of 61 stands. Stands were selected that were at least 0.5 ha and were relatively undisturbed since the last disturbance. A single plot was located near the center of each stand in an area that was representative of the stand composition and soil conditions. In each plot, trees (woody stems ≥ 1 m tall) were sampled as described in Roberts and Christensen (1988). Importance values for tree species (IVT) were calculated as relative density + relative basal area. Percent cover of species in the herbaceous layer (all vascular plants < 1 m in height) was visually estimated in twenty-five 0.5 × 2.0 m contiguous quadrats extending along the plot center line. Percent cover and frequency (proportion of quadrats in a plot in which a given species was found) were combined to generate importance values for herb layer species (IVH; relative cover + relative frequency). Species that occurred within the 0.1-ha plot but not sampled in the 0.5 × 2.0 m quadrats were assigned an IVH of 0.001.

Four replicate soil samples were taken from the A1 horizons, one sample from a soil pit adjacent to each plot and the other three at random points along the plot center line. Soil variables (see below) were calculated as mean values (n = 4) for each plot. Details of sample preparation and laboratory analysis can be found in Roberts and Christensen (1988). Briefly, after drying and sieving (2-mm screen), soil samples were analyzed for bulk density, water availability (% moisture content, by weight, between − 0.033 and − 1.5 MPa moisture potential), pH (1:1 soil: H_2O, glass electrode), and organic carbon. After extraction with a dilute acid solution (0.05 M HCl with 0.0125 M H_2SO_4), extractable PO_4 was determined by molybdenum blue colorimetry and extractable calcium, potassium, and magnesium were determined by atomic absorption spectrophotometry.

Data Analysis

Fernow Experimental Forest

Data for overstory and herbaceous layer species have been reported previously for each watershed separately (Gilliam et al. 1995). Because of minor differences between the two watersheds of each stand age and because of the focus of this paper on stand age-related comparisons, we combined data into two stand age classes: young (~20 years, even age) and mature (>90 years, mixed age). Stand age means of environmental variables were subjected to t tests between the two age classes.

Gradient lengths of the environmental variables shown in table 8.3 were determined for the herbaceous and overstory strata separately in each age class using CCA. CCA was performed with the computer program CANOCO version 3.10 (ter Braak 1990), a version of the Cornell Ecology program

DECORANA (Hill 1979, ter Braak 1987); all program defaults were used. Unlike DCA, CCA focuses on the relationships between plant species and measured environmental variables; thus, it provides direct interpretation of axes in the ordination (ter Braak 1986). All data were ln-transformed before ordination, according to suggestions of Palmer (1993).

Of importance in this chapter is CCA output in the form of trajectories of environmental gradients. CCA yields one trajectory for each environmental factor, the length of which is indicative of the importance of that environmental gradient in explaining species' patterns. We determined gradient lengths for all environmental factors by measuring the length of the lines in the trajectory figure. The focus of this paper is on the relative importance, as measured by vector length, of environmental factors in influencing species' patterns in young and mature stands. Gradient lengths for herbaceous layer versus overstory ordinations were subjected to Spearman rank correlation for each stand age class (Zar 1996).

University of Michigan Biological Station

Site-dependent changes in soil nutrients and vegetation in these plots have been described elsewhere (Roberts and Gilliam 1995a, Roberts and Christensen 1988). The dry-mesic and mesic sites sampled in this study represent the typical variation in site conditions and vegetation composition found in upland forests of the region; thus, we combined stands from both site types for this analysis. Stands were then subdivided into two age classes: Young (1–20 years, 25 stands) or mature (55–167 years, 38 stands). These age classes correspond roughly to those used at FEF, although there is clearly a wider range of ages within each of the two age classes at UMBS than at FEF.

Data from UMBS were subjected to CCA, using the same defaults and ln-transformations as used at FEF. The *t* test (PROC TTEST; SAS Institute Inc. 1985) was used for comparisons of stand and soil variables between stand ages. Gradient lengths of the environmental variables from the herbaceous layer and overstory were measured in a manner identical to that used at FEF.

Stand Age Comparisons of Environmental Variables Used in CCA

Fernow Experimental Forest

The two stand structural variables measured at FEF (tree basal area and stem density) varied significantly ($p<.05$) with stand age (table 8.3). Basal area for the mature stands was nearly twice that of the basal area for the young stands. Conversely, stem density for the young stands was nearly 2.5 times that for the mature stands. Such differences are typical of young versus mature forests: numerous, small stems early in succession followed by competitive thinning, which gives rise to fewer, but much larger, stems later in succession (Yoda et al. 1963; Christensen and Peet 1984). The other environmental data used in CCA for FEF were soil variables, including texture, organic matter, cation exchange capacity (CEC), pH, and extractable (exchangeable) nutrients, none of which varied significantly ($p<.05$) with stand age (table 8.3), confirming conclusions of earlier studies that general soil characteristics varied little across watersheds of sharply contrasting stand ages and histories (Gilliam 2002).

University of Michigan Biological Station

Although overstory basal area was lower at UMBS than at FEF, stand-age contrasts were similar among the two study sites. Basal area of the mature stands was just more than twice that of the young stands at UMBS (table 8.3), consistent with the successional thinning process that has been described for this forest type (Roberts and Richardson 1985). In contrast to the results of stand-age comparisons for FEF, several soil variables varied significantly with stand age at UMBS. In general, soils of mature stands were more acidic and less fertile than young stands, with soil fertility being largely a function of soil organic matter, which was significantly lower ($p<.05$) in mature stands than in young stands (table 8.3).

Canonical Correspondence Analysis of Vegetation Data

Individual CCA ordinations were run for both of the vegetation strata (herb layer and overstory) in both stand ages (young and mature) separately, resulting in four ordinations for each site. Original ordination figures (i.e., X-Y graphs of axis 1 by axis 2 scores for sample plots) for these analyses are not given; rather, the axis scores were used make correlations between strata. In all cases, axis 1 explained the greatest amount of variability in species data; thus, axis 1 scores were used in comparisons between strata. The correlation diagrams produced give a measure of the similarities in the ordering of plots along the first axis in the herbaceous and overstory vegetation ordinations.

Fernow Experimental Forest

Correlations between overstory and herbaceous layer varied with stand age at FEF. The correlation for young stands was not significant at ($p<.05$), whereas that for mature stands was highly significant ($p<.001$; fig. 8.1). Accordingly, we conclude that, although the two strata are not linked in the young stands, they are linked in the mature stands.

Another way of viewing this conceptually is to consider the proximity of

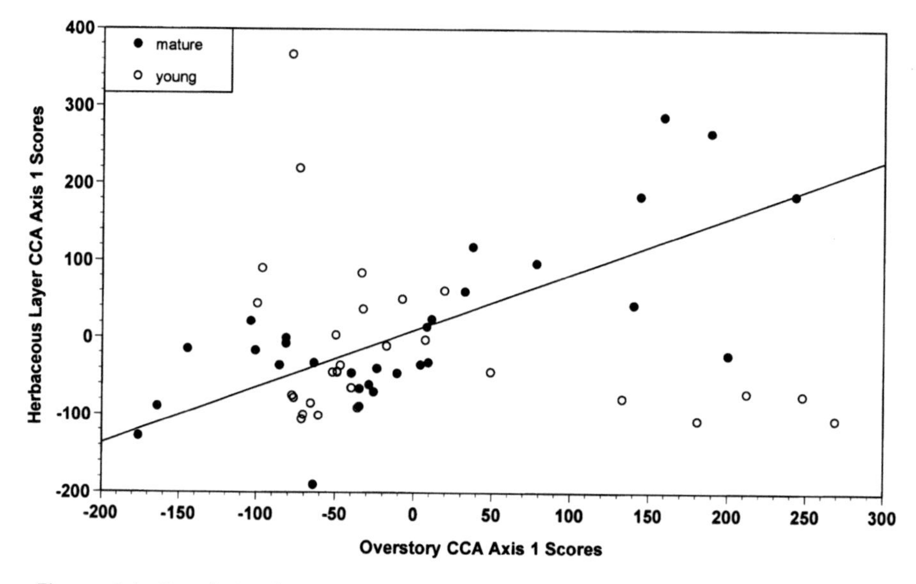

Figure 8.1. Correlation between overstory and herbaceous layer species composition of young stands (open symbols) and mature stands (filled symbols) for Fernow Experimental Forest, West Virginia. Overstory is represented on the X-axis as axis 1 scores from canonical correspondence analysis (CCA) of overstory species data. Herbaceous layer is represented on the Y-axis as axis 1 scores from CCA of herbaceous layer species data. Correlation for young stand plots was not significant ($p<.05$). The line shown is for mature stand plots only ($y = 9.51 + 0.73x$; $r^2 = .54$; $p<.001$).

points in the ordination diagrams. Points that are close to each other in ordination space represent sample plots that have similar species composition; conversely, points distant from each other represent plots that are dissimilar. Thus, a significant correlation between axis 1 scores from each stratum should indicate close similarity in the pattern and degree of spatial variation of species composition (sometimes referred to as *species turnover*) of the two strata. Once again, this degree of linkage appears to be related to stand age at FEF.

University of Michigan Biological Station

In contrast to results for FEF, overstory/herb layer correlations were highly significant ($p<.0001$) for both young and mature stands at UMBS (fig. 8.2). In fact, regression lines calculated to fit the data are nearly coincidental, with slopes of 0.85 and 0.88 for young and mature stands, respectively. Accordingly, we may conclude that the two strata are linked in these successional aspen forests, but that this occurs in a way that is independent of stand age (or certainly less dependent) than was found in stands at FEF.

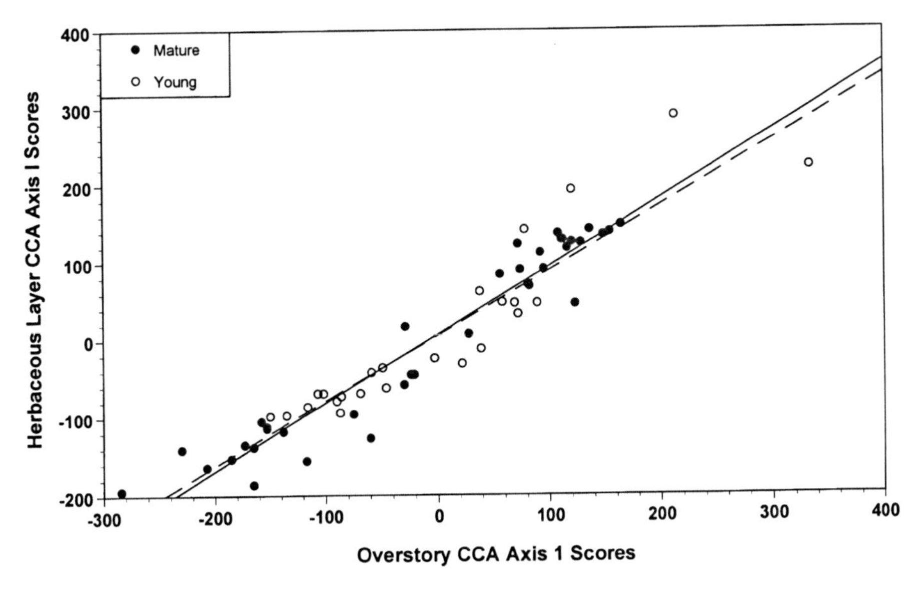

Figure 8.2. Correlation between overstory and herbaceous layer species composition of young stands (open symbols) and mature stands (filled symbols) for University of Michigan Biological Station. Overstory is represented on the X-axis as axis 1 scores from CCA of overstory species data. Herbaceous layer is represented on the y-axis as axis 1 scores from canonical correspondence analysis (CCA) of herbaceous layer species data. Solid line represents mature stands ($y = 7.02 + 0.88x$; $r^2 = .93$; $p<.001$). Dashed line represents young stands ($y = 5.21 + 0.85x$; $r^2 = .86$; $p<.001$).

Environmental Trajectories

As discussed previously, CCA output presents environmental data in the form of trajectories or vectors, one per environmental variable, the length of which is proportional to the importance of that factor (variable) in explaining species' patterns. Each line, along with the arrow, indicates the positive direction of the vector (i.e., increasing values).

CCA ordinations for this part of our discussion (e.g., figs. 8.3–8.10) are presented with vectors only (i.e., without the plot data discussed previously) because (1) the result is an ordination diagram with less clutter, and (2) the environmental trajectories are of primary importance in testing the mechanistic hypothesis for linkage of forest strata. Finally, because these are true vectors (ter Braak 1987, 1990; Palmer 1993), the direction of each line (originating from the center of ordination space) is also important in interpreting the meaning of CCA ordinations. Accordingly, vectors that point in similar directions represent environmental factors that are closely related to each other.

Fernow Experimental Forest

Some of the more important environmental factors for the overstory in young stands at FEF were elevation, stand density, and extractable NO_3 (which was of equal importance with sand content and extractable PO_4); some of the less important factors were cation exchange capacity and extractable calcium and magnesium (fig. 8.3). In addition, elevation and extractable NO_3 were closely related, along with soil organic matter (OM). This suggests that extractable NO_3 and soil OM increase with elevation in a way that is significant in influencing tree species' patterns in young stands. Furthermore, it links OM with the production of NO_3 in the soil, a relationship that has been demonstrated for this site using in situ incubations (Gilliam et al. 1996, 2001).

Environmental trajectories for the herb layer in young stands appeared to contrast sharply with those for the overstory in young stands, especially for extractable PO_4 (fig. 8.4). The most important environmental factors were, in descending order, extractable NO_3, soil OM, and extractable potassium, whereas the least important were, in ascending order, extractable PO_4, pH, and extractable NH_4 (fig. 8.4). The relative importance of both extractable NO_3 and potassium (and lack of importance of extractable NH_4) for the herb layer in young stands supports conclusions of Gilliam et al. (1996) that NO_3 is the predominant form of nitrogen taken up by plants of the herb layer at

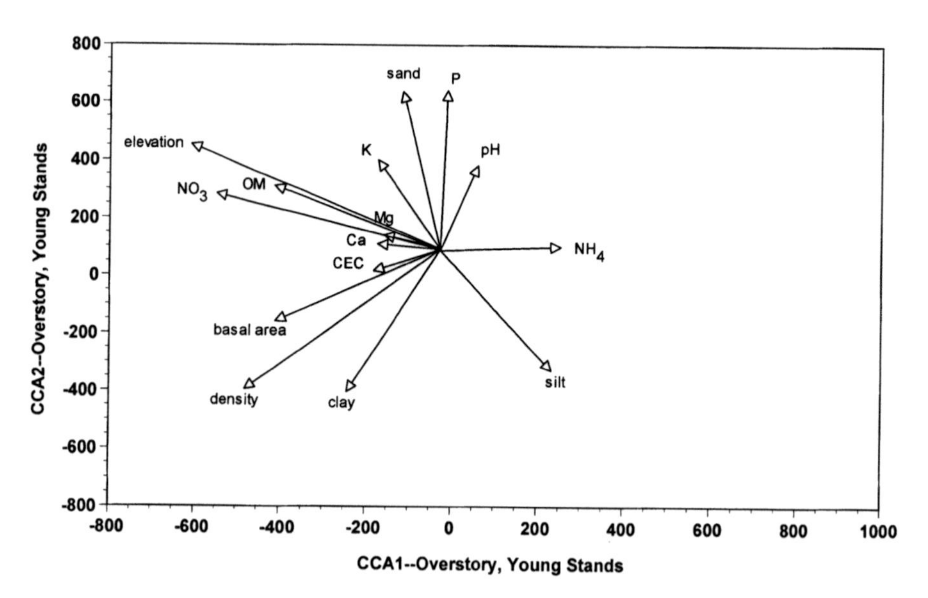

Figure 8.3. Environmental trajectories for the overstory of young stands for Fernow Experimental Forest, West Virginia, resulting from canonical correspondence analysis (CCA). The length of each line is proportional to the importance of the associated environmental factor in explaining species' patterns among sample plots within stands. CEC, cation exchange capacity; OM, organic matter.

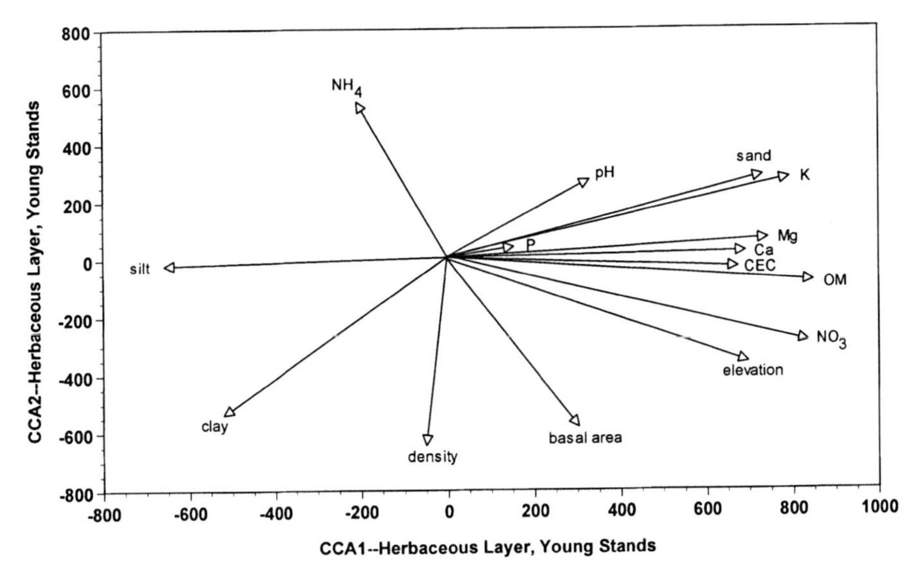

Figure 8.4. Environmental trajectories for the herbaceous layer of young stands for Fernow Experimental Forest, West Virginia, resulting from canonical correspondence analysis (CCA). The length of each line is proportional to the importance of the associated environmental factor in explaining species patterns among sample plots within stands. CEC, cation exchange capacity; OM, organic matter.

FEF and that NO_3 is taken up with potassium from these soils. This also confirms findings of numerous studies demonstrating the close correlation between nitrogen uptake and potassium availability (Barber 1995; Marschner 1995).

Environmental factors of importance for the overstory of mature stands differed somewhat from those for the overstory of young stands, suggesting a temporal (successional) shift in responses of tree species to variables such as soil resource availability. In descending order, the more important factors were stand density, soil clay content, soil OM, silt content, and extractable calcium, whereas the less important factors, in ascending order, were CEC, sand content, stand basal area, and extractable magnesium (fig. 8.5).

In contrast to comparisons of overstory versus herb layer for young stands, environmental vectors for the herb layer in mature stands appeared to be quite similar to those for the overstory in mature stands (fig. 8.6). The important factors were, in descending order, extractable NO_3, clay, stand density, and the unimportant factors were, in ascending order, CEC, stand basal area, sand, and extractable magnesium. In short, the two strata in mature stands showed considerable overlap for environmental factors that were both important and unimportant in explaining species' patterns. Finally, patterns of correlation among environmental variables were similar between strata for mature stands, in contrast to those between strata for young stands. For

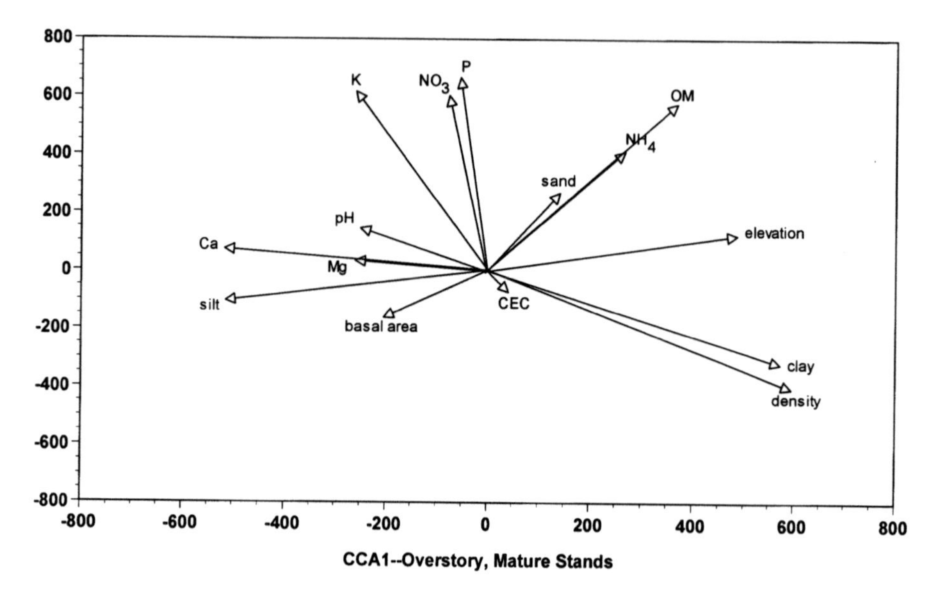

Figure 8.5. Environmental trajectories for the overstory of mature stands for Fernow Experimental Forest, West Virginia, resulting from canonical correspondence analysis (CCA). The length of each line is proportional to the importance of the associated environmental factor in explaining species patterns among sample plots within stands. CEC, cation exchange capacity; OM, organic matter.

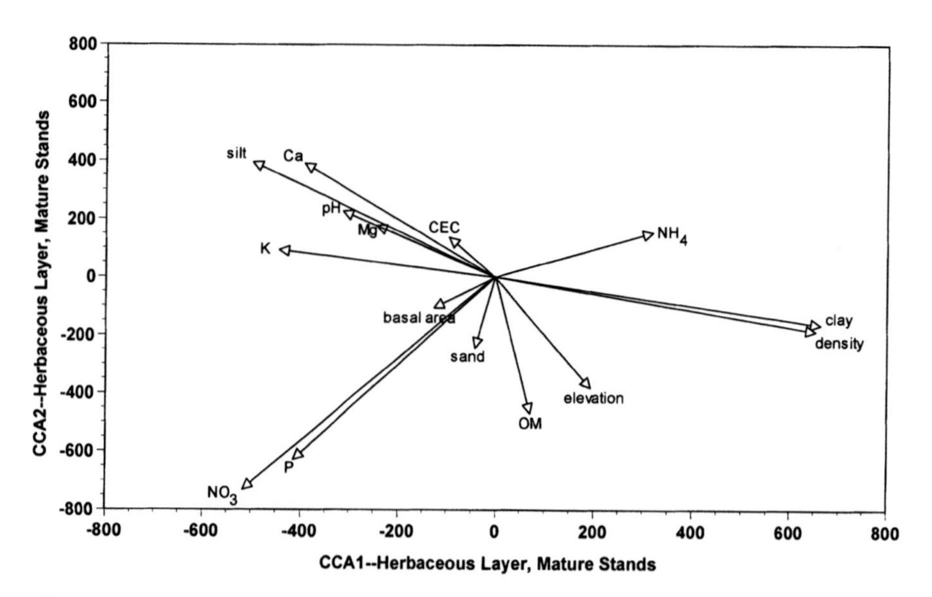

Figure 8.6. Environmental trajectories for the herbaceous layer of young stands for Fernow Experimental Forest, West Virginia, resulting from canonical correspondence analysis (CCA). The length of each line is proportional to the importance of the associated environmental factor in explaining species patterns among sample plots within stands. CEC, cation exchange capacity; OM, organic matter.

example, vectors for clay content and stand density showed close overlap, as did extractable NO_3 and phosphorus, for both strata in mature stands (figs. 8.5 and 8.6). Few such similarities were found in young stands (figs. 8.3 and 8.4).

University of Michigan Biological Station

The more important environmental factors for the overstory in young stands at UMBS were soil pH and extractable soil calcium, with less important factors being extractable soil potassium and tree basal area (fig. 8.7). In fact, pH and calcium were closely related, each with vectors essentially superimposed on one another, indicating that soil pH is determined in large part by levels of calcium in these soils. Also highly correlated were water availability index and OM, suggesting that OM is important in determining water availability in these coarse-textured, well-drained sandy soils.

Similar to the results for the overstory in young stands at UMBS, the more important environmental factors for the herb layer in young stands were pH and calcium, whereas soil potassium was of lesser importance; however, unlike the overstory, PO_4 was of equally high importance as calcium for the herb layer, and soil bulk density was the least important of all environmental variables (fig. 8.8). Thus, early in succession in these aspen stands, spatial

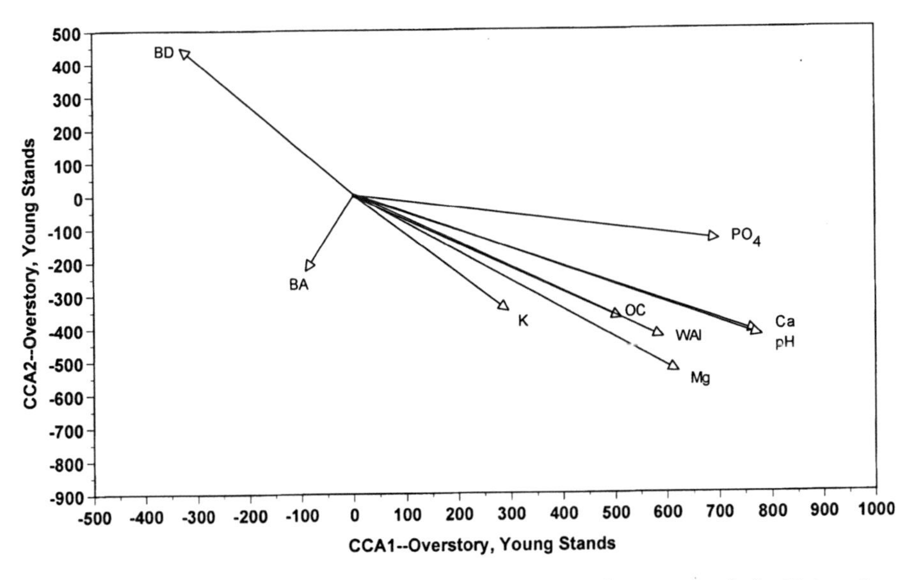

Figure 8.7. Environmental trajectories for the overstory of young stands for University of Michigan Biological Station, resulting from canonical correspondence analysis (CCA). The length of each line is proportional to the importance of the associated environmental factor in explaining species patterns among sample plots within stands. BA, basal area; BD, bulk density; OC, organic carbon; WAI, water availability index.

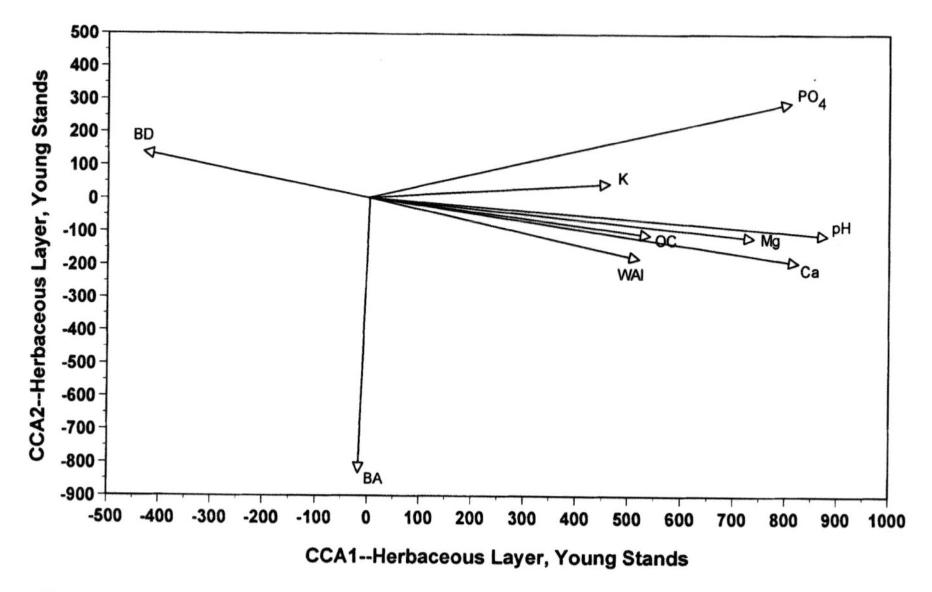

Figure 8.8. Environmental trajectories for the herbaceous layer of young stands for University of Michigan Biological Station, resulting from canonical correspondence analysis (CCA). The length of each line is proportional to the importance of the associated environmental factor in explaining species patterns among sample plots within stands. BA, basal area; BD, bulk density; OC, organic carbon; WAI, water availability index.

patterns of soil pH, as influenced by soil calcium, appears to exert a pronounced influence on the spatial patterns of composition of both overstory and herb layer species.

Soil calcium was by far the dominant environmental factor influencing species composition of the overstory in mature stands at UMBS, with extractable PO_4 and magnesium of secondary importance (fig. 8.9). Of minor importance were tree basal area and soil potassium, similar to results found for the overstory of young stands (fig. 8.7). As described previously (see "Study Sites"), these soils are generally acidic and low in fertility and OM, with available nutrients supplied largely from organic constituents, which is seen in the close coincidence of the OM vector with vectors of all extractable nutrients except potassium (fig. 8.9).

Finally, the more important environmental factors for the herb layer in mature stands at UMBS were soil calcium and magnesium, with extractable soil potassium and tree basal area being of minor importance (fig. 8.10), similar to the results found for the overstory of mature stands. In addition, the clustering of OM along with extractable nutrients (other than potassium) for the herb layer in mature stands was similar to that for the overstory in mature stands (fig. 8.9).

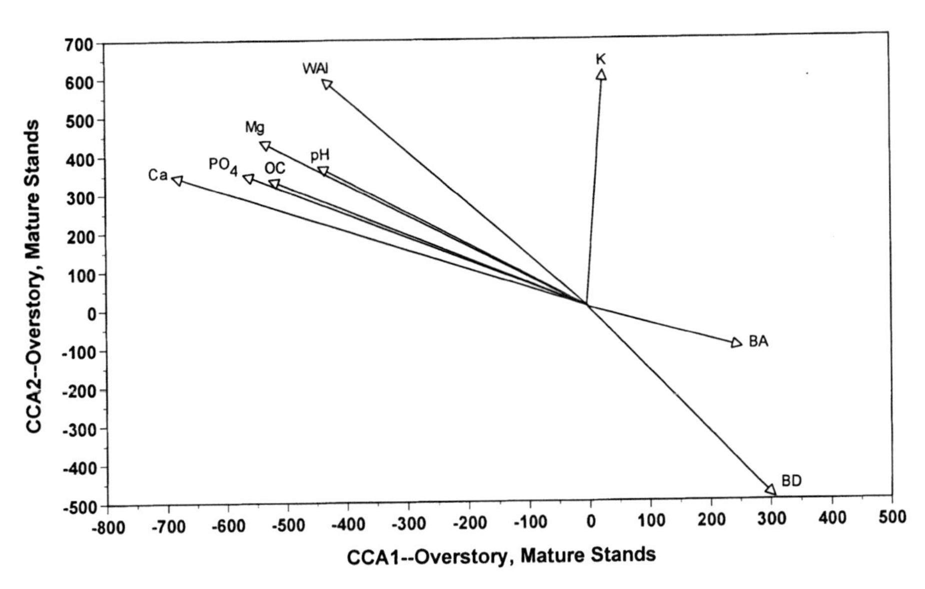

Figure 8.9. Environmental trajectories for the overstory of mature stands for University of Michigan Biological Station, resulting from canonical correspondence analysis (CCA). The length of each line is proportional to the importance of the associated environmental factor in explaining species patterns among sample plots within stands. BA, basal area; BD,bulk density; OC, organic carbon; WAI, water availability index.

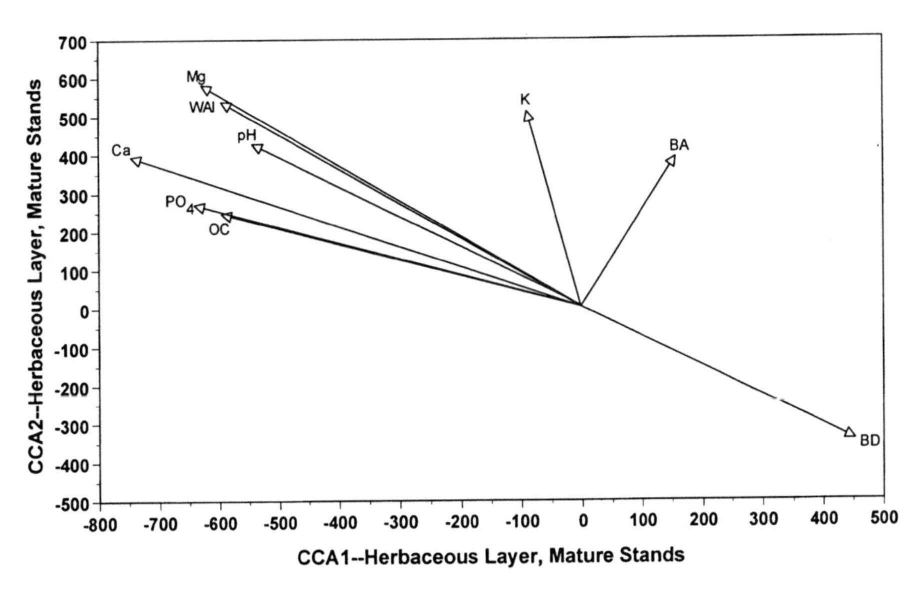

Figure 8.10. Environmental trajectories for the herbaceous layer of young stands for University of Michigan Biological Station, resulting from canonical correspondence analysis (CCA). The length of each line is proportional to the importance of the associated environmental factor in explaining species patterns among sample plots within stands. BA, basal area; BD, bulk density; OC, organic carbon; WAI, water availability index.

Correlations Between Strata

The preceding discussion essentially has been a visual inspection of environmental vector lengths to determine a gradient of importance of specific factors in explaining species patterns for each strata and stand-age combination. Quantitative evidence relevant to our mechanistic explanation of linkage can be provided by statistically determining whether the herbaceous layer and overstory are responding to environmental gradients in a parallel fashion. As described earlier in this chapter, because CCA performs a least-squares regression of species-based plot scores as dependent variables onto environmental factors as independent variables (Palmer 1993), it provides a direct quantitative assessment of these responses.

Accordingly, we can examine the evidence by comparing the rank order of environmental variables (from the most to the least important as determined by vector length) of herb layer versus overstory. Significant correlations in vector lengths between the herb layer and overstory would indicate that the two strata are responding to the same environmental factors to a similar degree. Consequently, considering our previous conclusions about stand-age–related linkage at FEF (i.e., strata are not linked early in succession, but become linked later in succession), our mechanistic explanation would be supported only if rank-order correlation is not significant for young stands at FEF but is significant for mature stands. Regarding UMBS, our explanation would be supported only if rank-order correlation is significant for both stand-age classes.

Fernow Experimental Forest

The herb layer–overstory correlation for young stands was not significant ($r = .02$, $p<.95$). In contrast, it was highly significant ($r = .92$, $p<.001$) for mature stands (table 8.6). Such stand-age–related contrasts in these comparisons support our hypothesis of linkage between forest strata. For FEF there appears to be a temporal (successional) shift in the degree of linkage. The lack of significant correlation between environmental vector lengths for herb layer and overstory in young stands indicates that the lack of linkage shown in figure 8.1 is the result of differences in response of these strata to environmental gradients at this point in successional time. The overstory appeared to respond to elevation and stand density, whereas the herb layer responded to factors related to soil fertility, such as soil OM and extractable NO_3 and potassium. Such results for the herb layer are consistent with conclusions of Gilliam and Turrill (1993) at FEF and those of other studies (e.g., Morris et al. 1993; Wilson and Shure 1993) that herb layer development early in forest succession is influenced strongly by competition for soil nutrients.

The significant correlation between environmental vector lengths for herb layer and overstory in mature stands indicates that the high degree of linkage shown in figure 8.1 is the result of the similar responses of these strata to

Table 8.6. Spearman rank correlation coefficients for importance of environmental factors for the herbaceous layer versus overstory of young and mature stands at Fernow Experimental Forest (FEF), West Virginia, and University of Michigan Biological Station (UMBS).

Site	Age class[a]	*r*	*p*
FEF	Young	0.02	.95
FEF	Mature	0.92	.001
UMBS	Young	0.90	.01
UMBS	Mature	0.97	.0001

Importance was measured as length of environmental vector for canonical correspondence analysis of each stratum and stand age for the two study sites.

[a]Young = 20 years (FEF), 1–20 years (UMBS); Mature >80 years (FEF), 55–167 years (UMBS).

environmental gradients, including factors related to stand characteristics, such as stand density, and to soil conditions, such as texture and nutrient availability. This supports conclusions of the herb layer study by Gilliam and Turrill (1993) at FEF that herb layer development may become more influenced by stand characteristics in later stages of forest succession.

University of Michigan Biological Station

In contrast to results for FEF, the correlation of herb layer versus overstory vector length for UMBS was highly significant for young stands ($r = .90$, $p<.01$); this correlation was even more significant ($r = .97$, $p<.0001$) for mature stands (table 8.6). As with results for FEF, these results for UMBS support our hypothesis that linkage arises from parallel responses of vegetation strata to environmental factors. Although it is not immediately clear why the two forest types contrasted with respect to stand-age–related changes in linkage, we suspect that it may be related to (1) the distinctness of the ecology of aspen-dominated successional stands of northern lower Michigan compared to the mixed-dominance of central Appalachian hardwood stands, and (2) a greater degree of soil heterogeneity at UMBS relative to FEF. That is, although soils of forest stands studied at UMBS are all Spodosols, they are derived from parent materials of widely varying glacial origin, including glacial outwash deposits (coarse-textured, sandy materials) and glacial till (mixtures of sand, silt, and clay). As we noted previously, there is a tendency to conclude that linkage occurs in studies that include a broader range of environments and β diversity. Thus, at UMBS, even the young stands showed linkage (although weaker than the mature stands). To adequately test the influence of diversity on linkage would require running ordinations of the total data set, along with ordinations of each site class separately. Our data sets were not large enough to allow this type of analysis.

Soil calcium is clearly an overriding environmental variable determining spatial patterns of species composition of both the herbaceous layer and the overstory in both young and mature stands at UMBS. Although additional factors, such as pH or soil OM, vary in importance among strata and stand ages, calcium is consistently the factor of greatest importance. In this region, calcareous substrates (indicated by high calcium concentrations) are associated with clay lenses or clay till within or underlying the sandy outwash surface deposits (Spurr and Zumberge 1956; Roberts and Christensen 1988). Van Breemen et al. (1997) found significant correlation between soil calcium and overstory species composition in southern New England forests. Thus, soil calcium may serve as an indicator for a suite of co-related soil nutrient and moisture variables at UMBS.

Summary

We have provided evidence that supports our mechanistic explanation of linkage in forest communities. Linkage among forest strata appears to arise from similarities among strata of forest vegetation in the responses of their respective species to environmental gradients. In central Appalachian hardwood stands of West Virginia, these responses may change through secondary forest succession and thus may be a function of stand age. Early in succession, spatial variation in species composition of the overstory appears largely related to the density of the stand, whereas variation in herb composition is related more to soil fertility. This changes later in succession when variation in herb composition responds more to stand structure (i.e., tree density), a time when overstory variability is also related to density.

In successional aspen forests of northern lower Michigan, where correlations between strata were significant for both young and mature stands, these responses may be much less related to stand age. The degree to which linkage between strata is the result of higher β diversity in the Michigan sample is not clear. We believe, however, that it is not an artifact of simply sampling across two extremes of environmental conditions (see McCune and Antos 1981), given the continuum of points along axis 1 for both vegetation strata, as depicted in figure 8.2.

Although the concept of linkage among vegetation strata of forest communities likely will continue to be debated among vegetation scientists, we believe that it is a concept with a high degree of importance and application. It furthers our understanding of and appreciation for the complexities underlying the structure and function of forest ecosystems (e.g., responses to disturbance [chapters 11 and 13, this volume] and mechanisms of secondary succession [chapter 9, this volume]). It may also be applied toward landscape-level investigations of forest cover types and remote sensing.

The generality and utility of this explanation, however, will only be seen after it is tested in a more formal and structured way. We invite further testing of this mechanism as a predictive hypothesis in other forest types and over a wider range of stand ages. Studies that examine linkage across different breadths of β diversity and spatial scales would be particularly useful. Given the widespread popularity and use of CCA among vegetation scientists (Barbour et al. 1999), this might be done with reasonable ease.

9

Temporal and Spatial Patterns of Herbaceous Layer Communities in the North Carolina Piedmont

Norman L. Christensen
Frank S. Gilliam

The questions ecologists, especially those who study plant communities, choose to ask are necessarily influenced by the character of landscapes immediately available for study. Thus, it is not surprising that researchers at mid-Atlantic universities in the first half of the twentieth century—individuals such as B. W. Wells, H. J. Oosting, W. D. Billings, and M. F. Buell, who established some of the first doctoral training programs in plant ecology—chose to focus much of their attention on the process of vegetation change that derives from the abandonment of farmland. Land degradation and a welter of economic factors resulted in abandonment of agricultural land across the mid-Atlantic region through the six decades after the Civil War, setting in motion a process of reforestation on an unprecedented regional scale. The region was awash in old fields and early successional forests. A unified theory of succession envisioned by many ecologists (e.g., Clements 1916, 1936) as a central element in our understanding of the distribution and structure of plant communities provided additional incentive to study these landscapes. The revegetation of old fields was destined to become the community ecologists' equivalent of the fruit fly or *E. coli*—in a sense, the "model organism" for the study of succession.

Although abandoned agricultural fields are now comparatively rare, and human actions associated with urban development have begun to reverse decades of forest spread, much of the landscape available to those of us who have succeeded those ecological pioneers retains the unmistakable imprint of old-field succession. Furthermore, we are the beneficiaries of research carried out over a period that begins to have real relevance to successional processes. Although ecologists no longer see succession as the integrated, unifying process envisioned by Clements, disturbance and the processes that derive from it continue to shape our understanding of the distribution and abundance of organisms on landscapes. Ecologists today are no less opportunistic than were their predecessors. This body of work that has accumulated over nearly a century provides unique opportunities to explore the nature of change on a rapidly changing landscape.

Succession on many forested landscapes often is depicted as a process that progresses from plant assemblages dominated by herbaceous species, to even-aged forests, and finally to uneven-aged, late successional or old-growth forests, with an almost-implied decline in the ecological significance of herb species. Certainly, herbs in old fields have received more study than their counterparts at other stages of succession. The herbaceous component is important in each successional stage; indeed, its diversity generally increases at most spatial scales as succession proceeds. Although most would agree that the factors that structure herb communities change with time, there has been little systematic consideration of the nature and mechanisms of such changes.

Our primary goal in this chapter is to explore variations in and dynamics of herb layer assemblages across the Piedmont landscape. We are particularly interested two questions: what are the key environmental factors influencing herb distribution at various stages in the succession from land abandonment? What do these patterns tell us regarding the mechanisms that underlie the dynamics of herb populations? To address these questions, we draw on several strands of research carried out over the last 70 years to provide a more comprehensive picture of the successional patterns and mechanisms of change in herb layer communities of forested landscapes of the Piedmont region of North Carolina. The importance of these variations for management of herb diversity is also considered.

A Succession of Successional Ideas and Their Relevance to Herbs

Early studies of old-field succession were concerned more with describing the specific patterns of change in species composition over time they were with determining the processes (i.e., mechanisms) that caused such change. After nearly a century of study, debate continues regarding the nature of those processes. Because of this, we provide here a brief historical chronology of some of the more influential hypotheses regarding successional change. We then apply these to the specific case of old-field succession.

The pioneering work of Henry Chandler Cowles on dune succession (Cowles 1899, 1901, 1911) was described by Tansley (1935) as "the first thorough working out of a strikingly complete and beautiful successional series" (p. 284). It was Cowles who led the way in "chronosequence" or "space-for-time" approaches to the study of succession. Cowles also recognized that early-invading organisms modified their environment in ways that affected establishment of successors. Although Cowles reckoned that this led to generally predictable patterns of change, he was also cognizant of the dynamic nature of plant communities in response to a variable abiotic environment. This awareness is captured in his characterization of succession as "a variable approaching a variable, not a constant" (Olson 1958).

Clements (1916, 1928) saw succession as much more predictable and directional, converging (regardless of starting conditions) inexorably to a stable climax community determined largely by climate. In Clements' view, the biotic reactions of dominant pioneer species determined the sequence of vegetational seres leading to that climax—a process later dubbed "relay floristics" (Egler 1954).

In his then-controversial paper, Gleason (1926) rejected virtually all of Clements' ideas on succession and, especially, on the nature of plant communities. Gleason reasoned that the principal mechanism driving succession was an interaction between migration of plant species (rates and modes of which varied greatly among species) and environmental selection (of a variable set of abiotic conditions, coupled with species-specific responses to those conditions). Acknowledging that successional change might be described as occurring in stages dominated by physiognomic types (e.g., herbs being replaced by shrubs and then trees), he argued that the sequence of invasion of species was highly individualistic and determined by their ability to disperse to a site and subsequently compete. This theme was reiterated by Drury and Nisbet (1973), who emphasized the importance of life-history traits of successional species. They suggested that most of what happens during succession is best understood as a consequence of differential growth, survival, and colonizing ability of species adapted to conditions along environmental gradients.

Egler (1954) argued in his initial floristic composition model of succession that, in many circumstances, succession was driven by patterns of early establishment and that subsequent change was largely a matter of differential longevity. Because early establishment would be variable owing to potentially random variations (e.g., in climate and seed rain), he posited that successional change is neither fixed nor predictable, as would be expected from Clements' relay floristics model. Mechanisms driving succession included stochastic migration of propagules to the disturbed site and differential longevity of plants. All pioneer species, many seral species, and some climax species are initially present after disturbance. Some of these germinate, becoming established quickly, whereas others germinate quickly but grow more slowly and for a longer period; still others become established later. Major changes in community dominance occur when larger, longer lived, and slower growing species outcompete smaller pioneer species.

In their often-cited review and syntheses of mechanisms of successional change, Connell and Slatyer (1977) concluded that succession is driven by one of three overriding mechanisms, proposed in their paper as alternative models. The facilitation model most closely fits the Clementsian vision of succession in which early invaders alter the environment in such a way as to make it more habitable for successors than themselves. The tolerance model captured elements of both Gleason and Egler in proposing that succession was largely determined by patterns of dispersal and differences in life history. The inhibition model depicts succession as a process in which early invaders establish, usurp resources, and thereby competitively exclude would-be successors. In this model, succession only proceeds when the populations of the current occupants decline due to their inability to reestablish.

Peet and Christensen (1980) argued that population-based approaches were a prerequisite to distinguish among the Connell-Slatyer models, and much work over the past two decades has focused at that level. Grime (1977, 1979) proposed that spatial and temporal patterns could be understood in terms of three primary plant strategies: competitor, stress tolerant, and ruderal (ephemeral), reflecting basic trade-offs in life-history traits. Competitors are continuously abundant but depressed under low resource levels, as might exist early in some successional processes. Stress-tolerant species are continuously rare, being displaced spatially to resource-poor sites, they would be more characteristic of later stages of succession on unique sites. Ruderal species are generally easily dispersed but are not effective competitors. They are temporarily abundant and displaced in time by more competitive species. Grime argued that plants could be classified into one of these three categories and that the relative abundance of species among the categories would change through succession.

Tilman (1985) proposed that temporal and spatial variations in species composition through succession could be understood in terms of variations in relative growth (and reproductive) rates of species in response to varying proportions of resources. Succession results from a gradient through time in the relative availabilities of limiting resources. Tilman argued that succession should thus be repeatable or directional only to the extent that the resource supply trajectories are repeatable or directional. He proposed that succession often involves a gradient change from habitats with resource-poor soils but high available light at the soil surface to habitats with resource-rich soils but low light availability.

T. M. Smith and Huston (1989) drew on ideas from both Grime and Tilman and proposed that life-history strategies entail compromises that determine their competitive abilities under varying resource levels. Generally, plants can be competitive under high levels of resource availability or tolerant of low levels (i.e., stress tolerant), but cannot be both. For example, a plant that is competitive under high light conditions or high available soil resources cannot also be shade or drought tolerant. Focusing particularly on light and water availability, Smith and Huston (1989) then simulated species change based on whether species were competitive or stress tolerant for light and/or water. For their simulations, they assumed that successional responses were due largely to changing light levels and that spatial variation at any time was largely due to water availability (other resources might be substituted for this sort of simulation). Their results suggested that, with respect to light, species composition for herbs should shift from competitive to stress-tolerant during succession and that at any particular time in succession competitors should be more prevalent on favorable sites with regard to soil resources. Correlations among species' abilities to compete for light and water might also explain observed differences in successional trajectories with respect to site. For example, some early successional species (e.g., pines) persist on dry sites because shade-intolerant (light competitors) species are often tolerant of drought stress.

Old-Field Herbs

In his 1932 description of the vegetation of North Carolina, Wells described the mosaic of old-fields as the "melting pot where foreigners and natives mingle" (p. 140). Wells described the general sequence of change, noting that the "foreigners" are relatively more important in the early years; indeed, many were weeds during cultivation, e.g., *Digitaria sanguinalis* (L.) Scop. (crabgrass) and *Cynodon dactylon* (L.) Pers. (Bermuda grass). Wells (1932; Crafton and Wells 1934) observed that the sequence of change in herbaceous dominance during the first 3–4 years was highly predictable, interpreting this as a facilitation process.

Oosting (1942) provided a much more quantitative sampling of abandoned fields of varying age (i.e., years after abandonment), paying particular attention to replicating each age with fields as similar as possible in physical features (e.g., soil type/texture, slope, aspect). Keever's (1950) classic study of the early stages of old-field development provided even more detail. One-year-old fields sampled early in the summer after abandonment had a total of 35 annual and perennial herbaceous species. Although these fields were not identical in species composition, they consistently had two species with highest density and frequency: *D. sanguinalis* and *Conyza canadensis* (L.) Cronq. (horseweed). Almost all first-year species, including *D. sanguinalis* and *C. canadensis*, were also found in 2-year-old fields. The Sorensen community coefficient of similarity for first- versus second-year fields was high (0.63), despite 26 new species appearing in the second year. At this time, however, there was a pronounced shift in dominance to *Aster ericoides* L., which was absent from first-year fields, and *Ambrosia artemisifolia* L., a minor component of first-year fields. Species richness dropped sharply in the third year after abandonment, corresponding to a rapid increase in dominance of the perennial grass, *Andropogon virginicus* L., which maintained dominance for several years.

Keever's (1950) experimental work showed little evidence for facilitation. Rather, she showed that the changes in species dominance could be understood in terms of modes of species dispersal and life histories (annuals replaced by biennials subsequently replaced by longer lived perennials). She rejected the hypothesis that such long-term dominance resulted from allelopathic compounds produced by *A. virginicus*, implicating competition for resources as a more likely mechanism—an implication more consistent with Connell and Slatyer's (1977) inhibition hypothesis. Keever (1983) later marveled at the uniqueness of these changes compared to the process succession on abandoned land elsewhere: "the sequence of species and the timing of these changes in old-field succession in the Piedmont of the Southeast are not typical of such succession elsewhere. Nowhere else is there such a fast and distinct change in species dominance. In most places there is a gradual overlapping of species dominance often extended over a much longer time" (p. 402).

Much of the attention on old-field vegetation has been devoted to understanding general patterns of change, with little study of the variability in those patterns. Schafale and Christensen (1986) examined variations among herb communities among 3- to 7-year-old fields (fig. 9.1). They found that species richness varied widely among such fields (14 to >50 species/0.1 ha) and was positively correlated with soil pH and cation availability. These same factors were also highly correlated with trends in species composition indicated by first-axis DCA ordination scores. Standing crop and productivity, however, were more highly correlated with soil organic matter and may reflect other variables associated with the conditions of field abandonment or water availability. This suggests that, at least to some extent, factors influencing variations in species composition are independent of those that influence production.

It was during these early years that seedlings of wind-dispersed pine (*Pinus taeda* L. and *P. echinata* Miller) and hardwood species, such as *Fraxinus americana* L., *Ulmus alata* Michx., *Liquidambar styraciflua* L., *Liriodendron tulipifera* L., and *Acer rubrum* L., became established. Pines are generally favored in an environment of high light availability and nutrient and water stress, and typically form a closed canopy by year 10. On moister sites, *Liquidambar styraciflua* and *Liriodendron tulipifera* may share dominance with pine. The suite of herb species typical of old fields is virtually absent thereafter. There is ample evidence that the dominant old-field herbs significantly influence the patterns of early tree establishment. For example, most pine seedlings become established during the first 3 years after abandonment, before the development of an herbaceous thatch, which inhibits pine seedling growth (Oosting 1942). This explains the even-aged structure of most old-field pine stands.

De Steven (1991a, 1991b) experimentally evaluated the role of dominant old-field herbs (and the animals that associate with them) with regard to the invasion of loblolly pine and five early-successional hardwood species common in the North Carolina Piedmont: *Liriodendron tulipifera, F. americana, Liquidambar styraciflua, U. alata,* and *A. rubrum*. De Steven found that competition from herbs and rodent herbivory had significant effects on seedling emergence and the growth of many, though not all, of these species. Loblolly pine exhibited the highest levels of seedling emergence, seedling survival, and seedling height growth in all treatment combinations. Accordingly, her data help explain why, although all of these species produce wind-dispersed seeds (i.e., seed rain) that arrive at old-field sites in potentially large numbers, loblolly pine initiates the woody species stage of old-field succession. The success of hardwood species that become established initially with pines is

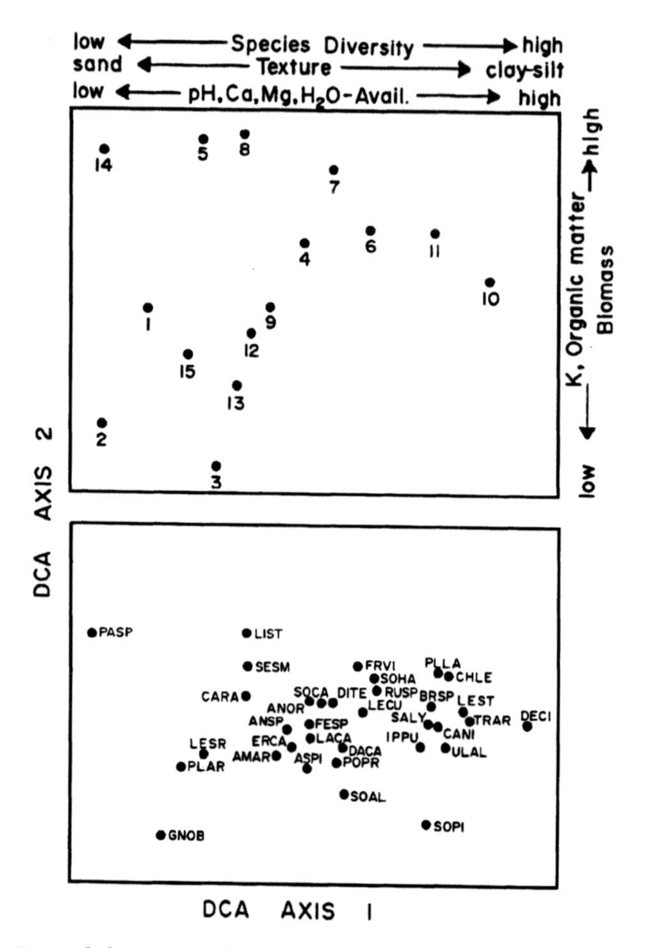

Figure 9.1. Detrended correspondence analysis (DCA) of herb community composition of 15 old fields 3–7 years after abandonment. Species codes are as follows: AMAR = *Ambrosia artemisiifolia* L.; ANOR = *Anthoxanthum odoratum* L.; ANSP = *Antenaria* spp., ASPI = *Aster pilosus* Willd.; BRSP = *Bromus* sp.; CANI = *Cassia nictitans* L.; CARA = *Campsis radicans* (L.) Seemann; CHLE = *Chrysanthemum leucanthemum* L; DACA = *Daucus carota* L.; DECI = *Desmodium ciliare* (Muhl. ex Willd. DC); DITE = *Diodia teres* Walter; ERCA = *Erigeron canadensis* L.; FESP = *Festuca* spp.; FRVI = *Fragaria virginiana* Duchesne; GNOB = *Gnaphalium obtusifolium* L.; IPPU = *Ipomoea purpurea* (L.) Roth; LACA = *Lactuca canadensis* L.; LECU = *Lespedeza cuneata* (Dumont) G. Don; LESR = *Lespedeza striata* (Thunberg) H. & A.; LEST = *Lespedeza stipulacea* Maxim.; LIST = *Liquidambar styraciflua*; PASP = *Panicum* spp.; PLAR =*Plantago aristata* Michaux; PLLA = *Plantago lanceolata* L.; POPR = *Polypremum procumbens* L.; RUSP = *Rubus* sp.; SALY = *Salvia lyrata* L.; SESM = *Senecio smallii* Britton; SOAL = *Solidago altissima* L.; SOCA = *Solanum caroliniense* L.; SOPI = *Solidago pinetorum* Small; SOHA = *Sorghum halepense* (L.) Persoon; TRAR = *Trifolium arvense* L.; ULAL=*Ulmus alata*.

determined as much by their amount of seed rain as by factors affecting seedling performance (De Steven 1991a, 1991b). McDonnell (1986), studying similar fields in New Jersey, found that the structure and spatial distribution of herbaceous and shrubby species influenced the behavior of animals that disperse woody plant seeds, thereby affecting the future structure of forest communities.

The Herbaceous Layer in Successional Forests

Although we focus primarily on temporal patterns of change in the herbaceous layer following abandonment, we first provide an overview of successional change in woody species. The general pattern of change among woody species during succession is captured in the model proposed by Oliver (1982) and Peet and Christensen (1988). The variety of changes from old-field abandonment to the formation of an even-aged, closed-canopy forest constitutes the *establishment phase* of development. During this period species patterns and composition are heavily affected by dispersal and spatial patterns affecting seed availability. The duration of the establishment phase varies from less than 10 years to several decades in some cases. Peet and Christensen (1988) suggested that the actual length of time appears to be directly related to factors that influence the initial stocking of woody stems (e.g., seed rain and early seedling survivorship) and site productivity (i.e., tree growth is faster and canopy closure occurs earlier on productive sites).

Canopy closure marks the initiation of the *thinning phase*. During this period, vegetation patterns and composition are affected heavily by limited light and intense competition for soil resources. The duration of this phase is also inversely related to tree density and site productivity. Thinning progresses more rapidly on productive sites and where tree density is highest.

The *transition phase* begins as tree mortality begins to create obvious canopy gaps (i.e., surviving trees are no longer able to close the canopy after a mortality event). This phase is marked by considerable small-scale (1–50 m) spatial heterogeneity. It is during this period that shade-intolerant pioneer trees (e.g., pines) are replaced by more shade tolerant successors. This stage can begin as early as 30–40 years after field abandonment in very densely stocked stands, but 70–80 years is more typical of average densities.

The *equilibrium* (mature) *phase* is characterized by an uneven-aged stand structure. Vegetation patterns are determined by key environmental variables (e.g., gradients of moisture and soil properties) and gap-phase transients (Watt 1947a). It is presumed that the broad-leaved hardwood stands of the Piedmont are typical of this phase and the likely end point of the old-field successional process (Billings 1938; Oosting 1942). However, Peet and Christensen (1987, 1988) provide clear evidence that most of these forests are experiencing considerable thinning as a consequence of a variety of historic disturbances (e.g., timber high-grading and livestock grazing). Furthermore, few of these stands were actually initiated as old fields (Christensen and Peet 1984).

Oosting (1942) compared the herb layer communities of forest stands representative of these various stages in upland and lowland areas. Generally, he found that herb species richness increased with stand successional age, and that, at any particular age, herb richness was highest in stands on moister sites. Like most ecologists of his time, Oosting paid little attention to variability among stands within broad age categories or within these broadly defined site types. However, the notion of successional convergence—that variations in initial conditions result in intersite variation in species composition that diminishes as communities mature—is all about the nature of changes in variance among sites through time.

Christensen and Peet (1984) sampled the herbaceous layer in 0.1-ha plots located in more than 200 forest stands representing a wide range of stand ages, site conditions and other landscape variables (e.g., tree density). They also measured a suite of more than 20 environmental features in each sample plot. They then divided the data set into 20–40, 40–60, 60–80, and 80+ year-old pine stands and hardwood stands for comparison. The range of environmental variation represented within each age class was very nearly equivalent.*

At the scale of 0.1 ha, herb layer diversity varied considerably among these stands, with some stands having only 20 species and other stands having well in excess of 100 species in the herb layer. On average, the number of species was remarkably constant among stand age classes at 50 species/ha. That said, only 203 species were sampled among all of the pine stands ($N = 111$), compared to 328 species among all hardwood stands ($N = 78$). As these results suggest, the range of variability in species composition among stands (Whittaker's β-diversity) was greater among hardwood than among pine stands.

Both species composition (indicated by first detrended correspondence analysis axis-ordination scores) and herb layer richness were highly correlated with a number of soil variables, most notably pH and soil cations (Fig. 9.2). Recall that Schafale and Christensen (1986) observed similar trends among old-fields. These relationships were strongest among the early and intermediate age pine stands and hardwoods and weakest (indeed, nonsignificant) in the 80+ year-old pines. These weak correlations in the older, transition phase pine stands suggest significant differences in the factors affecting diversity and species composition. Figure 9.3 shows that diversity varies most among stand ages (always greatest in oldest stands) at the lowest (least fertile) and highest (most fertile) soil cation sites.

These results suggest significant changes in the factors affecting the distribution of herb layer species during succession. Certainly, they indicate a

*Twenty deciduous forest stands on sites not represented among the pine stands (e.g., bottomlands and rocky outcrops) were excluded from this analysis.

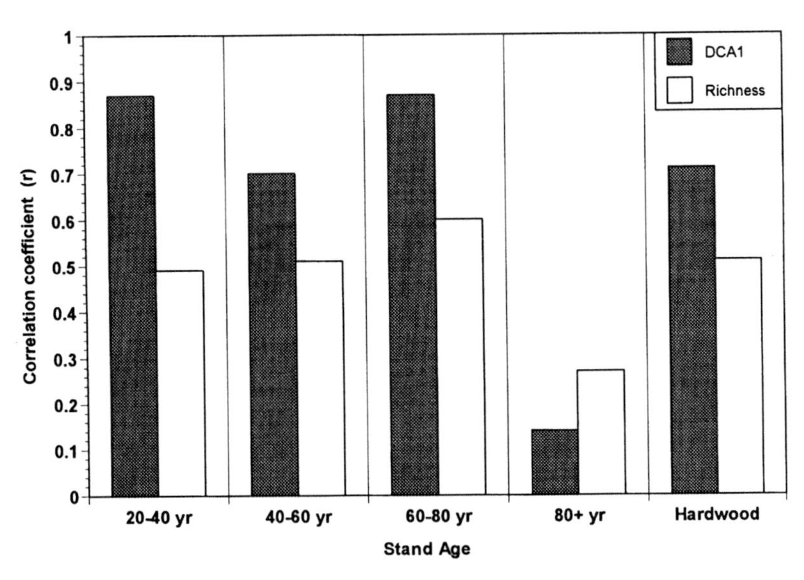

Figure 9.2. Spearman rank correlation comparisons of stand first-axis detrended correspondence analysis (DCA 1) ordination scores and species richness (number of species/0.1 ha) with soil pH.

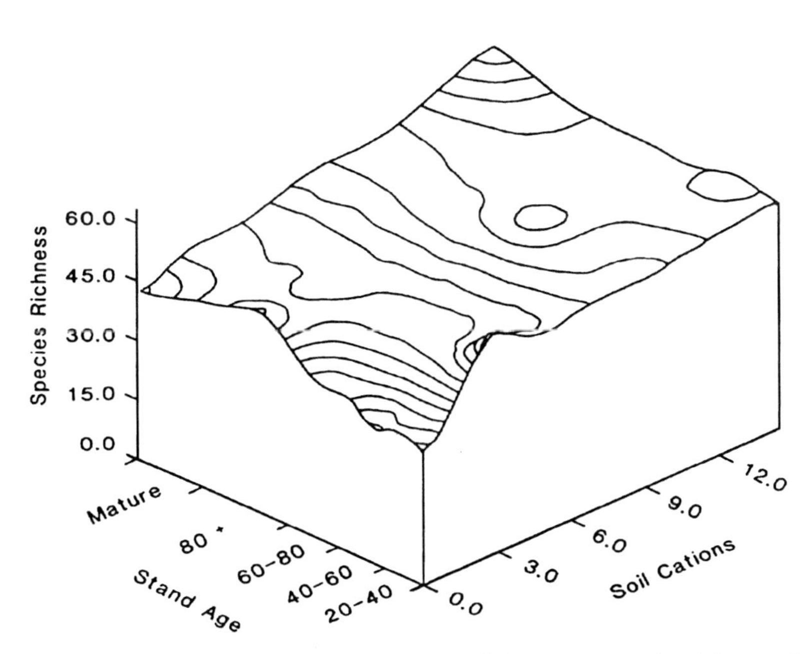

Figure 9.3. A three-dimensional representation of changes in species richness, stand age, and soil cations (highly correlated with pH) among Piedmont forests.

significant change in the relationship between species distributions and soil chemistry in transitional pine stands. High correlation between compositional variation and a factor such as pH implies that the range of conditions over which a typical species occurs is more constrained than where the correlation is lower. To test this, Christensen and Peet (1984) calculated the habitat breadth (abundance-weighted standard deviation of pH for stands within which a species occurs) for all species occurring in more than five stands with respect to soil pH. The results indicate that habitat breadth with regard to soil pH is greatest in the youngest pine stands, diminishes in intermediate-age stands, and increases slightly among transitional pine stands (fig. 9.4). It is lowest for species in hardwood stands.

If convergence in species composition were to occur as conceived by Clements (1936), one might predict that the variance in composition among stands (β-diversity) should diminish through time. Christensen and Peet (1984) found just the opposite, as β-diversity was highest in hardwood stands compared to pine stands of varying age. A more sophisticated approach to the convergence question is implied in Whittaker (1956) as the extent to which vegetation gradients (e.g., species composition—site relationships) in successional stands resemble those in hardwood stands. Christensen and Peet (1984) pursued this by comparing (using canonical correspondence analysis; CCA) the first three detrended correspondence analysis (DCA) axis scores of species in common between a particular pine age class and hardwood stands (fig. 9.5). The similarity generally increases with increasing pine stand age, but abruptly decreases in the transition-stage pines. This suggests that the distribution of species relative to one another becomes more like the distribution in hardwoods, but those relationships are altered considerably in transition-stage forests.

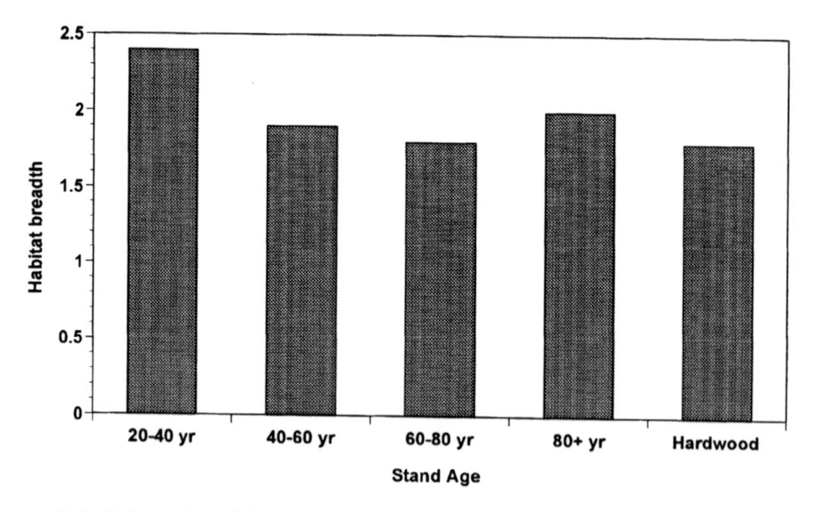

Figure 9.4. Habitat breadth (i.e., importance value weighted species distributions in relation to pH) among different-age forest stands.

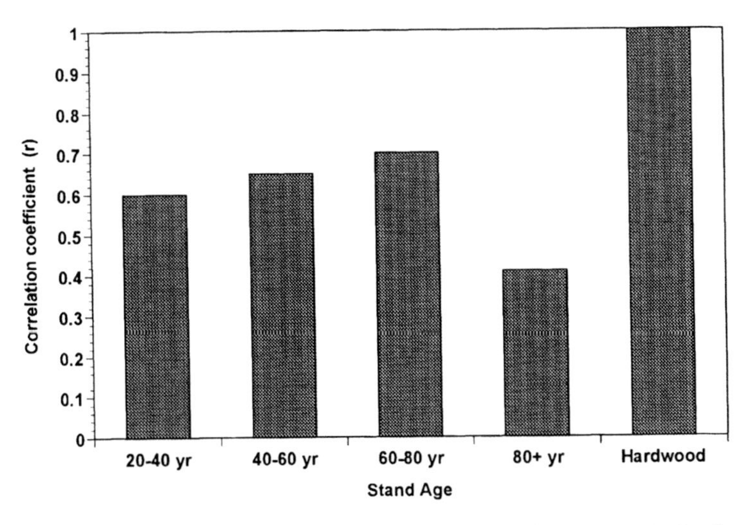

Figure 9.5. Canonical correlation analysis comparisons between axis 1, 2, and 3 detrended correspondence analysis species scores in pine stand age-classes and hardwoods. Hardwoods compared among themselves have a score of 1.0.

Conclusions

To conclude, we return to the questions posed in the introduction. First, what are the key environmental factors influencing herb distribution at various stages in the succession following land abandonment? Although there are major changes in species dominance and community composition during succession and across environmental gradients, these patterns are best understood as a consequence of individual responses of species related to their ability and opportunities to disperse to and compete at particular sites. A few of the individuals of woody plant species that will dominate late in succession may arrive early and simply outlive pioneers, supporting Egler's initial composition hypothesis; however, this is most certainly not the case for herbs. Virtually none of the herb species common in pine and hardwood stands is found in old fields.

At every stage of succession, a significant proportion of the variability in herbaceous species composition was correlated with soil site-variables. Across the entire landscape, soil moisture conditions (determined by topography and proximity to streams) accounted for much variation. However, when the very moist sites (e.g., riparian zones and swampy areas) were eliminated from consideration, soil chemical features such as pH and cation concentrations were most highly correlated with herb species distributions. Furthermore, these same variables were highly correlated with variations in overall species richness.

That soil resources explain variability throughout succession is certainly consistent with the assumptions of T. M. Smith and Huston's (1989) simulation models. It is important to note, however, that the strength of the correlations described above varied considerably among successional ages, as does the relative importance of other factors. Whereas soil factors were correlated within variation vegetation among old fields, the strength of those correlations (e.g., r^2, the proportion of total variance explained by a comparison) was not nearly as strong as among mid-aged pines or hardwoods. Although difficult to quantify, it was clear that old-field composition was heavily influenced by "landscape" effects such as the proximity of other fields and disturbed areas that provided important sources of seeds. Much of the variation among old-field soils was due to variations in farming practices (e.g., liming and fertilizer applications), whereas such variation among forest stands was tied to the chemical character of the parent rock. Thus, sites that might be considered to be rich in soil resources due to practices before land abandonment may become less so as those historical effects diminish through time.

The significant decline in correlation between soil site-variables and species composition among late-stage pine stands is noteworthy. Christensen and Peet (1984) suggested that this was a consequence of changes in the physical structure of these stands such as the creation of canopy gaps that produce considerable variance in the light environment. Put another way, understory light availability is uniformly low among pine stands at earlier stages and becomes highly variable at this stage. The change in the relative importance of explanatory variables in these transitional pine stands is certainly consistent with predictions from Tilman's (1985) resource ratio hypothesis and with T. M. Smith and Huston's (1989) simulations.

Variations in species richness through this successional process are considerably more complex than Odum's (1969) expectation that richness increases with increasing successional age. At the spatial scale of a 0.1-ha sample, the average number of herbaceous species encountered was remarkably similar (i.e., ~50 species) among successional ages. However, when comparing the total list of species encountered across the full range of environments among ages, Odum's prediction holds. Schafale and Christensen (1986) encountered fewer than 100 species among all their old fields, and Peet and Christensen (1988) encountered 203 and 328 in pine and hardwood stands, respectively. The key difference among these ages is in the numbers of species occurring in only a few stands at relatively low abundance. This coincides with Grime's (1979) prediction that low-abundance, stress-tolerant species should become more abundant in later stages of succession.

Our second question was, what do these patterns tell us regarding the mechanisms that underlie the dynamics of herb populations? Among the successional mechanisms proposed by Connell and Slatyer (1977), the tolerance and inhibition models seem to provide the most explanatory power for old-field succession in the North Carolina Piedmont. Keever's (1950) observations and experiments demonstrated clearly that variations in life history account for much of the change observed early in the successional process (e.g., the succession from annuals to biennials to perennials). Longer-lived herbaceous dominants such as broomsedge severely limit light to the soil surface and thereby prevent the subsequent invasion of many other herb species (inhibition). Similarly, it appears that usurpation of resources by pines prevents invasion of later successional species until the pine canopy begins to deteriorate.

In its most simplistic form (i.e., a species directly prepares the way for another species) there is no evidence for facilitation. However, when one considers the full array of interspecies interactions (e.g., competition, dispersal, herbivory) occurring during succession, invasion of particular species is often facilitated by other species. Pines may limit light and soil nutrients (inhibition), but in doing so influence the outcome of competition among potential successors. Thus, the successful competitors depend on the pines (Connell and Slatyer's operational definition of facilitation). Similarly, that structural features in old fields influence the dispersal of later successional species could be viewed as a form of facilitation.

It is clear that changes in the relative availability of resources are important in explaining successional patterns and that they are probably responsible for shifts in the relative abundance of species fitting into Grime's three strategic categories. That said, it is also true that designations such as "competitive," "stress-tolerant," and "ruderal" are comparative and relativistic and cannot be defined operationally in a rigid fashion. Furthermore, as Tilman (1985) suggests, shifts in relative importance of resources need not be linear or directional; when they are not (as suggested in some of the data presented here), then species shifts may be complex as well.

Much effort (certainly much of this discussion) to explain variability in the distribution of herbs on successional landscapes is focused on the importance of competitive interactions in the context of changes in resource availability. If such interactions were the only factors shaping the distributions of herbs in time and space, we should expect to see very high correlations between compositional variations and patterns of environmental variation. Although such correlations do explain significant amounts of variation in composition, they leave large amounts (often the majority) of such variation unexplained. It is fitting in closing this chapter to note that other mechanisms are likely equally important and that they may defy our attempts to create a unified theory of change. Christensen (1989) described the array of historical effects that influence vegetation composition including pre- and postabandonment land use and landscape effects, such as context and patch spatial scale. Grubb (1977) presented compelling arguments that the distributions of many species are best understood in regard to the conditions required for their establishment. When this is the case, competition among mature plants may explain little in the way of distribution. Such problems provide ample opportunity for future research.

10

Composition and Dynamics of the Understory Vegetation in the Boreal Forests of Quebec

Louis De Grandpré
Yves Bergeron
Thuy Nguyen
Catherine Boudreault
Pierre Grondin

The boreal forest of Quebec extends from the 48th to the 58th northern parallels (fig. 10.1). It can be subdivided into four forest domains (from south to north): the *Abies balsamea* (Mill.) and *Betula papyrifera* (Marsh.) domain, the *Picea mariana* (Mill.) and feathermoss domain, the *Picea mariana* lichen woodland, and the forest-tundra domain (Grondin et al. 1996; Saucier et al. 1998). This chapter will be restricted to the *Abies balsamea-Betula papyrifera* and the *Picea mariana*-feathermoss domains, which extend from 48° to 52° north latitude and include much of the closed-canopy, commercial boreal forest in Quebec. In this zone, average annual temperatures range between 1° and −2.5°C, and average annual precipitation is between 600 and 1400 mm. Temperature and precipitation decrease toward the north, and precipitation increases from west to east. The main tree species are *Picea mariana, Picea glauca* (Moench), *Abies balsamea, Pinus banksiana* (Lamb.), *Larix laricina* (Du Roi), *Populus tremuloides* (Michx.), and *Betula papyrifera*.

In addition to the south–north transition from *Abies balsamea*-dominated forests to *Picea mariana* forests, there is a strong east–west gradient that mainly relates to topography and forest fire frequency. In the east, a low fire frequency associated with higher altitude and a more pronounced topography allows for the dominance of late successional stands of *Abies balsamea* and *Picea mariana*, while a relatively flat topography and frequent fires are responsible for the abundance of post-fire stands dominated by deciduous species and *Pinus banksiana* in the west. The boreal ecosystem is strongly controlled by disturbances such as fire, insect outbreaks, and windthrow (Bergeron et al. 1998). Although fire is present in all of the boreal forest, there is a decrease in fire activity toward its eastern part. Since *Abies balsamea* is a late succes-

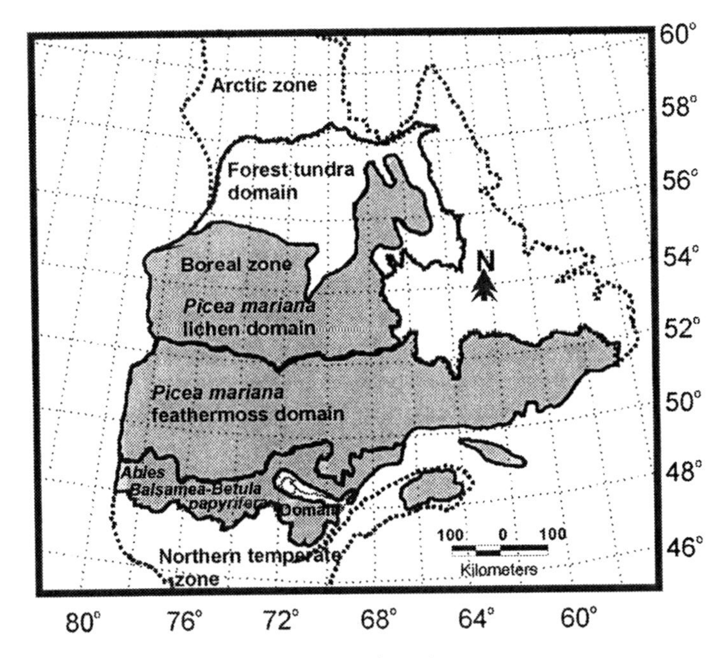

Figure 10.1. Boreal forest zones of Quebec.

sional species, the impact of spruce budworm outbreaks, which cause significant mortality of *Abies balsamea*, increases toward the east (Grondin et al. 1996; Bergeron and Leduc 1998). Major windstorms are also reported in the eastern part of the boreal forest (Ruel 2000). Forest canopy composition is also strongly associated with local site conditions. With the exception of its western part, which belongs to the northern Clay Belt where clay and organic soils are abundant, the area is located on the Canadian Shield and is characterized by tills of various depths and fluvio-glacial coarse deposits. The northern Clay Belt of Ontario and Quebec south of James Bay is a vast physiographic region created by lacustrine deposits left after maximal extension of proglacial lakes (Vincent and Hardy 1977).

The total vascular flora is estimated to contain less than 1000 species, including only 22 tree species. It is mainly represented by species with boreal affinities, with few intrusions from temperate-deciduous or arctic regions. Forest species compose a small part of this diversity with less than 300 common species. Although it is species poor, the boreal forest is characterized by understory dominants that vary from forbs and ericaceous shrubs to feathermosses, or *Sphagnum*. The variability in the composition of the understory layers has been recognized by many boreal ecologists and is the foundation of many classification systems using indicator species (Cajander and Ilvessalo 1921; Jurdant et al. 1977; Bergeron and Bouchard 1984).

The composition of the understory can be explained by several nonexclusive factors. It may be controlled by species' individualistic responses to regional climate, local site conditions (Whittaker and Levin 1975), and canopy cover composition (chapter 8, this volume). Disturbances also exert a strong influence on understory community composition and species distribution. Boreal plant species present various adaptive strategies to persist in the context of disturbance. The species can be classified into different groups according to their ability to tolerate disturbance, their successional status, their reproductive strategy, and their competitive ability (Noble and Slatyer 1980; Rowe 1983; Halpern 1989; Noble and Gitay 1996). Rowe (1983) identified five major groups of species in boreal forest ecosystems based on their reproductive strategies: pioneer species with high seed production and highly dispersed (*invaders*), species able to resprout following fire from buried vegetative organs (*endurers*), species surviving fire from protective organs such as thick bark (*resisters*), species evading fire by storing seeds in the humus or mineral layers (*evaders*), and late successional species more common in unburned areas and not adapted to maintain themselves in the context of recurrent fires (*avoiders*). The abundance of each of these groups over the landscape or in a plant community is largely determined by the fire frequency (Johnson 1979). At one end of the spectrum with frequent fires, invaders are more abundant, while at the other end avoiders dominate the landscape. Thus, a regional variation in the fire regime characteristics would have repercussions on the composition and diversity of the understory. In addition, forest management has recently contributed to changing understory composition and consequently has to be considered as a factor responsible for species distribution (chapter 13, this volume). In this chapter, we discuss the factors that control the composition and dynamics of the understory community of the Quebec boreal forest. In the first section, the influence of fire recurrence, site type, and cover type on understory community composition is analyzed throughout the range of the commercial boreal forest. The second section focuses on the role of forest management in understory community dynamics.

Influence of Fire Recurrence, Site, and Cover Type on Understory Community Composition and Dynamics

Methods

We used data from the ecological survey of the Ministère des Ressources naturelles du Québec to study the distribution of understory species (Bergeron et al. 1992). The ecological survey is based on an extensive uniform sampling including 15,000 plots well distributed in the southern boreal zone (Saucier et al. 1994). Sampling was designed to be representative of the distribution of physical features (soil type and hydric regime) and cover types present in each ecological region of the boreal forest. Trees, shrubs, herbs, ferns, and nonvascular species were sampled in circular plots of 400 m². Only forests with a dominant canopy of 7 m and more were sampled. Early postfire communities with an opened canopy were not sampled. The percent cover of all species was estimated over the whole area based on cover classes ("+" for the presence of one individual, 1–5%, 6–25%, 26–40%, 41–60%, 61–80%, 81–100%). The environment of each plot was described in terms of hydric regime, type of soil deposit, geographical position, altitude, and many more variables (Ministère des Ressources naturelles du Québec 1994).

We estimated fire recurrence using the forest fire data of the last 30 years and the forest age-class distribution based on the last provincial forest surveys. Fire recurrence was calculated for each ecological region (characterized by uniform association of dominant vegetation under uniform regional climate; fig. 10.2). For the analysis, five qualitative fire recurrence classes were considered: high (50–100 years), moderate (101–150 years), low (151–200 years), very low (201–250 years), and extremely low (> 250 years, fig. 10.2). Fire recurrence is thus based on the current forest age-class distribution over an ecological region. The site type is a combination of the surficial deposit and the hydric regime of each plot. To simplify the interpretations, we created five site type classes based on the soil descriptions of each plot: organic deposit with hydric moisture regime, fine-textured with subhydric moisture regime, medium-textured deposits with mesic hydric regime, coarse-textured deposit with mesic-xeric moisture regime, and thin soils with xeric-mesic moisture regime. Finally, the cover type for each plot was assigned based on the dominant tree species (one or two) using the species cover data. We identified

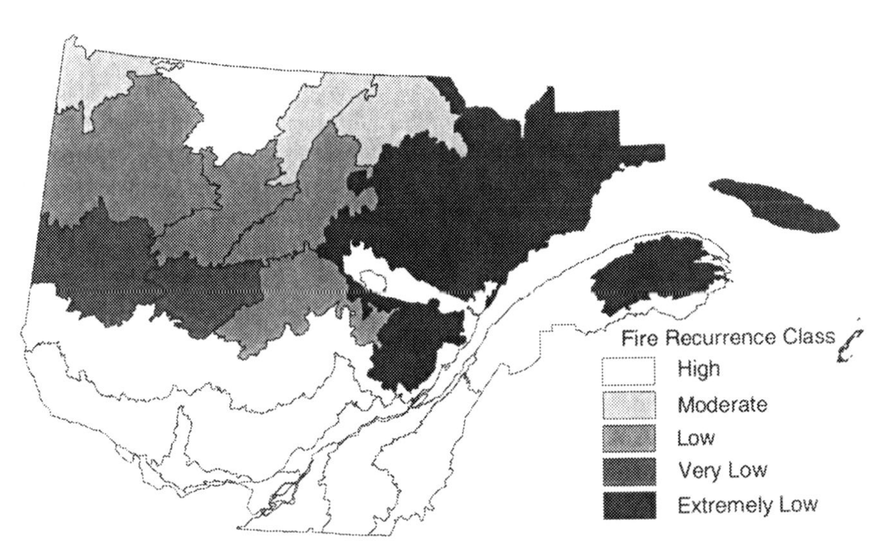

Figure 10.2. Fire recurrence map for the commercial boreal forest of Quebec. Fire recurrence was estimated within each ecological region. Ecological regions with similar fire recurrence were then grouped (high: 50–100 years, moderate: 101–150 years, low: 151–200 years, very low: 201–250 years, and extremely low: >250 years).

six dominant cover types: deciduous canopy, mixture of deciduous and coniferous species, canopy dominated by *Picea mariana*, mixture of *Picea mariana* and *Abies balsamea*, canopy dominated by *Abies balsamea*, and canopy dominated by *Pinus banksiana*.

Data Analyses

Data analyses were performed at two levels. First, we grouped plots with similar fire recurrence, site type, and cover type and used multivariate analyses to evaluate the effect of physical and ecological factors on understory species distribution and community composition. Second, understory species with similar life-history traits in the context of fire were grouped. The richness and cover patterns of these groups were interpreted along the fire recurrence gradient and in relation to cover type and site type.

Understory community dynamics were then described using a correspondence analysis (CA). To perform this analysis, the size of the data set had to be reduced, and plots were grouped. Each group represented a combination of fire recurrence, site type, and cover type ($5 \times 5 \times 6$ combinations). For each combination, we calculated the mean species cover based on the number of 400-m^2 plots. Although 150 groups were possible, only 108 were used for further analyses. Some combinations had no or too few observations (plots) to be used (only groups with more than five plots were used for further analyses). The CA was performed on the vegetation data set of 108 groups (considered as stands) and 189 species. The compositional gradient was interpreted using fire recurrence, site type, cover type, latitudinal and longitudinal gradients, altitude and richness of the different components of the understory (total richness, herbs [including ferns, *Lycopodium*, grasses, and sedges], shrubs, ericaceous shrubs, moss, lichens, and *Sphagnum* richness). These variables are shown in the ordination diagram (fig. 10.3). Quantitative variables are represented by vectors in the ordination diagram (Spearman correlation, $p < .05$), and class variables are represented by their centroids.

More than 13,500 plots were used for further analyses. To evaluate the effect of fire recurrence on understory community dynamics, we classified the understory species according to Rowe's (1983) reproductive strategies. Only the most common herb, fern, ericaceous, and shrub species were classified (species with a weight of 10 or more given by the CA were retained: 87 species out of 170 excluding lichen, mosses, and *Sphagnum*). Lichens, mosses, and *Sphagnum* species were not considered because they behave differently from vascular species along the successional sequence (see appendix 10.1 for complete list of species). This would have obscured the interpretations of the patterns of the different groups.

The invaders were characterized by 14 species, including *Epilobium angustifolium* (L.) and *Rubus idaeus* (L.). The most common reproductive strategy for boreal species is resprouting after fire from below-ground organs (endurers). The endurers composed close to 60% of the understory community and were separated in two subgroups: ericaceous endurers (7 species) and endurers (43 species). *Ledum groenlandicum* (Retzius), *Vaccinium myrtilloides* (Michx.), and *Kalmia angustifolia* (L.) are ericaceous endurers; *Aralia nudicaulis* (L.), *Betula glandulosa* (Michx.), and *Pteridium aquilinum* (L.) are typical endurers. The evaders were represented by seven species in the *Ribes* and *Viburnum* genus. The avoiders comprised 16 late successional species including *Taxus canadensis* (Marsh.), *Circaea alpina* (L.), and *Goodyera repens* (L.). None of the species was classified as resisters.

Fire Recurrence and Understory Community Composition

The first two axes of the CA explained 26% and 18%, respectively, of the variance in the species matrix (Eigenvalues: 0.364 and 0.241). The variation in understory composition along axis 1 is mainly related to a latitudinal gradient (fig. 10.3). As latitude increases, ericaceous and lichen species are more dominant in the understory (appendix 10.1), and the richness of these groups is higher. Both these species groups are associated with *Picea mariana* and *Pinus banksiana* cover types as well as high and moderate fire recurrence (fig. 10.3). As latitude decreases, on the right portion of axis 1, total richness of understory increases. Herbs and shrubs are more dominant (appendix 10.1) and are represented by more species. This higher richness is associated with the presence of deciduous and mixed dominated stands. At lower latitudes, a decrease in fire recurrence is observed, and *Abies balsamea* cover types are more frequently encountered (fig. 10.3). The second ordination axis is associated with hydric regime. On the upper part of the axis, organic-hydric sites are found, and they are characterized by a dominance of *Sphagnum* species (both in cover and richness; fig. 10.3 and appendix 10.1). Herb richness is higher on fine-textured soils with a subhydric regime, and lichen richness is highest on sites with xeric moisture regimes (fig. 10.3).

With increasing latitude, the successional sequence associated with *Picea mariana* (with or without *Pinus banksiana*) dominates the landscape. Although climate is related to the increased dominance of *Picea mariana*, fire recurrence also exerts a strong influence on species distribution. Regions with high and moderate fire frequency are dominated by *Picea mariana* and *Pinus banksiana* (table 10.1). Both species are adapted to fire, although *Pinus banksiana* is more restricted to regions that burn frequently. As fire recurrence decreases, *Abies balsamea* becomes more dominant in the landscape. The distribution and abundance of the different site types over the landscape (see table 10.1) will also contribute to influencing the distribution of tree species and probably affect fire recurrence (if the probability of burning decreases along an increasing moisture regime). *Pinus banksiana* is usually more abundant on sandy soils with a xeric moisture regime. *Picea mariana* can also be dominant on this type of soil but is more associated with organic soils with a hydric moisture regime (table 10.2). Deciduous species and *Abies balsamea* usually dominate on the mesic-xeric moisture regime (table 10.2). These factors will interact in space and time to influence understory community composition

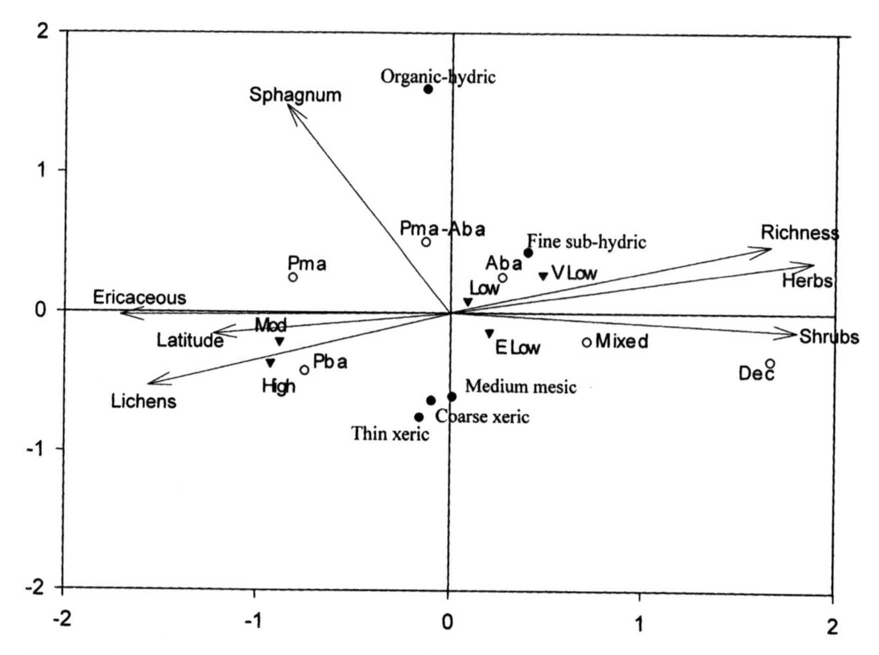

Figure 10.3. Scores of the environmental and ecological factors on the first two axes of the correspondence analysis of the 109 groups by 189 species. Centroids of fire recurrence classes, site and cover types with correlation coefficients of latitude and the richness of different species groups are shown on the diagram. Only significant correlations are shown on the graph. Fire recurrence classes: moderate, mod; very low, V Low; extremely low, E Low. Cover types: *Abies balsamea*, Aba; deciduous, Dec; *Picea mariana*, Pma; *Picea mariana-Abies balsamea*, Pma-Aba; *Pinus bankasiana*, Pba.

Table 10.1. Proportion of site types and cover types in the fire recurrence classes

	Site types (%)					Cover types (%)					
Recurrence	O	FS	MM	CX	TX	Dec	Mix	Aba	Pba	Pma	Pma-Aba
Frequent	10.5	3.7	67.2	16.6	2.0	5.1	9.1	3.1	37.5	44.9	0.3
Moderate	10.7	2.2	77.1	8.2	1.8	2.3	8.5	7.4	14.6	59.4	7.8
Low	20.0	16.6	47.9	11.6	3.9	9.8	18.4	6.1	15.0	46.5	4.2
Very low	20.2	16.5	40.1	17.7	5.5	18.9	25.4	8.4	18.0	25.5	3.8
Extremely low	7.9	17.7	57.7	7.6	9.1	10.2	13.1	29.7	2.9	29.6	14.5

*Cover types: deciduous canopy (Dec), mixture of deciduous and coniferous species (Mix), canopy dominated by *Picea mariana* (Pma), mixture of *Picea mariana* and *Abies balsamea* (Pma-Aba), canopy dominated by *Abies balsamea* (Aba), and canopy dominated by *Pinus Banksiana* (Pba).

Site types: organic deposit with hydric moisture regime (O), fine-textured with subhydric moisture regime (FS), medium-textured deposits with mesic hydric regime (MM), coarse-textured deposits with mesic-xeric moisture regime (CX), and thin soils with xeric-mesic moisture regime (TX).

Table 10.2. Proportion of cover types in the different site types

Site type	Cover types (%)					
	Dec	Mix	Aba	Pba	Pma	Pma-Aba
Organic-hydric	2.2	5.1	13.0	1.6	69.5	8.6
Fine subhydric	12.4	16.4	32.1	6.7	23.2	9.3
Medium mesic	13.0	18.5	17.9	7.8	32.3	10.5
Coarse xeric	7.8	12.6	8.8	36.2	29.4	5.2
Thin xeric	8.7	19.3	17.3	6.0	37.1	11.6

*Cover types: deciduous canopy (Dec), mixture of deciduous and coniferous species (Mix), canopy dominated by *Picea mariana* (Pma), mixture of *Picea mariana* and *Abies balsamea* (Pma-Aba), canopy dominated by *Abies balsamea* (Aba), and canopy dominated by *Pinus banksiana* (Pba).

and species distribution. Succession has not only to be interpreted along a time gradient (variation in cover type) but has to consider the physical setting of the landscape and the disturbance regime. In the following section, we address the impact of these factors on the understanding of boreal forest understory community dynamics using Rowe's (1983) reproductive strategies.

Successional Sequences

Understory Changes in the Abies balsamea-Betula papyrifera *Domain* In the *Abies balsamea-Betula papyrifera* domain (fig. 10.1), the most common successional sequence observed involves the early dominance of *Betula papyrifera* and *Populus tremuloides* after fire. In the absence of fire, the deciduous canopy will be replaced by a coniferous one and dominated by *Abies balsamea* (Bergeron and Charron 1994; Bergeron 2000; De Grandpré et al. 2000; Gauthier et al. 2000). As changes are observed in canopy composition with time elapsed since fire, the understory communities are also dynamic through time (De Grandpré et al. 1993). After canopy closure, pioneer species such as *Epilobium angusltifolium, Solidago rugosa* (L.), and *Rubus idaeus* disappear from the understory and are replaced by more shade-tolerant species including *Aster macrophyllus* (L.), *Acer spicatum* (Lam.), and *Aralia nudicaulis*. These species may persist for a long time in the understory by vegetative growth. Some changes in species composition do occur after canopy closure, but understory community dynamics are mostly characterized by shifts in community dominance, associated with changes in the composition of the canopy (Carleton and Maycock 1978). Species richness and overall abundance are higher under a deciduous canopy and decline with the increasing dominance of *Abies balsamea* in the overstory (Foster and King 1986; De Grandpré et al. 1993). As conifers increase their dominance in the canopy, light penetration to the forest floor decreases, and the understory is dominated by even more shade-tolerant species such as *Taxus canadensis*. These successional changes of the understory can be predicted from species life-history strategies, such as establishment ability and shade tolerance (Noble and Slatyer 1980; Halpern 1989). As *Abies balsamea* becomes a major component of the landscape, disturbances such as windthrow and spruce budworm outbreaks will affect community dynamics. Gap creation resulting from these disturbances will contribute to an increase in richness and allow many early successional understory species to remain longer in post-fire communities. Thus, understory community succession does not operate in a linear fashion with time elapsed since fire. De Grandpré and Bergeron (1997) hypothesized that openings of the canopy resulting from spruce budworm outbreaks allow for the reestablishment or the increase in cover of some early successional species, a situation that makes the understory more resistant to change after other disturbances.

Understory Changes in the Picea mariana*-Feathermoss Domain* In the *Picea mariana*-feathermoss domain (fig. 10.1), succession after fire is characterized by the establishment immediately after fire of a mixture of *Pinus banksiana* and *Picea mariana*. In the absence of fire for long time intervals, *Picea mariana* will dominate the canopy (St-Pierre et al. 1992). *Picea mariana* often dominates throughout the successional sequence, and few changes occur in canopy composition with time elapsed since last fire (Cogbill 1985; Foster 1985; Morneau and Payette 1989; Gauthier et al. 2000).

Several studies have shown that understory successional changes for vascular plants along the *Picea mariana* chronosequence occur before canopy closure (Dix and Swan 1971; Shafi and Yarranton 1973; Black and Bliss 1978; Foster 1985; Taylor et al. 1987; Sirois and Payette 1989). Most understory species in the *Picea mariana* boreal forest can survive fire and resprout afterward (Ahlgren 1960; Archibold 1979; Carleton and Maycock 1980; Rowe 1983). Foster (1985) observed that all species present before fire were present in the postfire community, with the exception of *Empetrum nigrum* (L.). Furthermore, only few invading species were part of the postfire community. Beside these minor changes in early postfire succession, few compositional changes will occur in later stages. After canopy closure, most of the changes are associated with bryophytes and lichens, while community composition of vascular plant species does not change much. The succession observed for the nonvascular understory would correspond to the facilitation model of Connell and Slatyer (1977). Taylor et al. (1987) hypothesized that these changes are related to the self-thinning of *Picea mariana* stands, which leads to an increase in soil moisture content. As evapotranspiration decreases, resulting from self-thinning, *Sphagnum* species slowly replace moss species (mainly *Pleurozium schreberi* (Brid.), (Taylor et al. 1987). The dominant component of the understory throughout the *Picea mariana* successional sequence is the ericaceous species.

Throughout the boreal forest range, variability is observed in the fire regime. Such variability may affect the distribution of species in the different regions of the boreal zone. For example, in southeastern Labrador, Foster

(1985) observed that only *Epilobium angustifolium* invaded newly burned sites, while all the other vascular species present in the early postfire community were resprouters, mainly ericaceous species. In the western boreal forest, along with ericaceous resprouters, species like *Aralia hispida* (Vent.), *Geranium bicknellii* (Britton), *Polygonum cilinode* (Michx.) and some grasses and sedges are only found immediately after fire (Ohmann and Grigal 1979; Viereck and Schandelmeier 1980). According to Rowe (1983), all of these species are shade intolerants belonging to the evader group. These species also produce a persistent seed bank (Archibold 1979; Rowe 1983; Mladenoff 1990). The absence of the evaders in southeastern Labrador may be explained by long intervals between successive fires, exceeding the longevity of seed bank.

Fire Recurrence and Cover Types

Along the *Abies balsamea* successional sequence, the variations in cover and richness of the functional groups are similar along the fire recurrence gradient (figs. 10.4 and 10.5). With the exception of the landscapes with high fire recurrence, where ericaceous endurers dominate the understory, typical endurers are dominant and tend to increase in dominance with decreasing fire recurrence. The richness pattern of the understory communities along the fire recurrence gradient is mainly associated with this group of species (fig. 10.5). Both the richness and the cover of the typical endurers group are highest under the deciduous cover type and decrease as *Abies balsamea* increases in the canopy. The species part of the invaders group increases in cover as fire becomes less common in the landscape. Although the relationship is strongest under deciduous canopy, it can still be observed under mixed and *Abies balsamea* cover types (fig. 10.4). Richness of this group also follows a similar pattern (fig. 10.5). Species belonging the evaders and avoiders groups are present with extremely low cover in landscapes with high to moderate fire recurrence, no matter what the cover type is (fig. 10.4). As fire becomes less frequent, these species groups increase in mean cover. Although present with low cover in the high to moderate fire recurrence landscapes, the richness of these groups is almost comparable to that observed in lower fire recurrence landscapes.

Along the fire recurrence gradient, understory vascular plant composition for the *Picea mariana* successional sequence is characterized by the dominance of endurers. Ericaceous endurers are dominant in terms of cover both in *Picea mariana* and *Pinus banksiana*-dominated stands (fig. 10.4). The richness of the ericaceous endurers group is not influenced by fire recurrence under both cover types (fig. 10.5). However, the mean cover of this group decreases almost linearly with decreasing fire recurrence under *Picea mariana* cover types, while it remains similar under *Pinus banksiana* cover type (fig. 10.4). In stands with a mixed composition of *Picea mariana* and *Abies balsamea*, typical endurers dominate the understory both in richness and cover along the recurrence gradient, with the exception of the landscapes subject to high

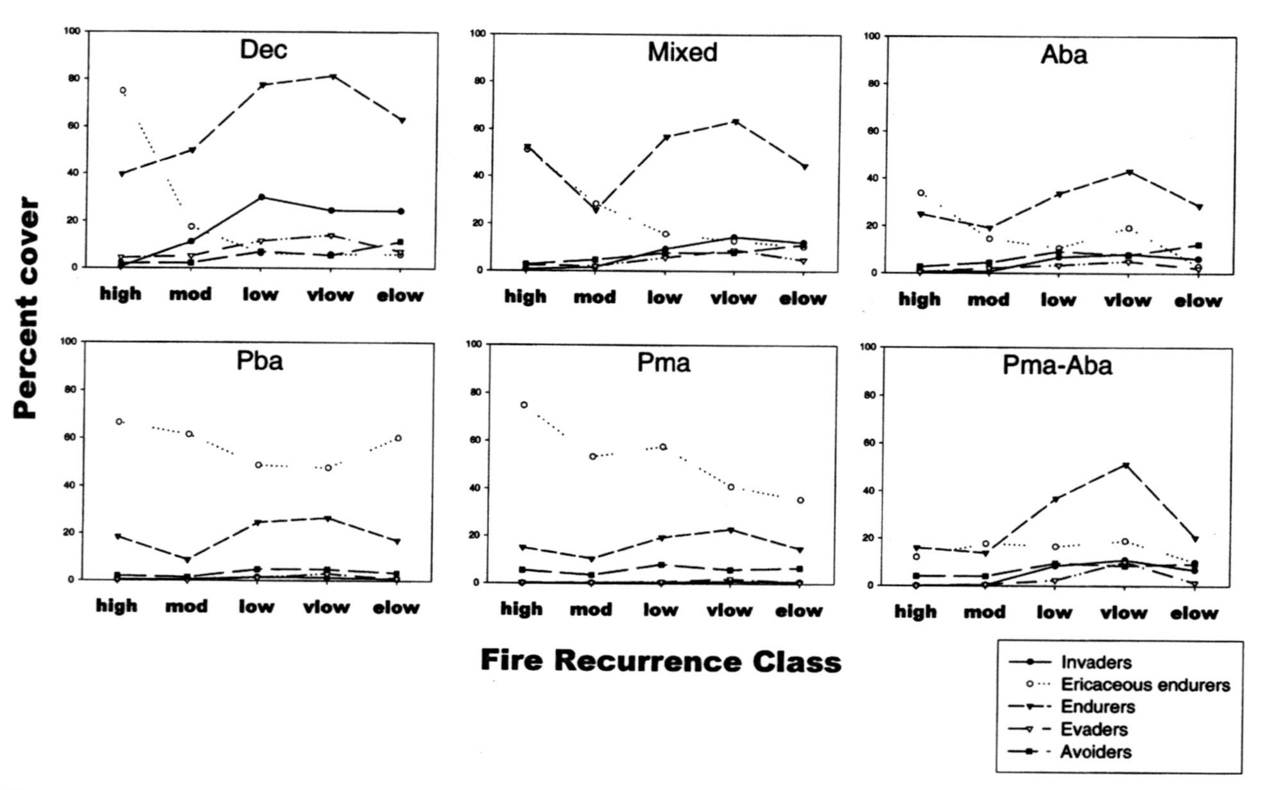

Figure 10.4. Percent cover patterns for the different reproductive strategies by cover type, along the fire recurrence gradient (high, moderate, low, very low, extremely low). Cover types: deciduous canopy (Dec); mixture of deciduous and coniferous species (Mix); canopy dominated by *Picea mariana* (Pma); mixture of *Picea mariana* and *Abies balsamea* (Pma-Aba); canopy dominated by *Abies balsamea* (Aba); and canopy dominated by *Pinus banksiana* (Pba).

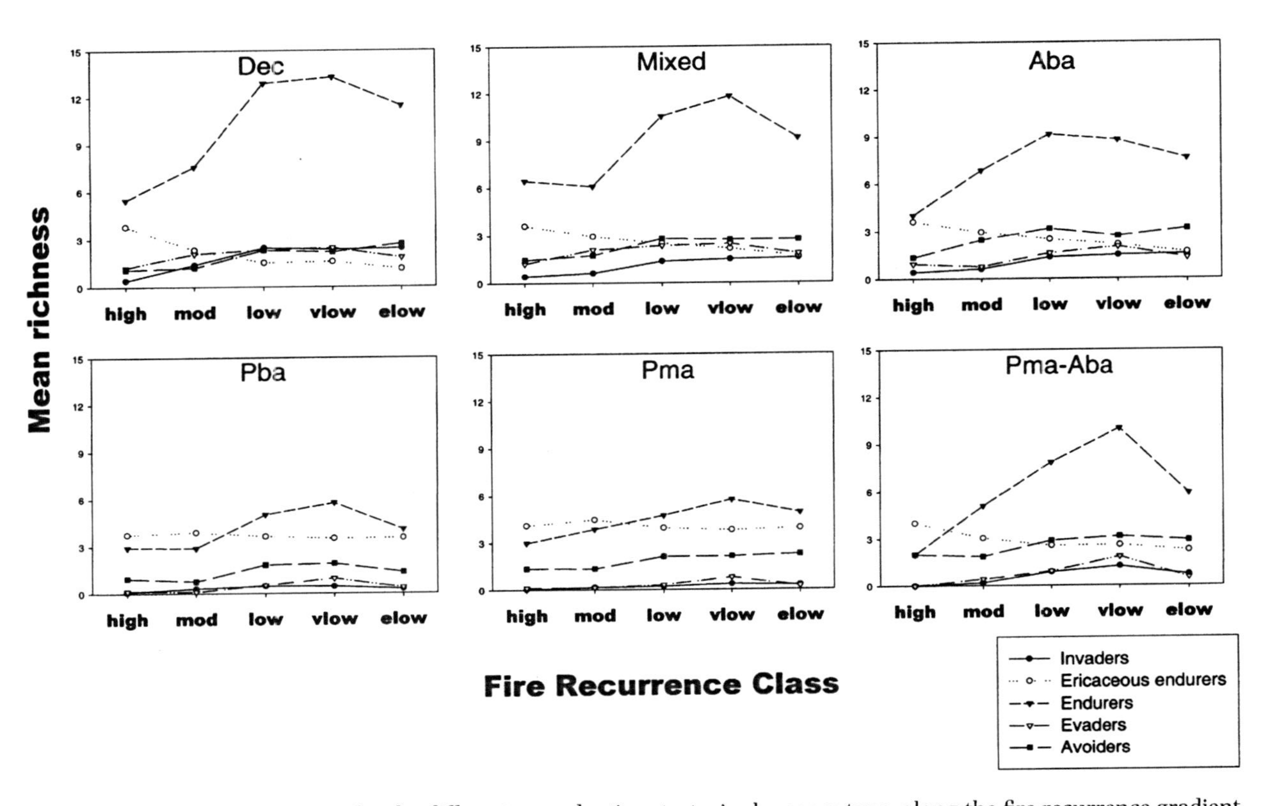

Figure 10.5. Richness patterns for the different reproductive strategies by cover type, along the fire recurrence gradient (high, moderate, low, very low, extremely low). Cover types: deciduous canopy (Dec); mixture of deciduous and coniferous species (Mix); canopy dominated by *Picea mariana* (Pma); mixture of *Picea mariana* and *Abies balsamea* (Pma-Aba); canopy dominated by *Abies balsamea* (Aba); and canopy dominated by *Pinus banksiana* (Pba).

fire recurrence, where ericaceous endurers have the highest richness. The typical endurers species group is characterized by an increase in both richness and cover as fire recurrence decreases. After reaching a peak where fire recurrence is very low, richness and cover decrease (figs. 10.4 and 10.5). The same pattern is observed under the *Picea mariana—Abies balsamea* cover type. The species parts of the invaders and evaders groups are almost absent along the fire recurrence gradient in *Picea mariana-* and *Pinus banksiana-*dominated stands (figs. 10.4 and 10.5). Some invaders and shade tolerant evaders are present under the *Picea mariana—Abies balsamea* cover type but only in regions with low to extremely low fire recurrences. Finally, avoiders are able to remain at comparable richness levels along the recurrence gradient, except under the *Pinus banksiana* cover type, where richness slightly increases as fire recurrence decreases. The cover of the avoiders increases as fire recurrence decreases (fig. 10.4), and this tendency is stronger in *Picea mariana–Abies balsamea*-dominated stands.

Fire Recurrence and Site Types

As they are dominant under the different cover types, endurers also dominate on the different site types. In landscapes with high to moderate fire recurrences, ericaceous endurers dominate the understory communities in terms of cover on all site types (fig. 10.6). As fire recurrence decreases, the cover of this group also decreases. The richness of ericaceous endurers decreases as fire becomes less frequent (fig. 10.7). The richness and cover patterns for typical endurers along the fire recurrence gradient show some differences depending on the site type. The richness and cover of this group is highest on fine-textured soils with subhydric moisture regimes. Avoiders do not vary much in richness and cover along the fire recurrence gradient on organic and fine subhydric site types. For site types with a more xeric hydric regime, avoiders show an increase in richness and cover with decreasing fire recurrence (figs. 10.6 and 10.7). As for invaders and evaders, the patterns vary according to site type, but in general there is a tendency for richness and cover to increase with decreasing fire recurrence (figs. 10.6 and 10.7).

Understory Community Composition and Species Distribution

Carleton and Maycock (1981) studied the affinities between understory and canopy. They concluded that the majority of the understory species (75%) did not show any specificity for single canopy classes. The lack of specificity between cover type and understory species has been associated with the nature of regeneration after fire (Carleton and Maycock 1981). Because most species have developed reproductive strategies to cope with fire, recurrence should have more influence on understory species distribution and community composition than cover type. Cover type distribution is also influenced

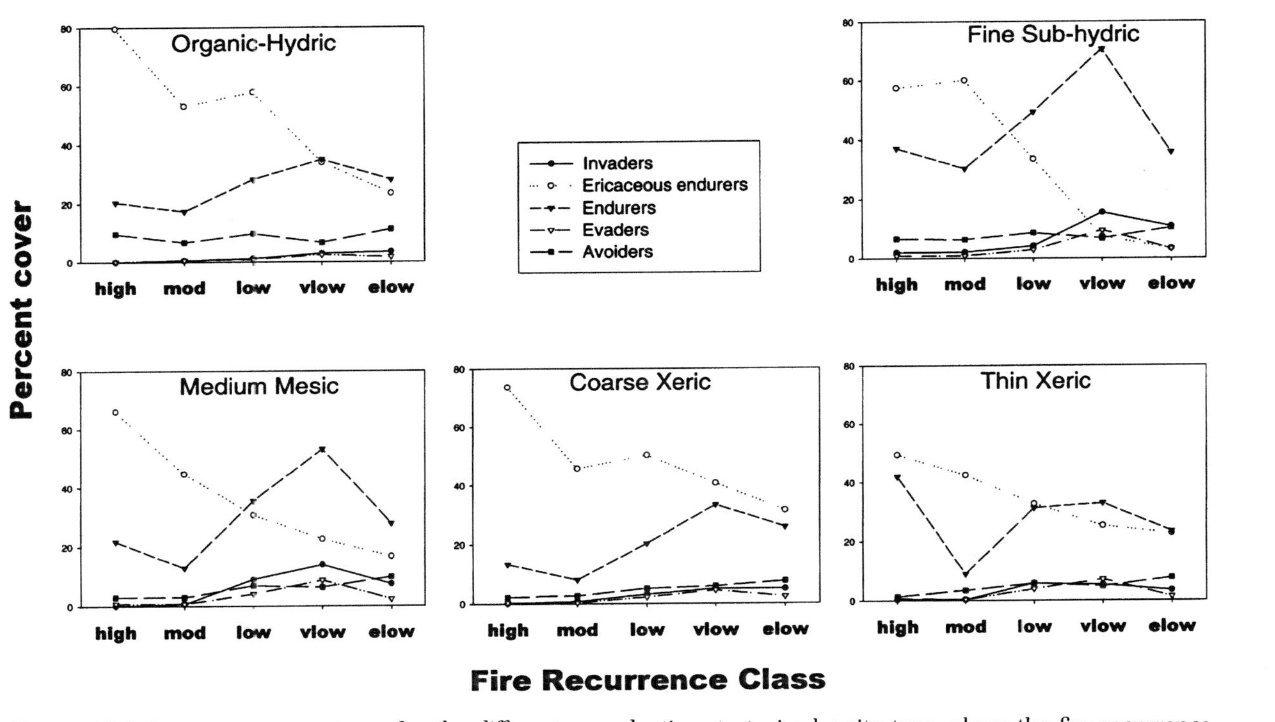

Figure 10.6. Percent cover patterns for the different reproductive strategies by site type, along the fire recurrence gradient (high, moderate, low, very low, extremely low). Site types: organic deposit with hydric moisture regime; fine textured with subhydric moisture regime; medium-textured deposits with mesic hydric regime; coarse-textured deposits with mesic-xeric moisture regime; and thin soils with xeric-mesic moisture regime.

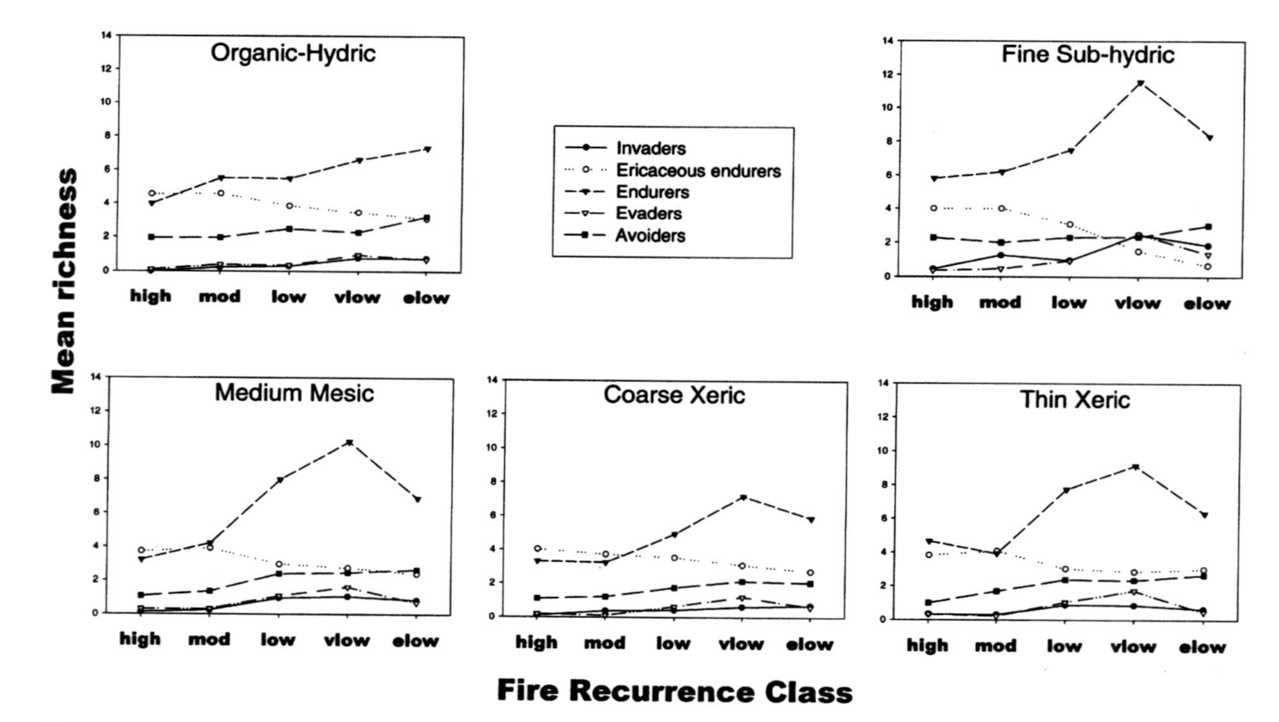

Figure 10.7. Richness patterns for the different reproductive strategies by site type, along the fire recurrence gradient (high, moderate, low, very low, extremely low). Site types: organic deposit with hydric moisture regime; fine textured with subhydric moisture regime; medium-textured deposits with mesic hydric regime; coarse-textured deposits with mesic-xeric moisture regime; and thin soils with xeric-mesic moisture regime.

by the fire recurrence gradient, as tree species also have strategies to cope with fire. Species distribution and understory community composition are largely influenced by site conditions (site type). Carleton et al. (1985) showed that 67% of the variance in understory species composition in the boreal forest of northern Ontario was explained by soils. Although site type influences the composition of the understory, our results showed that for the same site type, fire recurrence has an impact on the abundance and distribution of the species groups.

Only 17% of the stands sampled in the landscape subject to high fire recurrence are characteristic of the *Abies balsamea* successional sequence (table 10.1). In these stands, understory community composition does not correspond to the typical successional sequence described for *Abies balsamea* forests. Ericaceous endurers are dominant in deciduous, mixed, and *Abies balsamea* cover types in this landscape. The high fire recurrence contributes to the dominance of *Picea mariana* and *Pinus banksiana* cover types and of ericaceous species. Even on the different site types, ericaceous endurers dominate. This observation confirms the hypothesis of Carleton and Maycock (1981) that the lack of specificity between the canopy and the understory is related to the reproductive strategies of the understory species in relation to fire. Ericaceous endurers thrive as resprouters where fire recurrence is high (Rowe 1983). There is a definite impact of fire recurrence on the composition of the understory and also on species distribution. In contrast, the understories of deciduous, mixed, and *Abies balsamea* cover types support a higher mean species richness for the typical endurers group than in all other cover types where fire recurrence is high or moderate (figs. 10.4 and 10.5). In such landscapes, these stands could act as reservoirs for understory species that are usually found in the southern section of the commercial boreal forest, where *Abies balsamea* is dominant and fires are less frequent. In these stands, species like *Aralia nudicaulis*, *Clintonia borealis* (Ait.), *Cornus canadensis* (L.), and *Trientalis borealis* (Raf.) are found more frequently than in stands dominated by *Picea mariana* or *Pinus banksiana*. These species are typical endurers characteristic of the *Abies balsamea* forests of the south. Site type appears to play a similar role, and the presence of avoiders and some typical endurers in landscapes subject to high or moderate fire recurrence could be related to site types with hydric to subhydric moisture regimes. The mean richness of avoiders on organic and fine subhydric site types is almost twice that observed on other site types. Although these site types represent less than 15% of the landscapes, their contribution to the biodiversity over the landscape may well surpass their representation. A study in a 2-ha old-growth swamp forest in Sweden revealed that 33% of the bryophytes found throughout the country were found in this forest (Ohlson et al. 1997).

The richness patterns are in accordance with Connell's (1978) intermediate disturbance hypothesis, which predicts that diversity will be higher under intermediate disturbance. The argument behind this hypothesis is that intermediate frequency of disturbance will allow the coexistence of species with different competitive strategies (Huston 1979; Huston and Smith 1987). The increase in richness observed in all of the functional groups as fire recurrence decreases supports this hypothesis (fig. 10.5). Although the richness and cover patterns of the avoiders and the shade-tolerant evaders are easy to relate to the recurrence gradient (figs. 10.4 and 10.5), this is not the case for the invaders group. Richness and cover of invaders should have been higher in the landscape subject to high fire recurrence since this group is characterized by r-selected species such as *Epilobium angustifolium* and *Hieracium* sp. (Rowe 1983). The absence of the invaders where recurrence is high to moderate may be related to the fact that only closed forests were sampled. Understory successional changes in *Picea mariana*-dominated forests occur before canopy closure for vascular plant species; invaders could only be present in open forests. Foster (1985) observed that the only invading species to colonize the recently burned black spruce forests of southeastern Labrador was *Epilobium angustifolium*. In similar forests in northern Minnesota and western Canada, species such as *Geranium bicknelii*, *Aralia hispida*, and *Polygonum cilinode* were noted in recently burned areas among other shade intolerant species, which were absent from the prefire communities. These species have been identified as shade-intolerant evaders with a persistent seed bank (Rowe 1983). The absence of these species in the early postfire community in southeastern Labrador is probably related to the extremely low fire recurrence. A fire rotation of 500 years has been suggested for this part of the boreal forest (Heinselman 1981; Foster 1985).

Invaders are more abundant with decreasing fire frequency, and deciduous canopy could act as a reservoir from which these species could invade burned areas in regions where fire recurrence is low to extremely low. The increase in the abundance of invaders under lower fire recurrence could, however, be explained by the higher recurrence of other types of disturbances like spruce budworm outbreaks. De Grandpré and Bergeron (1997) and De Grandpré et al. (1993) have shown that in the *Abies balsamea-Betula papyrifera* domain, openings in the canopy resulting from spruce budworm outbreaks help maintain some early successional species.

Response of the Understory Layer to Forest Management

Until recently, the Canadian boreal forest has often been perceived as a simple and rather homogeneous system whose natural dynamics were mainly driven by the high recurrence of wildfires (Dix and Swan 1971; Carleton and Maycock 1978; Cogbill 1985). This idea was based on the frequent dominance of pioneer tree species such as *Populus tremuloides*, *Betula papyrifera*, *Pinus banksiana*, and *Picea mariana* in the overstory of many boreal stands. As a result, even-aged stand management is being applied over most of the territory covered by the boreal forest. However, recent studies have shown that stand structure of increasing complexity and/or replacement of pioneer species by more late successional species, such as *Abies balsamea* and *Thuja occidentalis* (L.), are also observed with the prolonged absence of major fire events (Bergeron and Dubuc 1989; Gauthier et al. 2000). With the increasing integration of sustainable forest management into forestry policies, the pertinence of a generalized application of the even-aged management strategy over most the boreal forest has also been questioned (Bergeron and Harvey 1997; Bergeron et al. 1999). We have shown in this chapter that significant spatial and temporal variation in composition and structure can be observed for the understory vegetation. It is also increasingly recognized that the understory vegetation greatly contributes to the diversity and nutrient dynamics of forest ecosystems of eastern North America (chapters 2 and 4, this volume). Given that a great proportion of the commercial timber that is now being harvested in eastern Canada originates from the boreal forest, the identification of the impact of generalized even-aged forest management on the understory layer of the boreal forest is important in assessing the sustainability of current forest practices.

In general terms, even-aged forest management in the boreal forests of Quebec consists of regeneration cuts that bring the forest stand back to an early seral stage, and it usually entails the complete removal of the overstory (Doucet et al. 1996). Adequate tree regeneration of the forest stand is ensured by the practice of careful logging on sites where advance tree regeneration is sufficient before harvesting, and by site preparation and/or planting on sites deficient in advance regeneration. As seen in chapters 2–4 (this volume) and in previous sections of this chapter, the composition and structure of the understory vegetation of forests are mainly determined by the light and soil environments encountered at this stand level. The changes observed in the composition and structure of the understory layer after harvesting mostly result from the modifications this disturbance brings about in terms of these two environments. The increased amount of light reaching the forest floor often leads to a rise in soil temperatures, favoring soil microbial activity and increasing decomposition rates and nutrient availability (Keenan and Kimmins 1993). The impact of harvesting on the soil's physical properties (compaction, structure, mineral soil exposure, residual humus depth) will vary with site type, season of harvesting, type of machinery used, machinery circulation patterns, and subsequent silvicultural treatments. The amount of change in the soil's structural characteristics thus depends on the intensity of harvesting activities and is best expressed in terms of severity (chapter 13, this volume). Given the predominance of cryptograms in the *Picea mariana*-feathermoss domain of the Quebec boreal forest and the resulting influence they exert through the buildup of a significant forest floor layer, the severity of forest floor disturbance will be more crucial for postharvest regeneration patterns of this domain compared to those of the *Abies balsamea-Betula papyrifera* domain.

The recurrent nature of wildfires in the boreal forest has led to the development of a variety of regeneration strategies by boreal understory plants. These strategies differ in their capacity to cope with different levels of forest floor disturbance severity (Rowe 1983) and to exploit the environmental resources (light and nutrients) available after the disturbance (Chapin and

Van Cleve 1981). Thus, many understory pioneer plant species (invaders) quickly establish on a site after a disturbance by seeding in from the surrounding areas or germinating from a dormant soil seed bank (evaders). These species generally prefer mineral soil or shallow humus seedbeds and are shade-intolerant plants with rapid growth rates that take advantage of the light- and nutrient-rich environment usually found after disturbance. These species are often observed after forest floor disturbances of moderate to high severity. Many understory species present in mature forests maintain their presence on a site after a disturbance by resprouting from undamaged underground organs (endurers). Most of these species are usually observed after forest floor disturbances of low to moderate severity and possess a certain degree of plasticity in their capability to exploit the light and nutrient resources of their environment. Finally, some species tend to be restricted to later seral stages of the boreal forest. These species often require establishment or growth conditions associated with older forests such as lower light availability, cooler soil temperatures, or higher soil moisture levels (avoiders). These plant species do not persist after disturbance or do so only if the severity of the forest floor disturbance is low. Although these different strategies have developed as a response to the natural variability in forest floor disturbance severity, they can also serve as a basis to explain the regeneration patterns observed after harvesting (Nguyen-Xuan et al. 2000).

Effects of Clearcutting on Diversity

As seen in the previous sections of this chapter, the understory vegetation of the forests of the *Abies balsamea—Betula papyrifera* domain is mainly characterized by various associations of herbs and shrubs. The majority of the species are endurers that take advantage of the improved light and soil conditions found after disturbance. Thus, clearcut harvesting with a low to moderate level of forest floor disturbance should not significantly affect the diversity of the understory vegetation. Indeed, a study conducted by Harvey et al. (1995) in recently harvested stands of the mixed-wood forest of western Quebec showed that low-intensity harvesting resulted in the release of pre-established understory herbs such as *Clintonia borealis*, *Cornus canadensis*, *Linnaea borealis* (L.), *Maianthemum canadensis* (Desf.), *Aralia nudicaulis*, and *Rubus pubescens* (Raf.). Harvesting of higher intensity favored the establishment and/or development of a number of grasses, sedges, and introduced species (e.g., *Calamagrostis canadensis* (Michx.), *Scirpus atrocinctus* (Fern.), *Carex stipata* (Mühl.), *Galeopsis Tetrahit* (L.), *Taraxacum officinale* (Weber), *Cirsium vulgare* (Savi), *Phleum pratense* (L.)). Some of these species, especially grasses and sedges, can spread aggressively over a significant portion of the disturbed area, resulting in a decrease in postdisturbance plant diversity (Abrams and Dickman 1982). Harvey et al. (1995) observed that disturbance severity was mostly determined by site type and season of harvesting. Finally, it has also been noted that in mixed-wood stands, clearcut harvesting tends to favor a conversion to a predominantly deciduous canopy cover (Harvey and Bergeron 1989). The conversion of mixed stands to deciduous stands over extensive areas may have profound implications for the diversity of understory species present at the landscape level.

The previous sections of this chapter have shown the importance of mosses, *Sphagnum*, lichens, and ericaceous endurers in the characterization of the understory plant associations observed in the *Picea mariana*-feathermoss domain of the Quebec boreal forest. The majority of low shrubs are able to withstand disturbances of low to moderate severity through the persistence of underground organs in the forest floor. The bryophytes and lichens usually persist only if the forest floor disturbance severity is low. In a study conducted by Brumelis and Carleton (1989) in the spruce-dominated forests of northeastern Ontario, low-intensity logging promoted the development of many preestablished ericaceous endurers such as *Cassandra calyculata* (L.), *Gaultheria hispidula* (L.), *Ledum groenlandicum*, and *Vaccinium myrtilloides* or the persistence of many bryophytes such as *Dicranum polysetum* (Sw.), *Hylocomium splendens* (Hedw.), *Pleurozium shreberi*, *Ptilium crista-castrensis* (Hedw.), *Sphagnum angustifolium* (C. Jens.), *S. magnellanicum* (Brid.), *S. nemoreum* (Braithw.), and *S. wulfianum* (Girg.). High-intensity logging favored the establishment of a number of grasses, sedges, herbs, and tall shrubs (*Calamagrostis canadensis*, *Carex canescens* (L.), *C. tenuiflora* (Wahl.), *C. trisperma* (Dewey), *Scirpus cyperinus* (L.), *Epilobium angustifolium*, *Hieracium aurantiacum* (L.), *Typha latifolia* (L.), *Salix discolor* (Mühl), all part of the invaders group). A conversion from conifer-dominated toward a mixed and deciduous cover type was also observed with increasing severity of forest floor disturbance in the *Picea mariana*–feathermoss domain (Carleton and MacLellan 1994). Thus, similar tendencies are observed for the impact of harvesting on species diversity for this forest domain as for the *Abies balsamea-Betula papyrifera* domain: the diversity of the post-disturbance vegetation tends to decrease with increasing forest floor disturbance severity. Johnston and Elliott (1996) also noted that, after harvesting, the persistence of species associated with older stands contributed to an increased postdisturbance diversity of *Picea mariana*-dominated stands.

Effects of Harvesting on Ecosystem Nutrient Cycling

The understory vegetation often exerts a significant influence on ecosystem-level nutrient cycling (chapters 2–4, this volume). For example, during early succession, many pioneer herbs and shrubs play an important role in the immobilization of the nutrients released after a disturbance (Marks and Bormann 1972). Similarly, the dominance of bryophytes and their significant contribution to the accumulation of a thick forest floor in older stands of the boreal forest also illustrates the important influence of this vegetation layer on ecosystem nutrient cycling (Heilman 1966, 1968; Van Cleve et al. 1981; Weber and Van Cleve 1981, 1984). However, in the former case, the nutrients immobilized in the biomass of the pioneer species tend to be returned to the system's pool of available nutrients as this vegetation gradually dies off with the closure of the overstory canopy and produces a readily decomposable litter. In the latter case, the nutrients immobilized by the bryophyte layer tend to be sequestered in the forest floor and removed from the system's pool of available nutrients as a consequence of the low decomposition rates observed in this compartment of the boreal forest ecosystem. Thus, through its influence on the composition of postdisturbance understory vegetation, harvesting exerts a certain influence on the ecosystem's nutrient cycling. In a study where we examined the relationships that exist between postdisturbance vegetation composition and soil fertility, we observed that postharvest stands with a greater proportion of graminoids, herb, and/or deciduous shrub species demonstrated greater levels of nutrient cycling (Nguyen-Xuan et al. 2000). Postharvest stands with a greater proportion of mosses and *Sphagnum* demonstrated lower levels of nutrient cycling. Consequently, the severity of the disturbance directly influences postharvest understory vegetation composition and indirectly affects postharvest nutrient cycling. This observation yields more profound implications for the *Picea mariana*–feathermoss domain than for the *Abies balsamea–Betula papyrifera* domain because of the predominance of bryophytes in the former domain.

Conclusion

The variability in the disturbance regime influences the distribution of understory species of the boreal forest. The richness and the abundance of the different group of species vary according to fire recurrence. High fire recurrence will favor ericaceous and terricolous lichen species, while a lower recurrence is associated with a higher richness of herbs and shrubs species. Because boreal understory and tree species have developed strategies to maintain in a context of fire, their regional distribution is closely linked to fire recurrence. Thus, the specificity between cover types and understory may only reflect similar responses to a particular disturbance regime (Carleton and Maycock 1981; Gilliam et al. 1995; chapter 8, this volume). As for fire recurrence, the proportion of the different site types over a landscape will affect both understory and tree species distribution. Although the canopy cover can modify the abiotic conditions of a stand, thus affecting the composition of the understory, many understory species are not restricted to a specific canopy, but rather to specific abiotic conditions.

Understory species composition and distribution are directly influenced by the disturbance regime. Understory diversity at the landscape level is highest with intermediate frequency of fire. This is true under any cover types and on all site types. The data fit Connell's (1978) intermediate disturbance hypothesis. Understory species have developed different reproductive strategies to maintain in a context of fire, and understory richness and abundance relationships between the different groups will directly be affected by disturbance characteristics. When we change the predominant disturbance regime from fire to forest harvesting, we are imposing a new disturbance for which species do not present any particular adaptive traits. Clearcutting alone is not a threat to the diversity of understory vascular plant species; however, it will change the relationships among species in the communities and contribute to change community composition. Brumelis and Carleton (1989) observed that under particular abiotic conditions, the effect of logging on community development had completely altered the successional processes. Understory communities are not only part of the biodiversity heritage, but they can also play important roles in boreal forest dynamics (De Grandpré and Bergeron 1997).

Acknowledgements We acknowledge the Ministère des Ressources naturelles du Québec for allowing us to use the ecological survey data. We are also grateful to Sylvie Gauthier for her comments on an earlier version of the manuscript. We thank Pamela Cheers for editing the English and Diane Paquet for correcting the manuscript.

Appendix 10.1. List of understory species, with their score on the first two axes of the ordination, and their reproductive strategy according to Rowe (1983)

Species name	Species code	Axis 1	Axis 2	Group	Reproductive strategy
Empetrum nigrum	EMN	−0.49	0.27	Ericaceous	Avoider
Circaea alpina	CIA	1.00	0.32	Herb	Avoider
Coptis groenlandica	COG	0.32	0.02	Herb	Avoider
Cypripedium acaule	CYA	0.04	−0.70	Herb	Avoider
Gaultheria hispidula	CHH	−0.19	0.17	Herb	Avoider
Goodyera repens	GOR	0.35	0.04	Herb	Avoider
Linaea borealis	LIB	0.44	−0.07	Herb	Avoider
Mitella nuda	MIN	0.88	0.62	Herb	Avoider
Monese uniflora	MOU	0.54	−0.04	Herb	Avoider
Monotropa uniflora	MON	0.70	−0.42	Herb	Avoider
Oxalis montana	OXM	0.86	0.03	Herb	Avoider
Pyrola elliptica	PYE	0.70	−0.15	Herb	Avoider
Pyrola secunda	PYR	0.22	0.06	Herb	Avoider
Rubus chamaemorus	RUC	−0.67	0.92	Herb	Avoider
Trilium undulatum	TRU	0.33	−0.50	Herb	Avoider
Cladina stellaris	CLT	−1.02	−0.69	Lichen	Avoider
Bazzania trilobata	BAT	0.12	−0.09	Moss	Avoider
Hylocomnium splendens	HYS	0.06	0.10	Moss	Avoider
Mnium punctatum	MNP	0.81	0.60	Moss	Avoider
Mnium sp.	MNS	0.70	0.51	Moss	Avoider
Pleurozium schreberi	PLS	−0.34	−0.24	Moss	Avoider
Ptilidium ciliare	PTI	−0.69	−0.16	Moss	Avoider
Ptilium crista-castrensis	PTC	−0.38	−0.15	Moss	Avoider
Rythidiadelphus triquetus	RYT	0.81	0.30	Moss	Avoider
Taxus canadensis	TAC	1.29	−0.20	Shrub	Avoider
Sphagnum fuscum	SPF	−0.59	0.65	*Sphagnum*	Avoider

(*continued*)

Appendix 10.1. *Continued*

Species name	Species code	Axis 1	Axis 2	Group	Reproductive strategy
Sphagnum squarrosum	SPQ	0.30	1.07	*Sphagnum*	Avoider
Sphagnum girgensohnii	SPG	−0.15	0.78	*Sphagnum*	Avoider
Sphagnum sp.	SPS	−0.27	0.91	*Sphagnum*	Avoider
Cassandra calyculata	CAL	−0.76	0.69	Ericaceous	Endurer
Kalmia angustifolia	KAA	−0.63	−0.48	Ericaceous	Endurer
Ledum groenlandicum	LEG	−0.63	0.16	Ericaceous	Endurer
Rhododendron canadense	RHC	−0.25	0.13	Ericaceous	Endurer
Vaccinium angustifolium	VAA	−0.36	−0.38	Ericaceous	Endurer
Vaccinium myrtilloides	VAM	−0.28	−0.36	Ericaceous	Endurer
Vaccinium uliginosum	VAU	−0.43	0.21	Ericaceous	Endurer
Actaea rubra	ACR	1.45	0.26	Herb	Endurer
Aralia nudicaulis	ARN	1.02	−0.39	Herb	Endurer
Aster macrophyllus	ASM	1.39	0.02	Herb	Endurer
Atirium filix-femina	ATF	0.90	0.56	Herb	Endurer
Clintonia borealis	CLB	0.74	−0.32	Herb	Endurer
Comandra livida	COL	−0.54	0.03	Herb	Endurer
Cornus canadensis	CON	0.45	−0.24	Herb	Endurer
Dryopteris disjuncta	DRD	0.68	0.41	Herb	Endurer
Dryopteris noveboracensis	DRN	1.14	0.30	Herb	Endurer
Dryopteris phegopteris	DRP	1.02	0.09	Herb	Endurer
Dryopteris spinulosa	DRS	1.07	−0.06	Herb	Endurer
Epigaea repens	EPI	−0.36	−0.45	Herb	Endurer
Equisetum sp.	EQY	−0.08	0.81	Herb	Endurer
Equisetum sylvaticum	EQS	0.27	1.32	Herb	Endurer
Gaultheria procumbens	GAP	−0.18	−0.61	Herb	Endurer
Kalmia polifolia	KAP	−0.75	0.53	Herb	Endurer
Lycopodium annontinum	LYA	0.44	0.27	Herb	Endurer
Lycopodium clavatum	LYC	0.98	−0.48	Herb	Endurer
Lycopodium complanatum	LYP	0.68	−0.53	Herb	Endurer
Lycopodium lucidulum	LYL	1.24	−0.37	Herb	Endurer
Lycopodium obscurum	LYO	0.97	−0.44	Herb	Endurer
Maianthemum canadense	MAC	0.64	−0.32	Herb	Endurer
Osmunda cinnamomea	OSC	1.05	0.18	Herb	Endurer
Petasites palmatus	PES	0.40	0.67	Herb	Endurer
Pteridium aquilinum	PTA	1.01	−0.82	Herb	Endurer
Rubus pubescens	RUP	1.00	0.68	Herb	Endurer
Smilacina trifolia	SMT	−0.39	1.20	Herb	Endurer
Streptopus amplexifolius	STA	0.76	−0.02	Herb	Endurer
Streptopus roseus	STR	1.23	−0.31	Herb	Endurer
Trientalis borealis	TRB	0.69	−0.10	Herb	Endurer
Vaccinium oxycoccos	VAO	−0.71	0.85	Herb	Endurer
Vaccinium vitis−idaea	VAV	−0.56	−0.22	Herb	Endurer
Viola spp.	VIS	1.13	0.39	Herb	Endurer
Alnus crispa	AUC	0.01	−0.45	Shrub	Endurer
Alnus rugosa	AUR	0.43	0.90	Shrub	Endurer
Amelanchier sp.	AME	0.28	−0.34	Shrub	Endurer
Betula glandulosa	BEG	−0.80	−0.30	Shrub	Endurer
Cornus stolonifera	COR	1.08	0.13	Shrub	Endurer
Corylus cornuta	COC	1.67	−0.56	Shrub	Endurer
Diervilla lonicera	DIE	1.31	−0.64	Shrub	Endurer

Species name	Species code	Axis 1	Axis 2	Group	Reproductive strategy
Lonicera canadensis	LON	1.37	−0.27	Shrub	Endurer
Lonicera vilosa	LOV	−0.66	0.54	Shrub	Endurer
Prunus pensylvanica	PRP	1.08	−0.47	Shrub	Evader
Ribes glandulosum	RIG	0.91	0.31	Shrub	Evader
Ribes lacustre	RIL	0.98	0.35	Shrub	Evader
Ribes triste	RIT	1.04	0.68	Shrub	Evader
Sambucus pubens	SAP	1.19	−0.09	Shrub	Evader
Viburnum cassinoides	VIC	0.68	−0.41	Shrub	Evader
Viburnum edule	VIE	0.80	0.18	Shrub	Evader
Vaccinium cespitosum	VAC	−0.84	−0.58	Ericaceous	Invader
Anaphalis margaritacea	ANM	1.19	−0.30	Herb	Invader
Aster accuminatus	ASA	1.25	0.22	Herb	Invader
Epilobium angustigolium	EPA	0.72	0.04	Herb	Invader
Fragaria spp.	FRG	1.07	0.61	Herb	Invader
Galium triflorum	GAT	0.86	0.96	Herb	Invader
Hieracium spp.	HIS	1.48	0.26	Herb	Invader
Melampyrum lineare	MEI	0.83	−0.71	Herb	Invader
Osmunda claytoniana	OSY	0.76	−0.09	Herb	Invader
Prenanthes spp.	PRS	1.17	0.06	Herb	Invader
Solidago macrophylla	SOM	0.80	−0.25	Herb	Invader
Thalictrum polygamum	THP	0.73	1.52	Herb	Invader
Cladina mitis	CLM	−0.47	−0.39	Lichen	Invader
Cladina rangiferina	CLR	−0.70	−0.34	Lichen	Invader
Cladonia sp.	CLS	−0.19	−0.18	Lichen	Invader
Acer spicatum	ERE	1.32	−0.47	Shrub	Invader
Prunus virginiana	PRV	1.29	0.66	Shrub	Invader
Rubus idaeus	RUI	0.98	0.35	Shrub	Invader
Sphagnum magellanicum	SPM	−0.14	1.54	*Sphagnum*	Invader

IV

Community Dynamics of the Herbaceous Layer and the Role of Disturbance

11

The Herbaceous Layer as a Filter Determining Spatial Pattern in Forest Tree Regeneration

Lisa O. George
Fakhri A. Bazzaz

Although small in stature and seemingly ephemeral in nature, herbaceous understory communities have the potential to exert a strong influence on the dynamics of forest tree regeneration. Through study of fern-dominated understory communities in central Massachusetts mixed-deciduous forests, we have found that this understory stratum can act as a selective filter of tree seedling species in their earliest life stages. All trees, first as seeds, must penetrate this layer to reach the forest floor and then germinate and grow under the "canopy" of the herbaceous stratum. The microenvironment at and near the forest floor is dictated largely by the nearby herbs and shrubs above. Tree seedlings that are able to germinate, survive, and grow in this environment will ultimately emerge from this stratum and become potential successors to the forest canopy. Tree seedling species performance may vary considerably under the particular rigors of a well-developed understory, and it is this variability that enables the understory to filter tree seedlings.

Because the tree seedling bank represents the starting capital for forest regeneration, predisturbance interactions between the understory and tree seedlings that shape the nature of the seedling bank will influence the rate and trajectory of postdisturbance forest regeneration (Brokaw and Scheiner 1989; Connell 1989). The presence of a well-developed understory stratum composed of herbs and/or shrubs generally reduces the density of tree seedlings and saplings beneath its canopy (Horsley 1977a, 1977b; Phillips and Murdy 1985; McWilliams et al. 1995; Hill 1996; George and Bazzaz 1999a, 1999b). A low density of advance regeneration may slow forest regeneration after canopy disturbance (Nakashizuka 1987; Taylor and Qin 1992; McWilliams et al. 1995). In stands that lack abundant advance regeneration, it may be possible for widely dispersed pioneer tree species to invade the stand, and with little competition from advance regeneration, become an important component of the next generation of forest (Veblen 1989; Taylor and Qin 1992). In extreme cases, understory interference with tree regeneration has prevented forest regeneration and has led to the formation of permanent shrub communities (Niering and Egler 1955). In the Allegheny Plateau region of northwestern Pennsylvania, competition from herbs has contributed to the formation of "savannahs" and "orchard stands," which were previously forested stands, logged heavily in the early 1900s, that have failed to regenerate for more than 50 years (Horsley 1977a, 1977b).

Although the invasion of tree seedlings into herbaceous and shrubby old-field communities has been well studied and provides insight into competitive interactions between tree seedlings and herbs (e.g., De Steven 1991a; 1991b; Gill and Marks 1991; Putz and Canham 1992; Burton and Bazzaz 1995; chapter 10, this volume), we restrict our discussion to tree regeneration in forest communities. Our studies in central New England at the Harvard Forest in Massachusetts have dealt primarily with tree regeneration in closed-canopy *Quercus rubra-Acer rubrum* forests with well-developed fern understories. Some of our work has included the study of shrubs in the understory stratum when present, such as *Viburnum* species and *Corylus cornuta* Marsh. (beaked hazel), a well-studied understory dominant in the western Lake States (Kurmis and Sucoff 1989). One fern species that we have focused on, *Dennstaedtia punctilobula* (Michx.) Moore (hayscented fern), has also been studied extensively in the Allegheny Plateau region of Pennsylvania and New York from the perspective of commercial forestry (Horsley 1977a, 1977b, 1988, 1993a, 1993b; Horsley and Marquis 1983; Bowersox and McCormick 1987; Drew 1988; Kolb et al. 1989, 1990; McWilliams et al. 1995). This body of literature concentrates on the link between dense herbaceous understory vegetation and regeneration failure, which is the failure of a forest stand to be restocked after commercial timber harvest with appropriate densities of tree seedlings of commercially valued species in an acceptable period of time (Grisez and Peace 1973). Other understory species that have been observed to interact with Appalachian hardwood regeneration include clonal ferns such as *Thelypteris noveboracensis* (L.) Nieuwl. (New York fern), *Pteridium aquilinum* (L.) Kuhn. (bracken fern), *Onoclea sensibilis* L. (sensitive fern), and *Osmunda claytoniana* L. (interrupted fern) (Horsley 1988). Several grass species such as *Brachyelytrum erectum* Schreb. (short husk grass), *Danthonia compressa* Aust. (wild oat grass), and *Danthonia spicata* (L.) Beauv. (poverty oat grass) are important constituents of the understory and seed bank in the Appalachian hardwood region and have also been reported to interfere with tree regeneration (Horsley 1977a, 1977b; Bowersox and McCormick 1987).

In the southern Appalachian region, the most prominent members of the understory stratum are evergreen shrubs *Rhododendron maximum* L. (rosebay rhododendron) and *Kalmia latifolia* L. (mountain laurel). These species have expanded since the chestnut blight to cover an estimated 2.5 million ha in the Appalachians (Monk et al. 1985) and have been implicated in changes in forest succession (Baker and van Lear 1998). Although the focus of this book is on herbs, we discuss these well-studied shrub species because they contribute to the general discussion of understory filtering of tree regeneration and because they provide a useful counterpoint to the effects of herbaceous understory vegetation on tree regeneration. Just as the concept of the understory filter is not confined to herbaceous species, it is not confined to forests of the eastern United States. The processes of understory filtering are being recognized in a diversity of temperate and tropical forests that are characterized by dense or aggressive understory vegetation. Well-studied examples include Costa Rican tropical rainforest where the understory is characterized by palms and cyclanths (Denslow et al. 1991), Chilean subalpine forests where the understory is characterized by *Chusquea* bamboo (Veblen 1982; 1989), and the giant panda reserves in China where the understory is characterized by dwarf bamboo species (Taylor and Qin 1992; Taylor et al. 1995).

Before becoming a tree, an individual must pass through several life stages under or within the herbaceous stratum, each of which is a potential winnowing point for the understory filter. We first discuss the influence of the understory filter on dispersal, germination, and survival of tree seeds and seedlings individually and then discuss the effects of the understory filter on tree seedling community attributes such as tree seedling distribution, species composition, and diversity. The influence of the understory filter on seedling growth will then be addressed, followed by a discussion of the role of the understory in determining the size structure of tree seedling communities. The filtering activity of the understory is not only complex because of its activity at multiple life stages of a tree, but also because the understory is a complex and changing mosaic of understory plants. In conclusion, we discuss the spatial and temporal distribution of understory plants and how the mosaic nature of the understory affects the complexity of the understory filter.

The Understory Microenvironment

Plant growth in the understory is commonly limited by the light environment, which is determined by a complex of variables including site topography, seasonal solar patterns, and overstory structure and composition (chapter 3, this volume). The herbaceous layer further reduces the quantity and quality of light reaching the forest floor. The fern understory in closed-canopy forests in Massachusetts reduces midsummer light levels approximately 70%—from an average of 3.5% full sun above ferns to 1% full sun below ferns (George and Bazzaz 1999a). The understory stratum of partially thinned Allegheny forest stands dominated by *Dennstaedtia punctilobula* reduces light from about 20% full sun above the fern stratum to less than 0.5% full sun below ferns; the fern understory also decreases the red/far-red ratio from 0.50–1.10 above ferns to 0.04–0.07 below ferns (Horsley 1993a). Forest understory communities dominated by *Rhododendron maximum* in the southern Appalachians have a similarly dramatic impact on the light environment. Midsummer light levels, which range from 4% to 8% of full sun above the *Rhododendron maximum* canopy, are reduced below the canopy to 1–2% of full sun (Clinton 1995; Beckage et al. 2000). In contrast to herbaceous ferns that senesce above ground at the end of the growing season, *Rhododendron maximum*, an evergreen shrub, dramatically reduces light levels even in the early spring when the deciduous overstory has not yet fully leafed out. Peak forest light levels (measured at 1 m above the forest floor) occur in the southern Appalachians during mid-April and can reach 50% full sun in the absence of *Rhododendron maximum* compared to a peak of only 15% of full sun below *Rhododendron maximum* (Clinton 1995).

In our Massachusetts study sites, average litter depth was greater below the fern understory, and a greater area of mineral soil was exposed where ferns were absent. This effect was particularly dramatic in sloped sites, which have the potential for high levels of soil exposure (George and Bazzaz 1999a). Similarly, in a deciduous forest in the Blue Ridge Mountains, litter biomass beneath a *Rhododendron maximum* understory was 20% higher than in areas without *Rhododendron maximum* cover (Beckage et al. 2000). In our Massachusetts sites, soil water content, soil organic material content, and pH did not differ between fern and fern-free areas (George and Bazzaz 1999a). Similarly, in an Allegheny forest with a dense *Dennstaedtia punctilobula* understory, no differences were detected between soil moisture, soil nitrogen or phosphorus concentrations, soil ammonium and nitrate concentrations, and net ammonium and nitrate production between fern and fern-removal treatment plots (Horsley 1993a). There is some evidence, however, that soil moisture is reduced below a *Rhododendron maximum* understory due to evapotranspirational loss and high interception (Clinton and Vose 1996; Beckage et al. 2000). Additionally, the potential importance of allelopathy of *Dennstaedtia punctilobula* and *Rhododendron maximum* leachates has been investigated and found not to be of importance in field environments (Horsley 1993b; Nilsen et al. 1999).

The Influence of the Understory Filter

Tree Seed Dispersal and Predation

In New England, fern fronds senesce by the time of autumn seed dispersal of most tree species and therefore do not represent a physical barrier for seeds being dispersed to the forest floor. We observed that spring-dispersed *Acer rubrum* L. seeds land on fern fronds, where they were held for several days before the seeds dropped to the forest floor. In dispersal trials in which *Acer rubrum* seeds were dropped over the canopies of *Osmunda claytoniana* and *Dennstaedtia punctilobula*, more seeds were caught by *Osmunda claytoniana* fronds (45%) than by the more fragile and vertically oriented fronds of *Dennstaedtia punctilobula* (31%) (George and Bazzaz unpublished data). All seeds were ultimately recovered, however, from the ground under the fern fronds. Whether seed catching by the understory plays an important role in subsequent predation rates or secondary dispersal has not been well studied. We do expect, however, that the understory stratum may be a more important barrier to seed dispersal in other systems where the understory is characterized by woody or evergreen species. In addition to leaves of understory plants being potential impediments to seed dispersal, the thick litter layer that develops below a *Rhododendron maximum* understory has been reported to act as a physical barrier preventing seeds from reaching the soil where moisture conditions are more favorable for germination (Clinton and Vose 1996).

Although we did not observe the fern understory to be a direct physical impediment to seed dispersal, we observed higher rates of seed predation under fern cover. In our experimental studies, acorns of *Quercus rubra* L. (northern red oak) experienced a 66% removal rate under ferns compared to a 45% removal rate in fern-removal plots (George and Bazzaz 1999a). Wada (1993) found even more dramatic acorn (*Quercus serrata* and *Quercus mongolica*) predation rates under the canopy of *Sasa* species (dwarf bamboos) in a Japanese temperate forest. Wada not only documented higher numbers of rodents under the bamboo understory, but he observed complete removal of all acorns experimentally placed below the bamboo understory. Higher rodent predation rates of seeds of *Acer rubrum* (red maple), *Pinus strobus* L. (white pine), *Rhamnus cathartica* L. (common buckthorn), and *Cornus racemosa* (gray dogwood) have also been observed under herb cover in New York old fields (Gill and Marks 1991). It appears in all of these cases that understory plants provide valuable cover for rodents from their own predators.

Seed Germination and Emergence

Several dominant understory species such as *Dennstaedtia punctilobula*, *Aster acuminatus* Michx. (whorled wood aster), and *Rhododendron maximum* decrease the germination of individual tree species (Drew 1988; Clinton and Vose 1996). We investigated the germination of a suite of tree species below the understory stratum to determine whether understory filtering at the germination stage varied among tree species. Figure 11.1 compares the patterns of spring seedling emergence in plots with and without a well-developed fern understory in mixed hardwood stands in Massachusetts. Response of seedling emergence to understory cover was either neutral or negative, and, more important, emergence patterns were variable among species. Spring emergence of *Acer rubrum* and *Fraxinus americana* L. (white ash) was not affected by fern cover, but emergence of *Pinus strobus*, *Quercus rubra*, and *Betula lenta* L. and *B. alleghaniensis* Britt. (black birch and yellow birch) was reduced under the fern understory to varying degrees. Although absolute values of seedling emergents varied annually and from site to site due to variability of seed rain, the qualitative response of each tree species to fern cover remained the same (George and Bazzaz 1999a).

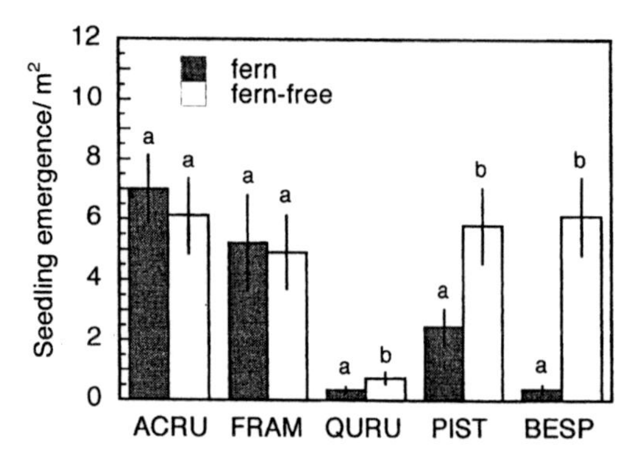

Figure 11.1. Density of emergent natural recruitment in fern and fern-free experimental plots. Each bar represents mean seedling emergence per square meter across six sites in central Massachusetts ($\pm$ 1 SE, n = 60). Letters indicate significantly different treatment means within seedling species (Fisher's Protected LSD, $p < .01$). Seedling species: ACRU = *Acer rubrum* (red maple); FRAM = *Fraxinus americana* (white ash); QURU = *Quercus rubra* (northern red oak); PIST = *Pinus strobus* (white pine); BESP = *Betula allegheniensis* and *B. lenta* (yellow and black birch, species combined). Figure modified from George and Bazzaz (1999a).

Not only does germination response to understory cover vary among tree species, but species responses seem to be mediated by different environmental factors. The aggregation of *Betula lenta* and *B. allegheniensis* germinants in areas of exposed soil, which were more common in fern-free areas, led to experiments that determined that the depth of litter and lack of exposed soil were the primary impediments to the germination of *Betula* species under ferns (George and Bazzaz 1999a). Further evidence indicated that low light under ferns limited summer germination of *A. rubrum* but did not affect spring germination of *A. rubrum* or *Fraxinus americana* because spring germination occurred before leaf out of the fern canopy. The emergence pattern of oak appeared to be a function of differential seed predation as discussed above, rather than of differential germination. Seedling recruitment below *Rhododendron maximum* understories has been reported to be limited primarily by the same means: light attenuation, inhibition by a thick litter layer, and higher seed predation rates under *R. maximum* (Beckage et al. 2000).

Tree Seedling Survival

The understory continues to selectively filter tree seedlings through its influence on seedling survival during establishment and early growth phases (Horsley and Marquis 1983; Phillips and Murdy 1985; George and Bazzaz 1999a, 1999b). Tree seedlings of five species were followed through their first full year of growth, and their survival rates are displayed in figure 11.2. Un-

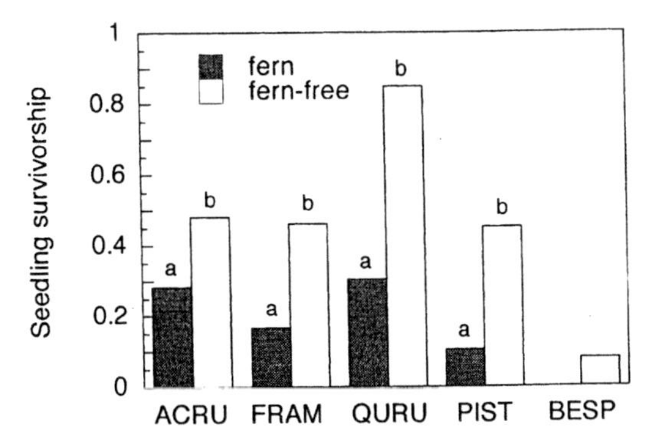

Figure 11.2. Proportion of seedlings cohorts from figure 11.1 surviving in fern and fern-free experimental plots for one year. Letters indicate whether survivorship of each species differed between fern manipulations (chi-square test, $p < .05$). No *Betula* seedlings survived for 1 year under ferns. Seedling species: ACRU = *Acer rubrum* (red maple); FRAM = *Fraxinus americana* (white ash); QURU = *Quercus rubra* (northern red oak); PIST = *Pinus strobus* (white pine); BESP = *Betula allegheniensis* and *B. lenta* (yellow and black birch, species combined). Figure modified from George and Bazzaz (1999a).

derstory cover negatively affected the survivorship of all species to varying degrees and precluded the establishment of any *Betula* seedlings below ferns.

Although understory interference with seed germination and emergence is mediated by several factors, the understory primarily influences seedling survival through above-ground interactions, particularly competition for light (Denslow et al. 1991; Pacala et al. 1994, 1996; George and Bazzaz 1999b). Herbivory is another important factor that can contribute to seedling mortality, although the identity and specific behavior of a herbivore will determine whether predation rates are higher or lower beneath the understory stratum. High levels of insect herbivory under a palm understory in a Costa Rican rainforest decreased seedling survival beneath the understory (Denslow et al. 1991); however, we found that fern cover protected *Quercus rubra* seedlings from high levels of insect herbivory in fern-free areas in Massachusetts (George and Bazzaz 1999b).

Tree Seedling Community Characteristics

Because the understory stratum can differentially affect predation, germination, and survival rates of tree seedling species beneath its canopy, the pattern of understory plant distribution will strongly influence patterns of tree seedling distribution. Even at the earliest stages in the life cycle of a tree, the understory has already begun a filtering process that leads to a characteristic distribution of new tree seedling germinants over the forest floor.

Because of the selective influence of the fern filter on seedling emergence in our New England study sites (fig. 11.1), *Betula lenta* and *B. allegheniensis* germinants were highly concentrated in fern-free areas, *Pinus strobus* and *Quercus rubra* germinants were more common in fern-free areas, and *Acer rubrum* germinants were more common in fern-free areas or nearly randomly distributed depending on the predominance of spring versus summer germination. *Fraxinus americana* germinants were distributed randomly over the forest floor with respect to understory cover.

Although survival patterns beneath the understory also varied among tree species (fig. 11.2), species did not necessarily follow the same patterns of response for survival as for emergence. The compounding of emergence patterns (fig. 11.1) and patterns of first-year mortality (fig. 11.2) produced the distribution patterns at the end of one year as displayed in figure 11.3. *Acer rubrum* seedlings were more common in fern-free areas by a 1.5:1 ratio, as were *Fraxinus americana* seedlings (2.6:1). *Quercus rubra* seedlings were strongly concentrated in fern-free areas (6:1) and *Pinus strobus* seedlings were very concentrated (10:1). *Betula* seedlings, at least in this sampling pool, had been eliminated under fern cover.

Because tree species distributions varied individually with respect to understory cover, the attributes of the tree seedling community in areas of high understory cover varied widely from the seedling community attributes of areas within the same stand with low understory cover. Because of the generally neutral or negative effect of the understory on tree seedlings, total seedling density is reduced beneath the understory stratum. The seedling bank below the understory also exhibits a shift in species composition when com-

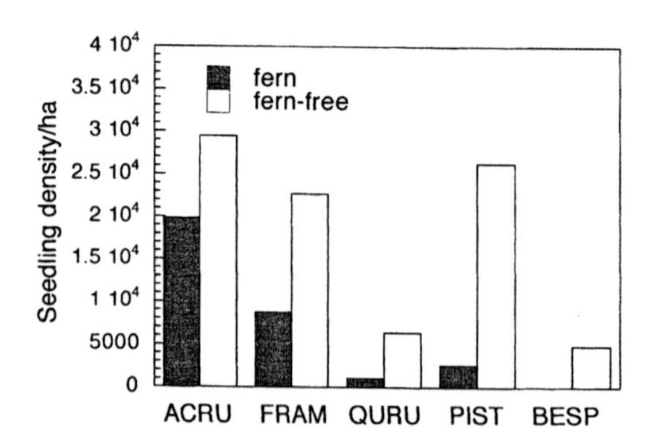

Figure 11.3. Density of surviving seedlings (from cohorts in fig. 11.1) in fern and fern-free experimental plots one year after seedling emergence. Data values are a combination of emergence values (fig. 11.1) and survivorship values (fig. 11.2) in fern and fern-free plots. Seedling species: ACRU = *Acer rubrum* (red maple); FRAM = *Fraxinus americana* (white ash); QURU = *Quercus rubra* (northern red oak); PIST = *Pinus strobus* (white pine); BESP = *Betula alleghaniensis* and *B. lenta* (yellow and black birch, species combined).

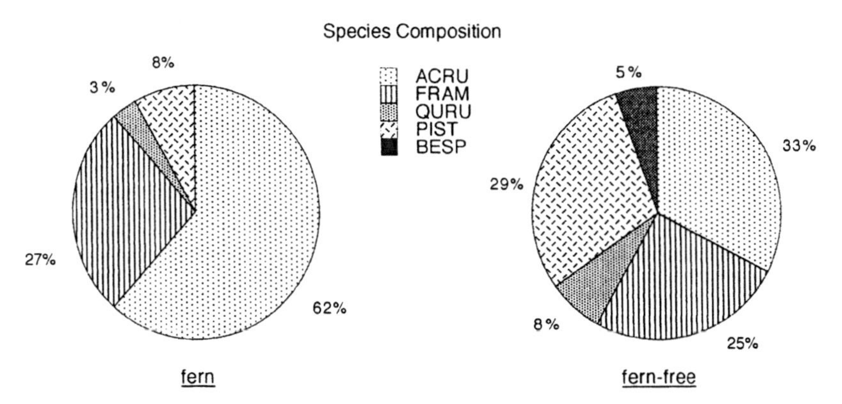

Figure 11.4. Percent species composition of 1 year-old seedlings in fern and fern-free experimental plots. Total cohort seedling density in fern plots = 32,150 seedlings/ha; total cohort seedling density in fern-free plots = 89,691 seedlings/ha (seedling densities as in fig. 11.3). Seedling species: ACRU = *Acer rubrum* (red maple); FRAM = *Fraxinus americana* (white ash); QURU = *Quercus rubra* (northern red oak); PIST = *Pinus strobus* (white pine); BESP = *Betula alleghenies* and *B. lenta* (yellow and black birch, species combined).

pared to areas with little understory cover. Figure 11.4 compares the species composition of the pool of 1-year-old seedlings between areas with and without a prominent fern understory in Massachusetts forests. Under ferns, *Acer rubrum* is the most dominant contributor to the seedling bank, followed by *Fraxinus americana*, and *Betula* species are absent. In fern-free areas *Acer rubrum*, *Fraxinus americana*, and *Pinus strobus* are codominant, and there is a strong representation of *Betula* species and *Quercus rubra*.

In the Blue Ridge Mountains of North Carolina, the selectivity of the *Rhododendron maximum* understory also causes a shift in the species composition of tree regeneration. Sapling density of *Quercus prinus* L. (chestnut oak) and *Quercus alba* L. (white oak) was depressed by high *Rhododendron maximum* cover, whereas regeneration of *Acer rubrum* was not affected by *R. maximum* cover. In contrast, greater regeneration of *Tsuga canadensis* (L.) Carr. (eastern hemlock) was associated with abundance of *R. maximum* (Phillips and Murdy 1985). This pattern of selective filtering resulted in a sapling bank beneath *R. maximum* in which *A. rubrum* and *Tsuga canadensis* were codominant, with a low representation of *Quercus* saplings. The sapling bank in areas of low *R. maximum* cover was dominated by *A. rubrum*, followed by *Quercus* species, with a low representation of *T. canadensis*.

In addition to reducing overall seedling density and shifting species composition of seedling communities, the understory filter may reduce the diversity of tree regeneration beneath its canopy (Drew 1988; Beckage et al. 2000). Where the understory of a stand is dominated by a single understory species of high density and aerial coverage such as *Rhododendron maximum*, stand-level diversity is generally reduced (Baker and van Lear 1998; Drew 1988). Diversity of herbaceous and woody regeneration may reach a fourfold difference between areas of scarce and high *R. maximum* cover (Baker and van Lear 1998). When the Shannon-Weiner index of diversity ($H' = -\Sigma p_i \ln p_i$) is calculated for the Massachusetts tree seedling communities (fig. 11.4), greater diversity is found in fern-free areas ($H' = 1.41$) compared to areas beneath the fern canopy ($H' = 0.96$). Not only is species richness higher in fern-free areas, but tree species are more evenly represented.

Seedling Growth

Growth rates of tree seedlings beneath the understory stratum are generally depressed and have been most strongly correlated with low light availability (Horsley 1993a; George and Bazzaz 1999b). Because tree seedling species vary in their growth response to light availability, the presence of understory cover can differentially influence seedling growth. Differential tree seedling growth in areas with and without dense understory cover will lead to different size structures in their respective seedling communities.

The selective influence of the understory on seedling growth has been well demonstrated in the Allegheny Plateau region of Pennsylvania. Understories of both *Danthonia spicata* (poverty oat grass) and *Dennstaedtia punctilobula* (hayscented fern) reduce seedling height growth of several timber species (Kolb et al. 1989, 1990). After 1 year of seedling growth, herbaceous interference had reduced the height of *Fraxinus americana* by 65%, *Liriodendron tulipifera* L. (yellow poplar) by 60%, and *Quercus rubra* by 29% but did not affect the height growth of *Pinus strobus*. After 2 years of growth in herb-free environments, seedlings of *F. americana* were taller (49 cm) than those of *Q. rubra* (40 cm), which were taller than *P. strobus* (10 cm). In contrast, in areas with herbaceous cover, seedlings of *Q. rubra* were taller (22 cm) than those of *F. americana* (11 cm) and *P. strobus* (8 cm) (Kolb et al. 1990).

The influence of the understory on the size structure of seedling communities will influence the outcome of tree–tree competition after release. For example, in the Allegheny community described above, *Quercus rubra* seedlings will have a size advantage upon release in areas with a well-developed understory, and *Fraxinus americana* seedlings will have a size advantage in areas lacking understory cover. Where the understory has a dampening effect on the growth of all seedling species (see Bowersox and McCormick 1987), the understory may tend to equalize competitive status among individuals and among species.

Whether a tree seedling emerges above the understory stratum depends on species growth rates tempered by their survivorship rates. In our studies in Massachusetts, the relative growth rate of *Betula alleghениensis* was higher than *Quercus rubra* and *Acer rubrum* beneath ferns and in fern-free areas (fig. 11.5). Although *Betula alleghaniensis* possessed the highest growth rate of all species in all treatments, it also had the lowest survivorship rates, particularly beneath ferns (see fig. 11.2). Conversely, the lower growth rates of *Q. rubra*

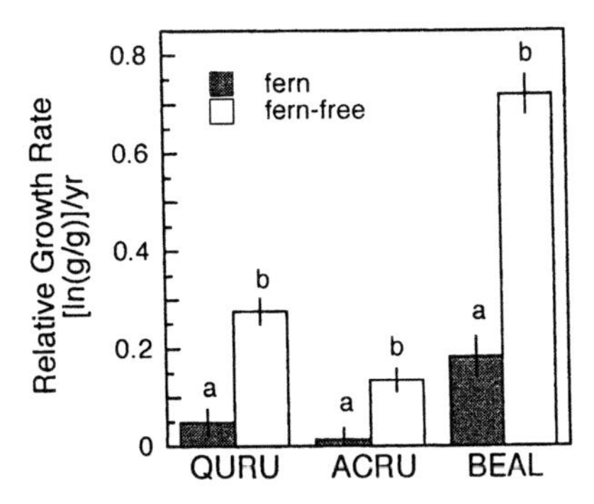

Figure 11.5. Relative growth rate (RGR) in terms of biomass, [ln(g/g)]/year of 3-year-old seedlings growing in fern and fern-free experimental plots. Values represent the average relative growth rates over the second and third years of growth. Each bar represents the mean of the manipulation (± 1 SE, n = 60). Means with different letters are significantly different among manipulations within seedling species (Fisher's Protected LSD, $p < .05$). Manipulations: fern = control plots where ferns were left intact; fern-free = plots where ferns were removed. Seedling species: ACRU = *Acer rubrum* (red maple); QURU = *Quercus rubra* (northern red oak); BEAL = *Betula alleghenensis* (yellow birch). Figure modified from George and Bazzaz (1999b).

and *A. rubrum* were balanced by higher rates of survivorship. In fern-free areas, the high growth rates of *B. allegheniensis* result in its seedlings quickly becoming the tallest of all species of a given cohort. In spite of the relatively high growth rates of *B. allegheniensis* under ferns, its seedlings generally do not emerge above the understory stratum because of low survivorship rates under ferns. They generally do not even survive long enough to overtop other seedling species which have an initial size advantage at emergence. Because of this well-recognized trade-off between traits that enhance persistence in low light and traits that promote maximum growth response to light (Pacala et al. 1994; Bazzaz 1996), it is often the slow growers that win the race to emerge above the understory canopy.

Spatial Heterogeneity of the Understory Stratum

Thus far, we have discussed only the influence of individual species of understory plants on the microenvironment and on tree seedling performance. The understory stratum, however, is a complex mosaic of many species that differ phenologically, morphologically, and physiologically. The influence of a single understory species can be very concentrated from the size perspective of a tree seedling, but from a stand-level perspective, tree seedling populations may be influenced by a number of different understory species of varying densities.

Typical understory communities in eastern forests can range in structure from nearly monospecific communities with uniformly dense cover to mixed-species communities with more patchy distributions. Many clonal understory plants form nearly monospecific patches that can range in size from a few meters to nearly 100 m in the case of *Rhododendron maximum* thickets (Baker and van Lear 1998). In our six Massachusetts study sites, the understory stratum was dominated by *Dennstaedtia punctilobula* and *Osmunda claytoniana* (interrupted fern), and commonly included *Osmunda cinnamomea* L. (cinnamon fern) and *Thelypteris noveboracensis* (New York fern). Because of their clonal growth form, these ferns formed relatively dense patches ranging in diameter from 2 to 20 m and in height from 30 to 70 cm. The ferns created an intergrading patchwork among which less densely distributed herbs and shrubs were interspersed. The patchwork was also punctuated with "holes" or areas with sparse understory cover. Although local cover could range up to 100%, average understory cover for individual stands was more typically around 60%.

Different understory species have the potential to alter the microenvironment in variable ways below their canopies (Burton and Bazzaz 1991; George 1996). For example, a species with a characteristically more open morphology or clonal density may permit greater light transmission through its canopy. Evergreen plants such as *Rhododendron maximum* and understory plants that leaf out in early spring will subject seedlings to a longer duration of reduced light levels than summer-green species. Differences in stature among understory species are also important to the activity of the understory filter because plant height will determine the length of time tree seedlings are subject to shading before a seedling is potentially able to overtop the understory canopy (Hill et al. 1994). Understory species drop litter of varying quantity and quality, and woody species may trap and hold more litter near their bases throughout the year when compared to deciduous herbaceous species (George and Bazzaz, pers. obs.). Understory species may also potentially capture or utilize below-ground resources differently.

Whereas there is evidence that tree seedlings respond to the specific identity and density of the understory, the presence or absence of understory cover of any species may be more important than understory species identity (Berkowitz et al. 1995; George and Bazzaz 1999a, 1999b). The presence of holes in the understory stratum, places that are favorable for tree seedling growth but have not yet been colonized by understory plants, are therefore an important spatial component of the understory mosaic.

Studies comparing seedling densities among different types of understory cover suggest that understory species can contribute to differential tree seedling performance within and among seedling species (Horsley 1977a, 1977b; Hill et al. 1994; George 1996). In a study of tree seedling density in gaps in the southern Appalachians, greater seedling densities were found in gaps with understory cover of *Kalmia latifolia* (1.6 seedlings/m^2) when compared to gaps with *Rhododendron maximum* cover (0.5 seedlings/m^2) (Clinton et al. 1994). These differences were ascribed to the higher leaf area of *R. maximum* compared to *K. latifolia*. After partial overstory removal in an Allegheny hardwood forest, greater seedling densities of *Prunus serotina* Ehrh. (black cherry) and *Acer saccharum* Marshall (sugar maple) were observed in areas with cover of *Brachyelytrum erectum* (short husk grass) compared to areas with *Dennstaedtia punctilobula* cover (Horsley and Marquis 1983). When grass or ferns were removed in experimental plots, *P. serotina* seedling growth and survival rates were similar in grass-removal and fern-removal plots, which indicates that *P. serotina* seedling performance varied due to the identity of understory plant species. The understory mosaic of patches of grass and ferns imposed a layer of environmental heterogeneity on initial site conditions to which *P. serotina* and other tree seedlings responded differentially.

Understory species may or may not segregate along easily detectable environmental gradients such as light or soil moisture (see Horsley 1993a). When they do, microsites occupied by different understory species may inherently have different potentials for tree seedling performance. In a New York northern hardwood forest, germination, growth, and survival of *Prunus serotina* seedlings differed between areas with *Dennstaedtia punctilobula* cover and areas with an understory of *Aster acuminatus* (Drew 1988). *P. serotina* height growth was lower under ferns compared to under aster but was also lower in fern-removal plots compared to aster-removal plots. Observed differences in seedling performance despite the removal of understory plants suggests that growth differences beneath the two types of understory were more associated with the microsites occupied by ferns and aster rather than with direct effects of the species of understory cover. Heterogeneity in site conditions were actually dampened or buffered in this case by the understory.

Heterogeneity in the local light environment beneath the overstory canopy may also be buffered by changes in local density of understory species. For example, *Dennstaedtia punctilobula* can cover large expanses of the forest floor, particularly in stands in which the overstory has been partially thinned (see Horsley and Marquis 1983; Drew 1988; Kolb et al. 1989). *D. punctilobula* density increases locally with light availability, and increased frond density reduces light transmission beneath the fern canopy (Hill 1996; Hill and Silander 2001). Therefore, *D. punctilobula* has been viewed as an equalizer of the variability in the spatial light environments of forests (Hill 1996).

The filtering influence of the understory can result in the aggregation of seedlings of the same species into certain areas defined by the type or density of understory cover. The grain of this aggregation will mirror the spatial structure of the understory stratum and will determine the future competitive and genetic neighborhood of trees during and after disturbance (Bazzaz 1996). The mosaic nature of the understory can provide different areas or niches in which regeneration of certain species is relatively more successful. In a Massachusetts forest with a structurally diverse understory stratum, we characterized the tallest seedling species of plots of four different understory types (fig. 11.6, George 1996). *Quercus rubra*, which has a tall initial seedling height, was well represented in all understory types. Successful *Acer rubrum*

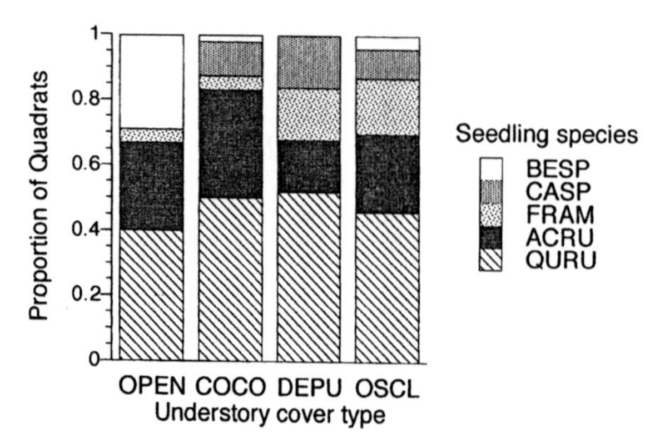

Figure 11.6. Proportion of quadrats in four understory cover types in which each seedling species is represented by the tallest seedling in the quadrat (n = 50 quadrats for each cover type). Understory cover types: OPEN = quadrats with 50% total understory cover; COCO = quadrats dominated by *Corylus cornuta* (beaked hazel); DEPU = quadrats dominated by *Dennstaedtia punctilobula* (hayscented fern); OSCL = quadrats dominated by *Osmunda claytoniana* (interrupted fern). Seedling species: BESP = *Betula alleghaniensis* and *B. lenta* (yellow and black birch, species combined); CASP = *Carya glabra* and *C. ovata* (pignut and shagbark hickory, species combined); FRAM = *Fraxinus americana* (white ash); ACRU = *Acer rubrum* (red maple); QURU = *Quercus rubra* (northern red oak). Figure from George (1996).

regeneration was proportionally best represented in *Corylus cornuta* plots, and regeneration of *Betula* species was nearly exclusively represented in plots with sparse understory cover. *Carya glabra* (Miller) Sweet and *C. ovata* (Miller) K. Koch (pignut hickory and shagbark hickory) seedlings were the tallest seedlings only in plots with well-developed understory cover, particularly cover of *Dennstaedtia punctilobula. Fraxinus americana* regeneration was well represented in areas of *Osmunda claytoniana* and *D. punctilobula* cover.

When seedling species are aggregated in certain patch types, intraspecific competition may predominate within patch types, and spatial aggregation into monospecific neighborhoods may lead to coexistence of species at the community level (Hibbs 1982; Pacala 1986; Silvertown and Law 1987; Pacala et al. 1993). The spatial structure of the understory stratum of this community appears to have contributed to the spatial aggregation of tree seedling species and promoted diversity of advance regeneration at the stand level. If this community lacked areas of open understory, successful *Betula* regeneration would have been drastically reduced. If the understory was entirely open, however, successful *Carya* regeneration would have been eliminated. The spatial segregation of advance regeneration by understory patches may be particularly important for preserving the representation of less common tree species such as *Carya, Betula,* and *Fraxinus americana* in the canopy.

Temporal Dynamics of the Understory Filter

The understory filter can be a consistent as well as powerful force that structures the community of advance regeneration. Many clonal understory plants, including those we have discussed as important components of the understory filter in their respective communities, have extremely long life spans, which may even exceed the life span of canopy trees (Watt 1947b; Knight 1964; McGee and Smith 1967; Oinonen 1967a, 1967b; J. P. Anderson and Egler 1988). Once established, they may persist many years in the understory in closed forest conditions (McGee and Smith 1967; Kurmis and Sucoff 1989; Gilliam and Turrill 1993; Gilliam et al. 1995; Hill and Silander 2001), and they may also persist through conditions of extreme canopy disturbance (Horsley 1988; Halpern 1989; Hughes and Fahey 1991). Clones of *Osmunda claytoniana* in one of our Massachusetts study sites are 150–200 years and spread only an average of 1.1 cm per year (Knight 1964). This temporal stability, coupled with low mobility, enable these plants to create an imprint on the microenvironment that will be experienced year after year by the same tree seedling individuals.

In deciduous forests of the eastern United States, some degree of canopy disturbance is necessary for most species of advance regeneration to grow to canopy height (Canham 1985, 1988). To understand tree seedling and sapling response to canopy disturbance, we must understand the response of understory plants and how the activity of the understory filter will change in the new context of altered resource flux. In response to overstory canopy disturbance, established understory plants have often been observed to increase in density (Huenneke 1983; Moore and Vankat 1986; Mladenoff 1990), compensating for canopy opening by reducing resource flux to tree seedlings below the understory stratum. For example, light levels at the forest floor did not increase significantly after experimental canopy gap formation in mixed-oak forest of the southern Appalachians where *Rhododendron maximum* cover was present and undisturbed (Beckage et al. 2000). Small canopy gaps may promote increased understory plant density where plants are already present but do not necessarily provide opportunities for their establishment. Density of *Dennstaedtia punctilobula* increases with light availability due to canopy opening (Hill and Silander 2001); however, we have observed many areas in our study sites with favorable light availability where no ferns are present. Others have also found differences in the understory layer between gaps and intact forest to be primarily expressed as changes in plant abundance rather than as species presence (Ehrenfeld 1980; Mladenoff 1990).

Although many understory plants increase in density in response to canopy disturbance, this is by no means a universal response. Several understory species show little response to single- or multiple-canopy-tree gaps (Collins and Pickett 1987, 1988; Mladenoff 1990), and, less commonly, others may show a density decrease (Veblen 1982; Mladenoff 1990). In a New Jersey oak forest where single- and multiple-tree gaps had been formed by gypsy moth defoliation, Ehrenfeld (1980) reported an increase in relative density of *Cornus florida* L. (eastern flowering dogwood) but little change in the herbaceous species *Osmunda cinnamomea Onoclea sensibilis* (sensitive fern), and *Maianthemum canadense* (Canada mayflower). The mode of canopy gap formation will also affect the level of disturbance to the understory stratum. Standing tree death due to drought, disease, or herbivory will have much less of a destructive impact on the understory stratum than windthrow or commercial harvesting (see chapter 14, this volume).

More extensive canopy opening used in commercial thinning of the overstory in the Allegheny Plateau region stimulates growth of understory plants such as *Dennstaedtia punctilobula* and grasses (Horsley and Marquis 1983; Drew 1988; Horsley 1988). Concomitant disturbance to the litter layer may promote establishment of new individuals of understory species that require mineral soil for establishment (Groninger and McCormick 1992), and these plants may virtually cover the entire area of the forest floor after canopy thinning. Light beneath ferns in stands with a relatively open overstory can be lower than levels experienced beneath the understory in closed forests (Horsley 1993a). In clearcut areas, however, where high insolation reduces soil moisture, fern cover can improve the moisture environment of seeds and improve germination levels (Kolb et al. 1989).

When forest stands experience large-scale disturbance such as hurricanes or clearcutting, understory herbs and shrubs can persist from underground structures as well as from new recruitment. In areas of the Harvard Forest that have been subject to simulated hurricane, reestablished fern cover reduced light levels to approximately 6–10% of full sun (Carlton and Bazzaz 1998). In this case, these levels were a substantial improvement compared to the light environment tree seedlings experienced in adjacent closed-forest stands under fern cover (1% full sun) or even outside of fern cover (3% full sun).

In areas of large-scale disturbance, forest understory plants may persist but may decrease in density due to competition with other invading species. After partial thinning of the canopy, *Dennstaedtia punctilobula* has been reported to increase in cover value up to 90%, but after complete harvest of canopy trees *D. punctilobula* cover has been observed to decrease to 60% due to competition with *Rubus allegheniensis* Porter and *R. occidentalis* L. (Horsley and Marquis 1983). Although the density of *D. punctilobula* varies over time, it is a good example of a forest understory herb that can persist throughout the whole disturbance cycle. Because of its persistence, an understory of *D. punctilobula* will likely continue to filter tree seedlings throughout the post-disturbance regeneration phase and into the next generation of forest. The extent of change in the filter's activity before, during, and after a disturbance event will depend on the extent to which understory plant growth preempts, neutralizes, and/or modifies available resources.

Summary

Because the understory stratum can differentially influence seed predation, germination, and seedling growth and survival, the understory acts as an ecological filter that can structure the nature and composition of the tree seedling bank. The tree seedling bank represents the starting capital for future forest regeneration, so forces that structure this community will influence the rate and trajectory of regeneration and ultimately help determine the composition of future forests.

Understory communities that we expect to be the most important filters of tree seedling regeneration are those in which the understory plants commonly reach high densities, at least locally, and cover large areas, as well as those that are temporally persistent. Although the three study systems in eastern forests on which we have focused, New England forests characterized by mixed-fern understories (George and Bazzaz 1999a, 1999b); southern Appalachian forests characterized by evergreen shrub understories (Beckage et al. 2000); and Allegheny forests characterized by a dense understory of *Dennstaedtia punctilobula* (Horsley 1993a), differ in overstory and understory structure and composition, they all demonstrate the substantial impact of the understory stratum on the light environment near the forest floor that tree seeds and seedlings experience. Important effects of the understory stratum on quantity and quality of litter have been repeatedly demonstrated, in contrast to the absence of significant understory effects on soil nutrient resources. Understories of *Rhododendron maximum* may influence tree seedling performance by decreasing soil moisture, but soil moisture has not been demonstrated to be an important mechanism of understory filtering in other systems, except in cases where an intact understory stratum conserves soil moisture in areas of high insolation after canopy opening.

The mechanisms by which the understory filters tree seedlings vary from tree species to species and even within a tree species at different life stages. We have found little evidence that the understory interferes directly with seed dispersal of trees in eastern forests; however, the presence of understory cover may play a more important role in influencing the secondary dispersal or predation of seeds by animals. The understory canopy often provides cover for seed predators, which leads to higher seed predation rates below the understory, particularly of large-seeded species such as *Quercus* spp. Primarily through the understory's influence on litter structure and light attenuation, it is able to differentially influence tree seedling germination and emergence. Low light levels beneath the understory and the behavior of herbivores affect seedling survival and growth, and these factors further enable the understory to filter tree seedlings. The seedling bank below the understory may be lacking in seedlings of relatively shade-intolerant species, but because the activity of the understory filter involves so many additional factors, including seed predation, seedling herbivory, and alteration of the litter layer, it is not always possible to predict the composition of the seedling bank beneath a well-developed understory based on shade tolerance alone.

The generally neutral or negative activity of the understory filter results in decreased tree seedling density below the understory stratum. Depending on postdisturbance conditions, reduced density of advance regeneration may increase the time necessary for forest regeneration, or in extreme cases it may prevent forest regeneration. Selective filtering of tree regeneration by the understory during dispersal, germination, and early growth phases most commonly shapes the species composition and diversity of the tree seedling bank below the understory canopy by eliminating or decreasing representation of certain species. The species composition as well as the density of the seedling bank will influence the trajectory of regeneration directly by determining which species are available for postdisturbance release and indirectly by determining the competitive environment for species that seed in after disturbance.

The differential influence of the understory filter on tree seedling growth leads to a typical size structure among seedling species that will influence the outcome of future tree–tree competition. The compounded influence of the understory filter on tree seedlings over all of the life stages discussed leads to a characteristic distribution of tree seedlings over the forest floor with different species commonly aggregated to various degrees in patches defined by understory type or density. The degree of spatial structuring of the tree seedling community will mirror the spatial structure of the understory community, and the degree of seedling aggregation will define future competitive and genetic neighborhoods and may have important implications for future forest diversity.

The effects of the understory on future forest composition and structure must be interpreted in the context of both predisturbance and postdisturbance conditions. The understory, if intact, will continue to filter tree regeneration after a disturbance event, although the activity of the filter may be altered in the context of new resource levels. Understory plant populations may grow in size and density in response to disturbance and may neutralize canopy recruitment opportunities for trees by preempting available light resources. Where a well-developed understory stratum interferes strongly with tree regeneration, a relatively large-scale canopy disturbance may be required for significant release of advance regeneration. The presence of a well-developed understory stratum may require that certain types of disturbance that directly damage the understory, such as windthrow or fire, are necessary to regenerate the forest canopy. Because interactions between the understory filter and tree seeds and seedlings determine the density, species composition, diversity, and size structure of the pool of advance regeneration within a stand, exploring these predisturbance processes will expand our understanding of postdisturbance tree regeneration.

12

Invasions of Forests in the Eastern United States

James O. Luken

The story of conservation in the eastern United States has historically been one of forest preservation (Whitney 1994; M. Williams 1989). With this story came the dogma that indigenous trees beget a ground layer (plants < 1 m tall) and understory (plants 1–3 m tall) of indigenous herbs and shrubs. In many relatively undisturbed sites this dogma appeared to mesh well with data on the structure and composition of plant communities. To this day, one can find numerous forest preserves in the eastern United States where the ground layer and understory are diverse and composed of indigenous species expressing several functional roles. However, beginning in the early 1970s, ecologists began focusing their attention not just on pristine, remote forested communities but also on forests in populated areas or areas subjected to human disturbance. When asking questions about the effects of fragmentation and historical disturbance, ecologists generated descriptive data; the data revealed forest communities supporting many nonindigenous species. The presence of these nonindigenous species and their impacts is now being addressed under the heading of a new and rapidly advancing discipline: invasion ecology. From the perspective of forest ecology, nonindigenous species are noteworthy because they offer unique opportunities to understand how plants arrive, establish, and eventually persist vis-à-vis the filter of the tree layer. From the perspective of conservation, nonindigenous species are potentially problematic because the forest may change in structure and composition so that management goals are no longer met. Both ecologists and conservationists are interested in predicting how forest communities will be changed as urbanization and human impact spreads even farther across the eastern United States (Daehler and Strong 1993; Mack et al. 2000).

Previous research on temperate forests indicates a high degree of unexplained variation in the presence and importance of nonindigenous species (C. E. Williams 1996; Lonsdale 1999). Previous attempts to assess the effects of nonindigenous species in forests have been weakened by the assumption that nonindigenous species inevitably have negative impacts and by a lack of experimental data (see Woods 1997 for a review). At best, we know that disturbance contributes to colonization by nonindigenous species (Stapanian et al. 1998) and that temperate forests may be more susceptible to additions of nonindigenous species than are other types of biological communities (Lonsdale 1999). However, few researchers have attempted to systematically assess the existing data on forests of the eastern United States with the goals of understanding the distribution, relative abundance, and impacts of nonindigenous species in the ground layer and understory. In this chapter I address the following questions: what nonindigenous plant species are found in eastern forests? What do these nonindigenous plant species indicate about forest communities? How might the impacts of these nonindigenous species be interpreted relative to our current knowledge of the structure and function of the ground layer and understory?

The Terms of Invasion Ecology

Discussions of invasions have been obfuscated by the use of many poorly-defined terms (Luken 1994). Most of these terms refer either to the site of evolutionary origin or original range (e.g., indigenous vs. nonindigenous), the ability to establish in a disturbance or to enter intact vegetation (e.g., colonizer vs. invader) (Bazzaz 1986; Pysek 1995; Schwartz 1997), or the human perception of problems associated with a species (e.g., weed) (Randall 1997). Recognizing the need for standardization, Davis and Thompson (2000) proposed two types of events that can be considered as colonizations: species entering new regions and species occupying discrete disturbances. They developed a dichotomous classification scheme for plant species based on dispersal distance, uniqueness to the region, and impact. Their scheme yielded eight types of organisms, all of which could be component species of forests (table 12.1). The purpose of this exercise was to reinforce the point that both nonindigenous species and indigenous species function as colonizers and that these colonizations can be understood without invoking special status for nonindigenous species. Lessons learned from studying succession will likely serve as good starting points for understanding how the structure and composition of forest communities are changed through time (Luken 1997; Davis and Thompson 2000; Davis et al. 2000).

The scheme presented in table 12.1 raises some provocative issues. Uniqueness to a region has spatial and temporal components. Thus, nonindigenous species that are now locally common could be categorized with other common, indigenous species. Traditionally, the status of a plant species is determined by referring to published floras where information on origin (e.g.,

Table 12.1. A classification of colonizers based on dispersal distance, uniqueness to the region, and environmental impact as defined by Davis and Thompson (2000). Potential examples are provided for eastern deciduous forests.

Type	Dispersal distance	Uniqueness to region	Environmental impact	Example
1	Short	Common	Small	A indigenous annual herbaceous species present in the forest understory colonizes a gap and occupies the gap for only one season.
2	Short	Common	Great	A indigenous perennial herbaceous species present in the forest understory colonizes a gap and dominates the gap community for many years.
3	Short	Novel	Small	A nonindigenous annual herbaceous species persisting as a founder population at the forest edge colonizes a gap and occupies the gap for only one year.
4	Short	Novel	Great	A nonindigenous perennial herbaceous species persisting as a founder population at the forest edge colonizes a gap and dominates the gap community for many years.
5	Long	Common	Small	A indigenous annual herbaceous species present in a local old-field is wind-dispersed into a forest gap and occupies the gap for only one season.
6	Long	Common	Great	A indigenous perennial herbaceous species present in a local old-field is wind-dispersed into a forest gap and dominates the gap community for many years
7	Long	Novel	Small	A nonindigenous annual herbaceous species is dispersed into a forest by hiker's boots. It establishes a small population and persists for only one season.
8	Long	Novel	Great	A nonindigenous perennial herbaceous species is dispersed into a forest by hiker's boots. It establishes in a forest gap and dominates the gap community for many years.

introduced from Eurasia) is provided. This status is developed without consideration of whether the species is common. From the perspective of understanding the invasion processes, it is cogent to first know if a species is locally established, as this will determine propagule pressure and potential for colonizing disturbances in forest stands. However, it is unlikely that terms referring to plant status will be universally abandoned, even though plant status is a complex issue with many poorly examined nuances (Schwartz 1997). The idea of indigenous versus nonindigenous is firmly entrenched in the philosophy of conservation in the United States, and this alone argues for the continued use of some term to indicate uniqueness to the North American continent (Luken 1994; Schwartz 1997). In this chapter, I use the term *nonindigenous* to indicate species that have been introduced over relatively long distances to the North American continent. However, at the spatial scale of forests that now occur throughout the eastern United States, nonindigenous species may be novel or common; the same can be said for indigenous species. In keeping with the suggestion of Davis and Thompson (2000), the terms *invasion* or *invader* will be used in situations where species have large measured impacts. Species new to a forest patch but with high impacts are referred to as novel *invasive* colonizers. Species new to a forest patch but with minimal impacts are referred to as novel *noninvasive* colonizers. The terms *colonizer* or *colonization* are used generally to refer to situations where plants occupy new areas.

Nonindigenous Species in Eastern Forests

There have been few attempts to assess how the contribution of nonindigenous species to ground layer and understory communities changes across a range of eastern U.S. forests. Stapanian et al. (1998) described results of the Forest Health Monitoring Program, in which cover of ground layer and understory species was assessed on a national scale. They found that different regions of the eastern United States supported nonindigenous species of different origins. Southeastern forests had mostly nonindigenous species from eastern Asia, whereas northeastern forests had nonindigenous species mostly from Eurasia. The idea that nonindigenous species may sort relative to environmental gradients has not been tested, but it is suggested by the fact that different regions of the eastern United States report different nonindigenous invaders. For example, the shrub *Berberis thunbergii* DC. is widespread in forests of the New York metropolitan region (Ehrenfeld 1997), while the shrub *Lonicera maackii* (Rupr.) Herder is widespread in forests of Ohio, Tennessee, and Kentucky (Hutchinson and Vankat 1997). Both of these species are presumably invasive based on measurements showing high densities, biomass, or production (Luken 1988, Ehrenfeld 1997). Both of these species are widespread in the eastern United States, but they appear to emerge as invasive only in smaller subsections of their entire ranges.

As a first step toward understanding the distribution of nonindigenous species in forests of the eastern United States, I examined floristic surveys and used these data to provide a list of colonizers that are potential invaders. Floristic surveys are biased in that they are typically done in nature preserves. Previous research suggests that preserves are less invaded than other sites (Lonsdale 1999). Information from floristic surveys emerges as lists of all plant species found within specific areas, but with little or no quantitative information provided on importance of the species in a community. Typically, the site or area of interest is an entire state park or natural area, but various habitats within the park or preserve may be included in the survey. I located 10 relatively recent floristic surveys that included 16 distinct forest communities. The communities ranged from savannahs ($n = 3$) to mixed mesophytic forests ($n = 4$) to floodplain/bottomland forests ($n = 6$) to pine forests ($n = 3$). I included only those surveys where plant presence was recorded within a specific type of forest, a criterion that eliminated many floristic surveys. I recorded the number of indigenous and nonindigenous graminoids, herbs, vines, and shrubs that composed the ground layer and understory. Unfortunately, few floristic surveys provide areas of the study sites. At best, the area of the entire park or preserve would be included in the survey description, but this would include the entire range of habitats including lawns, old fields, meadows, and forests. Because the number of species is likely related to site area, it is problematic to use these data for comparisons of richness or to determine relationships between richness of nonindigenous species and richness of indigenous species (e.g., Lonsdale 1999). However, I provide a scatter plot to show relative numbers of species in the two status categories (fig. 12.1).

Floristic surveys recorded 144 nonindigenous forest species, of which 44% were perennial herbs, 23% annual herbs, 14% shrubs, and the remaining 19% were evenly distributed among biennial herbs, graminoids, and vines. Plant species occurring in three or more sites included *Alliaria petiolata* (Bieb.) Cavara & Grande (biennial herb), *Arthraxon hispidus* (Thunb.) Makino (annual grass), *Berberis thunbergii* (shrub), *Cardamine hirsuta* L. (annual herb), *Celastrus*

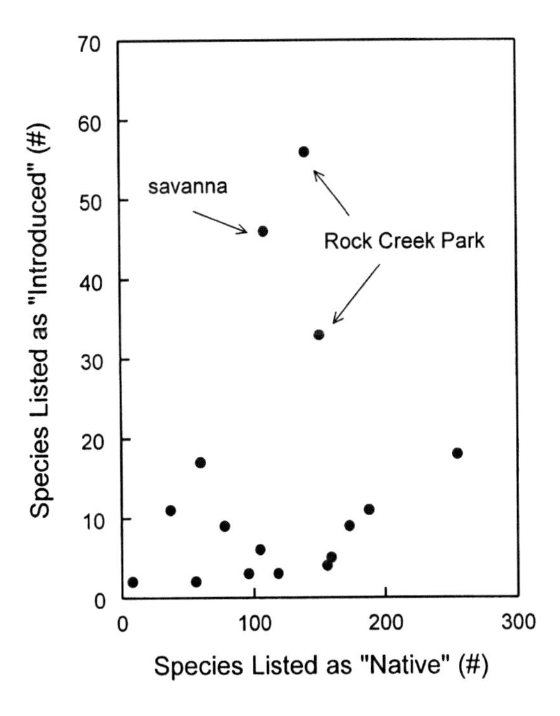

Figure 12.1. Numbers of indigenous and nonindigenous plant species in 16 eastern U.S. forest sites. Data were taken from the following floristic inventories: Basinger et al. (1997); Carpenter and Chester (1987); Crouch and Golden (1997); Easley and Judd (1990); Fleming and Kanal (1995); Herring and Judd (1995); Joyner and Chester (1994); Kearsley (1999); Reznicek and Catling (1989); Swanson and Vankat (2000). Rock Creek Park sites are urban mesic and floodplain forests.

orbiculatus Thunb. (vine/shrub), *Glechoma hederacea* L. (perennial herb), *Lonicera japonica* Thunb. (vine), *Lysimachia nummularia* L. (perennial herb), *Microstegium vimineum* (Trin.) A. Camess (annual grass), *Polygonum cespitosum* Blume (annual herb), *Polygonum persicaria* L. (annual herb), *Ranunculus ficaria* L. (perennial herb), *Rosa multiflora* Thunb. (shrub), *Rumex acetosella* L. (perennial herb), and *Stellaria media* (L.) Cyrillo (annual herb). Detrended correspondence analysis (DCA) of nonindigenous species (those occurring in at least 2 of the 16 forest sites) indicated both regional and community affinities. For example, forested sites from Ontario clustered at the low end of axis 1, and forested sites from Florida were located at the high ends of axes 1 and 2 (fig. 12.2). In contrast, floodplain forests, mesic forests, and a Tennessee pine forest clustered at the middle of axis 1. The distribution of species in ordination space (fig. 12.3) revealed a cluster of relatively common and well-known nonindigenous species (e.g., *L. japonica*) at midpoints of the axes that were associated with mesic and floodplain forests, and then relatively restricted nonindigenous species associated with the Ontario and Florida sites (figs. 12.2 and 12.3). Although I did not examine the ranges of these species in the native habitat, Panetta and Mitchell (1991) noted close correspondence between climate in the native habitat and climate in the area successfully colonized for several plant species.

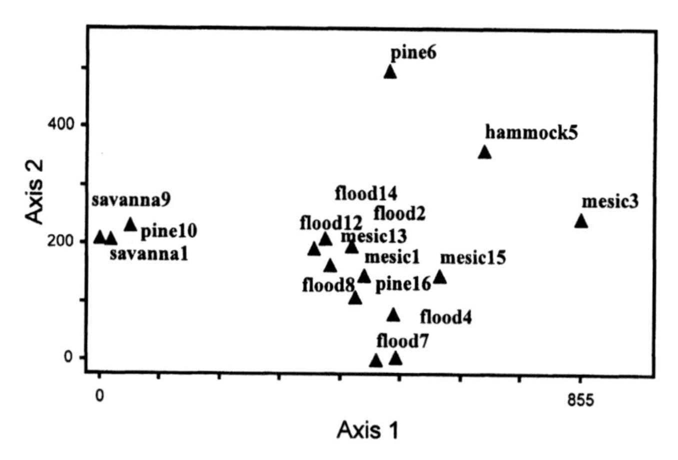

Figure 12.2. Detrended correspondence analysis (axis 1, eigenvalue = 0.247, length of gradient = 8.55; axis 2, Eigenvalue = 0.487, length of gradient = 4.97; cumulative r^2 = .405) of 16 forest sites based on the presence or absence of nonindigenous species. Flood = flood plain or bottomland forest; hammock = Florida hammock; mesic = mesic or upland forest; pine = pine-dominated forest; savannah = relatively open savannah.

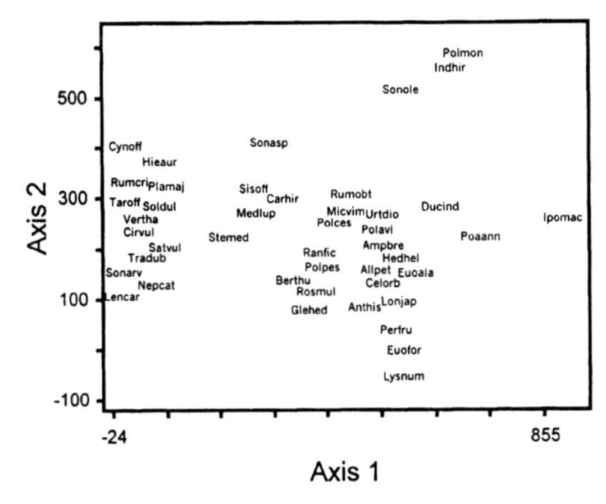

Figure 12.3. Detrended correspondence analysis of nonindigenous species found in 16 eastern U.S. forest sites. Species are as follows: Allpet = *Alliaria petiolata* Ampbre = *Ampelopsis brevipedunculata* (Maxim.) Trautv.; Anthis = *Arthraxon hispidus*; Berthu = *Berberis thunbergii*; Carhir = *Cardamine hirsuta*; C elorb = *Celastrus orbiculatus*; Cirvul = *Cirsium vulgare* (Savi) Tenore; Cynoff = *Cynoglossum officinale* L.; Ducind = *Duchesnea indica* (Andr.) Focke.; Euofor = *Euonymus fortunei* (Turcz.) Hand.-Maz.; Euoala = *Euonymus alatus* (Thunb.) Siebold; Glehed = *Glechoma hederacea*; Hedhel = *Hedera helix* L.; Hieaur = *Hieracium aurantincum* L.; Indhir = *Indigofera hirsuta* Harv.; Ipomac = *Ipomoea macrorhiza* Michx.; Lencar = *Leonurus cardiaca* L.; Lonjap = *Lonicera japonica*; Lysnum = *Lysimachia nummularia*; Medlup = *Medicago lupulina* L.; Micvim = *Microstegium vimineum*; Nepcat = *Nepeta cataria* L.; Perfru = *Perilla frutescens* (L.) Britt.; Plamaj = *Plantago major* L.; Poaann = *Poa annua* L.; Polavi = *Polygonum aviculare* L.; Polces = *Polygonum cespitosum*; Polpes = *Polygonum persicaria*; Polmon = *Polypogon monspeliensis* (L.) Desf.; Ranfic = *Ranunculus ficaria*; Rosmul = *Rosa multiflora*; Rumobt = *Rumex obtusifolius* L.; Rumcri = *Rumex crispus* L.; Satvul = *Satureja vulgaris* (L.) Fritsch; Sisoff = *Sisymbrium officinale* (L.) Scop.; Soldul = *Solanum dulcamara* L.; Sonarv = *Sonchus arvensis* L.; Sonasp = *Sonchus asper* (L.) Hill; Sonole = *Sonchus oleraceus* L.; Stemed = *Stellaria media*; Taroff = *Taraxacum officinale* Weber; Tradub = *Tragopogon dubius* Scop.; Urtdio = *Urtica dioica* L.; Verrtha = *Verbascum thapsus* L.

Which Colonizers Emerge as Invaders?

The data from floristic surveys indicate a relatively large pool of nonindigenous species that are potential invaders of forests. Furthermore, it appears as though this pool of nonindigenous species is sorted relative to environmental gradients across the eastern United States. However, floristic surveys typically do not include information on importance of individual species, except for an occasional qualitative assessment of frequency. Indeed, many nonindigenous species recorded in floristic surveys exist as single populations that have not spread much beyond the point of initial entry, while a smaller subset of the nonindigenous species are true invaders. (The small number of introduced species that actually becomes invasive has been examined by Williamson and Fitter [1996] with the goal of identifying critical traits that contribute to success in a new environment.) The distinction between invasive and noninvasive hinges on a determination of impact (Davis and Thompson 2000), but unfortunately, impacts have not yet been determined for most nonindigenous species. Thus, one must examine measured parameters that may indicate impact.

I reviewed some published literature on forests in an effort to extract those nonindigenous species shown to be invasive. There were some noteworthy biases regarding research questions and choices of research approaches. Most studies on the composition of forest communities have been done in relatively remote areas where nonindigenous species are not present. Presumably, this represents the bias of researchers in their attempt to capture understanding in the absence of human impact. In contrast, research focused on the impacts of nonindigenous species commonly targets restricted areas that are successfully invaded. Presumably, this represents the bias of researchers in their attempt to understand maximum invader impact. The problem of providing an unbiased sample of relative importance of nonindigenous species was addressed by Stapanian et al. (1998). In the absence of improved research methods, one can begin to determine which species are potentially invasive based on site-specific research and a clear definition of impact. I assume that species with large impacts in the forest community must usurp a relatively large share of the resource base, change the disturbance regimen, or modify biotic interactions so that loss of associated species occurs. Documenting these effects is difficult (Woods 1997). Published studies usually describe the ground layer and understory communities in terms of coverage, biomass, production, or importance, and thus it is necessary to make an inferential assessment of impact (e.g., a species with relatively high coverage must usurp a relatively large share of the resource base). Table 12.2 lists some invaders of forests as inferred from relatively high coverage or importance.

The list of species documented as invaders of forests (table 12.2) is brief when compared to the list of 144 nonindigenous species recorded in the floristic studies. The fact that few nonindigenous species have large impacts has been noted elsewhere with the suggestion that, in most situations, potential invaders and resident species are not fundamentally different (Levine and D'Antonio 1999). Gordon (1998), however, examined nonindigenous plant species in a variety of Florida habitats and concluded that many of these species do have large impacts. The factors that contribute to invasiveness of individual plant species are not well understood, but it is clear from the studies listed in table 12.2 that invasiveness in forests may be shared by more than one species and that even in invaded communities high impact is not solely by nonindigenous species. For example, as importance of *Alliaria petiolata* declines, the importance of *Stellaria media* increases (Luken and Shea 2000). Use of fire to manage *A. petiolata* was associated with more than a twofold increase in the importance of *Eupatorium rugosum* Houtt. and dominance of the community by this species (Nuzzo et al. 1996).

Table 12.2. Invasive plant species of eastern forests and selected characteristics of the invaded community.

Species	Community type	Criteria for invasiveness	Associated invasive species	Reference
Alliaria petiolata	Illinois sand forest	High percent cover	*Eupatorium rugosum* *Parthenocissus quinquefolia* *Rhus radicans*	Nuzzo et al. 1996
Alliaria petiolata	Maryland lowland forest	High percent cover	*Impatiens* spp.	McCarthy 1997
Alliaria petiolata	Kentucky mesic forest	High importance value	*Stellaria media* *Galium aparine*	Luken and Shea 2000
Glechoma hederacea	Maryland floodplain forest	High percent cover	*Alliaria petiolata*	Pyle 1995
Lonicera japonica	Eastern forests	High cover and frequency	*Ligustrum sinense*	Stapanian et al. 1998
Lonicera maackii	Kentucky mesic forest	High importance value	*Alliaria petiolata*	Luken et al. 1997
Lonicera maackii	Ohio mesic forest	High importance value	*Impatiens biflora*	Vankat and Snyder 1991
Lonicera tatarica	New England mesic forest	High percent cover	*Impatiens biflora* *Hydrophyllum virginianum*	Woods 1993
Stellaria media	Illinois mesic forest	High percent cover	*Carex jamesii*	Gibson et al. 2000

A further complicating factor for understanding why some species are invasive is the variation in invasiveness across the landscape. For example, *Alliaria petiolata* dominates the ground layer of forest communities in Illinois, Kentucky, and Maryland (Pyle 1995; Nuzzo et al. 1996; McCarthy 1997; Luken and Shea 2000), but is not an important invader in forests of the northeast (or at least it is relatively less invasive as indicated by the data presented in Woods [1993]). A possible explanation for variation in invasion success can be found in research where community changes are monitored over several years in the context of removal treatments. For example, when McCarthy (1997) removed *A. petiolata* from experimental plots, he facilitated a successful invasion of *Impatiens* spp. Interestingly, in the northeastern forests not heavily invaded by *A. petiolata*, *Impatiens biflora* Walt was extremely important (Woods 1993). The extensive database on forests invaded by *A. petiolata* suggests the presence of an invasion window (sensu Johnstone 1986; Davis et al. 2000) that can temporarily open for several species, not just for nonindigenous species. Again, nonindigenous species may not be fundamentally different from resident species in terms of ability to colonize and eventually invade. A more important factor may be how long species can persist after the initial colonization or how species can change the environment so that persistence is facilitated.

I noted some correspondence between nonindigenous species documented as invasive (table 12.2) and nonindigenous species documented in three or more floristic surveys. Those species most frequently encountered throughout the eastern United States will also be the ones most likely to emerge as invaders. In a search for characteristics that predict invasiveness of woody species, Reichard (1997) suggested that those with large ranges in the native region were more likely to invade a new region. Also, species that are invasive in one place are likely to be invasive in other places (Reichard and Hamilton 1997). Invaders listed in table 12.2 are also commonly found growing outside of forests, and in some instances they participate in early successional communities (Luken 1988). Thus, the ability to invade forests may not be linked to a single identifiable trait, but rather to the ability to express different traits across a wide range of environments (i.e., phenotypic plasticity).

Two of the invaders listed in table 12.2 have been well studied in terms of characteristics that contribute to success in the ground layer and understory. Anderson et al. (1996) reviewed the literature on *Alliaria petiolata* and maintained that the following traits are key to invasion success: an autogamous breeding system, high seed production, and rapid growth. Similar traits are important for successful invasion by *Lonicera maackii* (Luken and Mattimiro 1991). Forest invaders do, indeed, share many similarities with colonizers of successional communities (Bazzaz 1986). However, early-successional communities and forests differ dramatically in light availability. Successful colonizers of forests must overcome light limitation. *Lonicera maackii* has relatively high acclimation ability and is relatively plastic in terms of leaf morphology and biomass allocation (Luken et al. 1997b). This is apparently critical for growth and seed production across the full range of light environments, a characteristic not shared by many species adapted to low light conditions. High phenotypic plasticity apparently allows *A. petiolata* to successfully set seed in a variety of light environments and forest habitats (Byers and Quinn 1998; Meekins and McCarthy 2000). Finally, Schierenbeck et al. (1994) compared growth of *Lonicera sempervirens* L. and *Lonicera japonica* in response to herbivory. They concluded that the nonindigenous *L. japonica* was less susceptible to herbivore damage and was better able to compensate for lost tissue than the indigenous *L. sepervirens* (Schierenbeck et al. 1994).

Long-term studies of forest succession in the eastern United States demonstrate that nonindigenous species are sorted along resource gradients that emerge with the developing tree layer. Typically, the number of nonindigenous species declines with site age, but a few key invaders may persist in the forest environment. For example, Vankat and Snyder (1991) studied a chronosequence of sites in Ohio that included a 2-year-old field, a 10-year-old field, a 50-year-old field, a 90-year-old forest, and an old-growth forest. Analysis of the ground-layer flora indicated 30–35 nonindigenous species in the old fields and 3–8 nonindigenous species in the forests (fig. 12.4). During succession, annual and biennial species decreased as a percentage of the flora, while woody species increased as a percentage of the flora. Nonindigenous understory species included *Alliaria petiolata, Berberis thunbergii, Duchesnea*

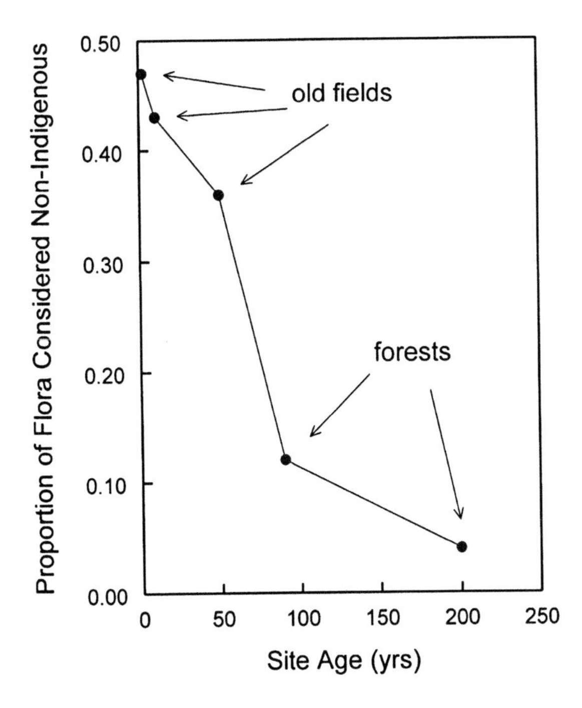

Figure. 12.4. Changes in proportion of the flora considered nonindigenous during forest succession in Ohio. Adapted from Vankat and Snyder (1991).

indica (Andr.) Focke, *Ligustrum vulgare* L., *Lonicera japonica*, *Lonicera maackii*, *Ornithogalum umbellatum* L., and *Rosa multiflora*. None of these species was common in the old-growth forest, but *L. maackii* shared dominance of the 90-year forest ground layer with the indigenous forest herb *Impatiens biflora*. Maximum species richness was found in the 50-year-old field, the site that also supported the greatest number of nonindigenous species. Gibson et al. 2000) monitored changes in the understory of an abandoned recreation area in a mesic Illinois forest. Their data also indicated a decline in importance of nonindigenous species through time, but *Stellaria media* persisted as an important invader and shared dominance of the forest understory with the indigenous sedge, *Carex jamesii* Schwein. A survey of Ontario woodlands suggested that more mature woodlands supported fewer nonindigenous species, but the few nonindigenous species persisting in mature forest were more invasive (Francis et al. 2000).

Occasionally, a nonindigenous species in eastern forests may express such strong dominance that forest development is diverted. Fike and Niering (1999) showed that invasion by the woody vine *Celastrus orbiculatus* had diverted the typical path of old field to forest succession in Connecticut. The data of Fike and Niering also showed two types of invasion patterns for other nonindigenous species. *Lonicera japonica* was present with low coverage in the old fields but gradually expanded in coverage over a period of 40 years; *Rosa multiflora* was absent during the first 20 years but gradually expanded in coverage during the last 20 years. Long-term studies of forest development provide, perhaps, the best information on how nonindigenous species enter and persist in forest communities. Several questions arising from previous research should be examined further. First, are soil disturbances necessary for invasions of forests? What are the unique traits of successful invaders that allow them to overcome light limitation? How does richness of the community influence the susceptibility of the forest to invasion?

Which Types of Forest Communities Are Invaded?

There have been no attempts to determine which forest communities in the eastern United States may be most susceptible to invasion. Research done in the Pacific Northwest suggests that riparian forests may be more strongly invaded than other forest communities (Deferrari and Naiman 1994; Planty-Tabacchi et al. 1995). However, this conclusion was based on landscape-level assessments of two parameters: numbers of species and coverage of only nonindigenous species. These parameters do not indicate what share of the community resource base is used by nonindigenous species. Furthermore, the conclusion that riparian forests are more heavily invaded is likely linked to the fact that richness of nonindigenous species increases with richness of the entire community (Levine and D'Antonio 1999, Levine 2000). The published literature dealing with species composition of some eastern forests indicates either an insignificant presence of nonindigenous species (Rogers 1981; Moore and Vankat 1986; Whitney and Foster 1988; Hughes and Fahey 1991; Gilliam and Turrill 1993; Roberts and Gilliam 1995b; Ebinger et al. 1997; Rooney and Dress 1997a; Goebel et al. 1999), a combination of indigenous and nonindigenous species, with nonindigenous species expressing an intermediate or low level of importance (Carleton and Taylor 1983; Brothers and Spingarn 1992; McCarthy and Bailey 1996; Bowles and McBride 1998; Hanlon et al. 1998; C. E. Williams et al. 1999a, 1999b; Laatsch and Anderson 2000), or communities where one or more nonindigenous species are successful invaders (see table 12.2).

Stapanian et al. (1998) attempted to address this systematic variation in invasion by sampling forests throughout North America. They presented estimates of mean proportional coverage of nonindigenous species in forests of the northeastern and southeastern United States as 1.5% and 13%, respectively. This study is remarkable for three reasons: the relatively low total coverage of nonindigenous species in eastern forests, the observed regional difference, and the unbiased nature of the sampling design. I examine each of these factors in the context of understanding why different forests in the eastern United States may be differentially invaded.

Relatively low total coverage of nonindigenous species in forest understories suggests that the forest environment is a strong filter of potential invaders. In contrast, nonindigenous species are universally more important in early successional communities. Filtering by the forest environment could be a limitation to growth, a limitation to dispersal, or some combination of the two factors. Light limitation is apparently due to the fact that most nonindigenous species in the United States were originally introduced into high-light environments, whereas movement into low-light environments was secondary (see Luken and Thieret 1996). Research indicates that forest edges support a rich assemblage of nonindigenous species but that most of these species do not invade forest interiors (Brothers and Spingarn 1992; Goldblum and Beatty 1999), presumably because of light limitation. These edge species are generalists and many function well in high-light environments. In contrast to high-light environments where soil disturbance is an important factor for invasion success, invasions of the forest understory appear to be linked to light availability and propagule availability. For example, canopy gaps may (Luken et al. 1997a; Goldblum and Beatty 1999) or may not (Moore and Vankat 1986; Hughes and Fahey 1991) lead to species additions, depending on whether potential colonizers are near the gaps. Hutchinson and Vankat (1997) found that high light availability and proximity to a seed source were important determinants of invasion success by *Lonicera maackii*. Group-selection harvests of Alabama bottomland forest created patches where the number of nonindigenous species increased from 6 to 34 (Crouch and Golden 1997), but it is unknown if this was the result of increased light, increased dispersal into the sites, or stimulation of buried seeds.

Considering that light limitation and availability of potential invaders are important factors in determining which forests will be invaded, it has been argued that riparian forests are more susceptible because of flood-induced canopy gaps, soil disturbance, and relatively greater availability of potential invaders (Pysek and Prach 1993; DeFerrari and Naiman 1994). However, results from eastern riparian forests are not conclusive. Pyle (1995) found that Maryland floodplain forests were heavily invaded, whereas C. E. Williams et al. (1999b) found that Allegheny Plateau riparian forests were not heavily invaded. The difference between these two studies may be due to differences in availability of nonindigenous species.

The observation that southeastern forests are more heavily invaded by nonindigenous species than northeastern forests was explained in terms of higher degrees of anthropogenic disturbance and conversion of large areas to *Pinus*-dominated communities (Stapanian et al. 1998). However, floristic studies suggest that pine forests in the southeast generally support fewer nonindigenous species than adjacent hardwood communities (Easley and Judd 1990; Herring and Judd 1995; Crouch and Golden 1997), and it may be that invasion susceptibility (more nonindigenous species or more importance of nonindigenous species) increases generally in warmer climates. Clearly, distance from populated areas is an important determinant of availability of nonindigenous species that could potentially invade eastern forests. The highest number of nonindigenous species (56) recorded for a floodplain forest was in Rock Creek Park in Washington, DC (Fleming and Kanal 1995), whereas floodplain forests in relatively isolated preserves of the southeast had few nonindigenous species (Carpenter and Chester 1987; Herring and Judd 1995; Basinger et al. 1997; fig. 12.1). Relatively open savanna—like forests may also be more susceptible to invasion due to higher light availability (Tester 1996; Laatsch and Anderson 2000). Several studies have suggested that hot spots of diversity are also more susceptible to invasions (Wiser et al. 1998; Levine and D'Antonio 1999; Stohlgren et al. 1999), although this has not been demonstrated for eastern forests; the data presented in figure 12.1 do not allow a valid test of this relationship.

Finally, Stapanian et al. (1998) designed a sampling scheme in which sample points were established on a nation wide grid system. In short, data were not collected in a biased, site-specific manner but were collected so that statistical analyses could be validly conducted at multiple scales. Furthermore, the data collected included two important aspects of invasion: numbers of species and relative coverage of the species. While this approach clearly has some utility for assessing forest characteristics over large areas, it may not reveal patterns of community change that emerge relative to the scale of dominant ecological processes. For example, Zampella and Laidig (1997) sampled riparian areas of the New Jersey pinelands in an effort to demonstrate whether watershed-level disturbances were reflected in community composition. Indeed, riparian invasions were facilitated by watershed-level disturbances. In addition, we see not only that disturbance is a component of invasion, but that disturbances change the availability of species and composition of communities far from the point of the actual disturbance (Zampella and Laidig 1997). An understanding of which forest communities are most susceptible to invasion will require a sampling protocol that allows for estimation of (1) changing composition of the pool of potential invaders as sorted by environmental factors, (2) changing numbers of potential invaders associated with urban/rural effects, (3) changing composition of forest communities associated with environmental factors, and (4) changing disturbance regimens that affect the degree of canopy closure and thus light availability.

Potential Interactions Between Invaders and Associated Species

Woods (1997) provided an excellent review of community-level effects associated with plant invasions and showed that effects of plant invaders range from dramatic to insignificant. He included few examples from eastern U.S. forests, suggesting that few such studies exist. Commonly, invader impacts are inferred by examining the relationship between invader density and species richness of the associated community (Woods 1993; Hutchinson and Vankat 1997) or by comparing species richness between narrowly circumscribed invaded and noninvaded sites (Gordon 1998; Lockhart et al. 1999). Conclusions from these types of studies are invariably confounded by a lack of information on site characteristics at the time when the invasion was initiated, and it is not possible to separate the effects of invaders from the potential effects of disturbances that likely initiated the invasions (Woods 1997).

Two studies on forests in the eastern United States, one in Florida and one in Ohio, demonstrate the types of research that are necessary to determine impacts of invaders (Horvitz et al. 1998; Gould and Gorchov 2000). Gould and Gorchov (2000) addressed the effects of *Lonicera maackii* on demography of three species of annual plants growing in Ohio forests. At two sites of differing invader densities (historically disturbed and relatively undisturbed), they established three treatments: *Lonicera* removal, *Lonicera* present, and *Lonicera* absent. They transplanted seedlings of the target species to the study plots and monitored plants for 1 year. The results indicated a significant increase in fitness (combination of survival and fecundity) for all transplanted species in the removal treatment. Plant responses were more dramatic in the heavily invaded site. Presumably greater degrees of invasion were associated with greater environmental changes associated with the treatments. The heavily invaded site also had less tree basal area, so presumably, when *Lonicera* was removed, light was more available than at the less-invaded site. Gould and Gorchov (2000) suggested that the direct effect of *Lonicera* was light limitation, although they pointed out that the results could differ in another year with different environmental conditions.

Horvitz et al. (1998) assessed a disturbance by Hurricane Andrew to determine effects of colonizers in forest preserves of Metro-Dade County, Florida. They studied three sites with differing amounts of hurricane damage over 2 years after the hurricane. Treatments included control areas allowed to recover in the absence of management and managed areas where nonindigenous vines were removed. The results clearly demonstrated that nonindigenous vines were competitively superior to many indigenous species. Rapid response of the nonindigenous vines was facilitated by the fact that the forests were invaded before the disturbance, and invaders simply responded by various regeneration mechanisms. Horvitz et al. (1998) concluded that nonindigenous species in these urban forests maintained the functional roles that they expressed in the native habitats. These functional roles were presumably matched to the disturbance regimen of the urban forest environment and were particularly matched to the occasional catastrophic disturbance associated with a hurricane.

The two previously described studies (Horvitz et al. 1998; Gould and Gorchov 2000) provide glimpses of the unique approaches and opportunities required to assess interactions between invaders and associated species. Specifically, one needs to be able to manipulate invader effects, and one needs to understand the interacting impact of disturbance. However, these approaches are primarily of value in understanding the interactions in retrospect and do not offer much for predicting community development beyond the dominance of the target invaders; both studies involved communities that had been invaded by nonindigenous species many years in the past, and it is unknown to what degree the invaders had already modified the invaded system. Because invaders may change community composition and ecosystem-level processes (Walker and Smith 1997; Lippincott 2000), it is important to consider how invaded systems may be changed so that future invasions are either facilitated or hindered. The degree of these changes will likely determine whether management approaches fail or are successful (Luken 1997). For example, Luken et al. (1997a) created gaps in dense thickets of *Lonicera maackii* to determine whether this might contribute to restoration of the ground layer in an urban Kentucky forest. Removal of the dense shrub canopy increased light availability, and richness of the ground layer increased. However, the responding species were generally not forest herbs, and some plots were quickly invaded and dominated by *Alliaria petiolata* and seedlings of *L. maackii*. The conclusion was that long-term dominance of the site by *L. maackii* had changed the composition of the bank of potential colonizers, and successful restoration required introductions of target species. Provencher et al. (2000) reached a similar conclusion in southeastern pine forests.

Should We Manage Nonindigenous Forest Species?

The presence of nonindigenous species in forest preserves poses a variety of dilemmas for resource managers and nature stewards. Invasion by nonindigenous species is inevitable, and this may even be facilitated by the mere visitation of a preserve, as visitors and their vehicles disperse seeds (Lonsdale 1999). Forest preserves located closer to urbanized regions will have relatively greater numbers of potential invaders established in forest edges and in surrounding landscapes (see, e.g., Fleming and Kanal 1995). This increases the number of nonindigenous species that may at some time in the future become invasive. Colonization of forests by nonindigenous species forests may be concurrent with local extirpations of other species (Robinson et al. 1994; Rooney and Dress 1997a). Thus, the potential flora of a preserve is dynamic and will not likely conform to any conservation plans that focus on maintaining a relatively stable plant community (Pickett et al. 1992).

Another dilemma is the paucity of information on impacts of nonindigenous species in forest understories. In response, resource managers have simply developed lists of problem species based on human perceptions of impacts, group consensus, and other means. For example, the Tennessee Chapter of the Exotic Pest Plant Council has placed species in categories ranging from severe threat to lesser threat. Heibert (1997) discussed the shortcomings of such an approach when developing resource management policies; seldom are indigenous species examined in terms of their abilities to function as invaders. Species placed in "severe threat" or similar categories will likely be targeted for some type of control. However, in forests, as in other types of plant communities, any effort to remove established plants is a type of disturbance that may create opportunities for new colonizations to occur (Hobbs and Huenneke 1992; Luken 1997; Provencher et al. 2000). Thus, resource managers may be placed in situations where they are managing species with unknown impacts and where the outcomes of the management process are highly unpredictable.

There are few examples of successful management of nonindigenous species in forests (Mack et al. 2000). Generally, management experiments elucidate the regeneration mechanisms that invaders possess (Horvitz et al. 1998) or they show that responses to management are species specific (Laatsch and Anderson 2000). Attempts to control *Alliaria petiolata* have been well documented. Luken and Shea (2000) assessed the effects of repeated fall burning and found that importance of *A. petiolata* actually increased in some sites. Nuzzo et al. (1996) and Schwartz and Heim (1996) found that burning maintained populations of *A. petiolata* at or near preburn levels. For this species, presence of seeds in the seed bank and reproduction of plants missed by control measures were the primary factors that maintained populations. R. C. Anderson et al. (1996) reviewed the literature on *A. petiolata* and concluded that "it may be unrealistic to expect to eliminate the plant from many habitats it has already invaded" (p. 181). Management or removal experiments, although generally showing that eradication of *A. petiolata* is not possible, do indicate the existence of other species that can quickly occupy the invader niche in forest understories when management activities create a pulse of resource availability (e.g., *Stellaria media*, Luken and Shea 2000; *Eupatorium rugosum*, Nuzzo et al. 1996; and *Impatiens* spp., McCarthy 1997). Perhaps the best approach for management of nonindigenous species is the elimination of founder populations (Hobbs and Humphries 1995). This is realistic in small forest preserves, but is likely not possible over large areas.

Research indicates that forest preserves through time have experienced both additions and losses of species (Reader and Bricker 1992b; Robinson et al. 1994; Rooney and Dress 1997a); nonindigenous species may be using a resource base that is currently available as a result of species loss. Furthermore, the emerging picture of plant invasion blurs the clear distinction between nonindigenous and indigenous species. Thus, in some situations it may be necessary to assume that additions of nonindigenous species are inevitable. Finally, assuming that nonindigenous species have negative impacts requires at least some explicit statement of the conservation goals for a preserve. Typically, this may involve attempting to recapture some historical condition of a forest (Laatsch and Anderson 2000), although there may be other goals depending on land ownership. There cannot be an impact unless the resources of concern have been quantified and unless the effects of the invader on these resources have been estimated. Exercises to quantify monetary losses associated with invasions (Pimental et al. 2000) are relatively straightforward for commodity production systems but are less reliable when resources of concern are not firmly identified (Hiebert 1997), as is the case with many preserved forests.

Should we attempt to control or eradicate forest invaders? The answer to this question obviously depends on the management goals of the forest, the resources of concern, the impacts of the invader, the long-term impacts of proposed management activities, and the system parameters that affect the long-term mix of species available to participate in ecological interactions. The process of determining these factors may lead to new perspectives on the functions and impacts of nonindigenous species (C. E. Williams 1997), a better understanding of how communities generally become invaded, and will surely lead to better understanding of how forest communities will change in the future.

Summary

Although it is clear that the structure and composition of the ground layer and understory are changing in many forests throughout the eastern United States as a result of additions and losses of species, the causes are not well understood. Human impacts are both overt and subtle. Specifically, humans add to the pool of potential forest invaders, and human-generated disturbance of forests apparently increases light availability and thus forest invasibility. Although the pool of nonindigenous species now present in forests is large, the number of species actually documented as invaders is relatively small. The proven invaders show a wide range of invasion success across the eastern United States that is apparently linked to limiting environmental factors. A systematic analysis of the relative contribution of nonindigenous species to ground layer and understory importance in the eastern United States is needed because of biases in the design of previous site-specific research. A single study completed to date suggests that the relative importance of nonindigenous species may be low; however, further work is needed to assess invasion along environmental gradients that are associated with human impact.

Low light availability is a key limitation to forest invasion. Many nonindigenous species can establish in forest edges but cannot invade forest interiors. Those species that do invade forest interiors show a high degree of phenotypic plasticity, can respond quickly to changes in light availability, and can produce seed under a wide range of light conditions. Most successful forest invaders are generalists that also grow well outside the forest environment. Although previous research suggests that riparian forests are more susceptible to invasion as a result of the disturbance regimen, this has not been fully tested with forests of the eastern United States. Forests located in urban areas will have a larger pool of potential invaders.

Impacts of invaders on community-and ecosystem-level processes in eastern U.S. forests are not well understood. Removal experiments indicate that once a site is invaded, resilience of invaders may be high due to various mechanisms for population regeneration. Removal experiments also suggest the existence of an invasion window that can open not just for nonindigenous species but for a larger subset of potential colonizers that includes indigenous species. It is now clear that most nonindigenous species are not fundamentally different from indigenous species, and thus a complete understanding of changes in the ground layer and understory will come from research where arrival, establishment, and persistence are assessed relative to changing environmental factors.

Although forests invaders are often managed, few studies can demonstrate the utility of this activity. Efforts aimed at preservation of the ground layer and understory should focus first on understanding how long-term forest changes are leading to both additions and deletions of species. The preservation context of the forest should be fully established before management, as this will determine whether it is realistic to initiate management actions.

13

Response of the Herbaceous Layer to Disturbance in Eastern Forests

Mark R. Roberts
Frank S. Gilliam

Various definitions of disturbance have been proposed, some restricted to discrete events (White and Pickett 1985) and others including long-term environmental fluctuations (Ryan 1991). We restrict our discussion to relatively discrete events because these are more readily tractable, and most types of direct management interventions (e.g., forest harvesting) are discrete in time and space. Accordingly, we adopt the definition of White and Pickett (1985) as our working definition of *disturbance: "any relatively discrete event in time that disrupts ecosystem, community, or population structure and changes resources, substrate availability, or the physical environment"* (p. 7). We recognize the admonition of Pickett et al. (1989) that disturbance affects all levels of organization differently, from individuals to ecosystems and landscapes, and that most empirical work focuses on effects of disturbance on community structure rather than on ecological processes.

Although ecologists generally recognize that disturbances of various kinds strongly influence structure and function of plant communities (e.g., White 1979, Pickett and White 1985) most attention has focused on disturbance size and frequency, with little explicit consideration of severity (Malanson 1984; Peterson and Pickett 1991; Peterson 2000). If more attention were given to the severity aspect of the disturbance regime, our understanding of ecosystem dynamics and our ability to predict consequences of management actions would be greatly refined. Thus, we emphasize disturbance severity but give due consideration to size and frequency where appropriate. We modify Oliver and Larson's (1996) definition of *disturbance severity* as the amount of forest overstory removed and the amount of understory vegetation, forest floor, and soil destroyed. With disturbance severity quantified in this manner, we are able to evaluate disturbances that vary both in the amount of overstory and understory destruction. Quantifying overstory and understory disturbance independently is particularly important with respect to forest harvesting because different degrees of overstory removal and different harvesting systems can be combined with a variety of understory treatments, creating numerous combinations of overstory and understory disturbance conditions.

The herbaceous layer simultaneously is exposed to and responds to many forest disturbances, ranging from microscale disturbances such as frost heaving and trampling by large vertebrates (McCarthy and Facelli 1990), to more extensive and intensive disturbances resulting from herbivory (Rooney and Dress 1997a, 1997b), tree mortality (Beatty 1984; Collins et al. 1985; Moore and Vankat 1986; Stone and Wolfe 1996; Goldblum 1997), inundation (Lyon and Sagers 1998; C. E. Williams et al. 1999b), periodic fire (Gilliam and Christensen 1986; De Granpré and Bergeron 1997), catastrophic wind damage (Peterson et al. 1990; Castelli et al. 1999; Peterson 2000), and forest management practices, including timber harvesting and use of herbicides (Duffy and Meier 1992; Roberts and Dong 1993; Gilliam et al. 1995; Halpern and Spies 1995; Hammond et al. 1998).

Principal Types of Disturbances in Eastern Forests

Natural Disturbances That Primarily Affect the Forest Canopy

Using the above definition of disturbance severity, the principal types of relatively discrete natural disturbances in eastern forests can be generally categorized into those that primarily affect the overstory and those that primarily affect the understory. Insect defoliation, wind, and ice storms are the most common types of disturbances affecting the forest canopy. Outbreaks of the defoliating insect spruce budworm (*Choristoneura fumiferana* (Clem.)), are common throughout the range of balsam fir (*Abies balsamea* (L.) Mill.) in eastern forests (Blais 1983). Balsam fir is the most susceptible species to damage by spruce budworm defoliation, although the spruces (*Picea* spp.) are the preferred host for the insect. Severity of defoliation can range from light (partial removal of current foliage) to severe (complete shoot destruction with widespread tree mortality). The spruce budworm prefers large, mature overstory trees, but at extremely high larval densities, the insect may also defoliate understory saplings and seedlings (Pardy 1997). Numerous studies have addressed growth responses of defoliated trees and stands (e.g., Peine 1989), tree regeneration (e.g., Osawa 1994) and succession of woody species (e.g., Ghent et al. 1957; Batzer and Popp 1985) after insect attack. Overstory mortality typically results in growth release of advance regeneration of balsam fir and spruces and establishment of shade-intolerant and mid-tolerant deciduous species such as *Betula papyrifera* Marsh., *B. alleghaniensis* Britt., *Po-*

pulus spp., and *Acer rubrum* L. (Ghent et al. 1957; Batzer and Popp 1985). Studies of herbaceous layer response to spruce budworm defoliation are lacking.

The gypsy moth (*Lymantria dispar* L.) is another important canopy-defoliating insect in eastern forests, affecting hardwood forests throughout the northeastern and mid-Atlantic states of the United States (USDA Forest Service 1993). Like the spruce budworm, the gypsy moth preferentially selects certain tree species over others such as *Quercus* spp., *Betula* spp., *Tilia* spp., and *Populus* spp. (Muzika and Liebhold 1999) and larger size classes of trees, principally those in the forest canopy (Campbell and Sloan 1977). The overall effect of moderate to severe defoliation by insect larvae is to release understory individuals, thereby altering species composition and initiating a new cohort of woody seedlings and saplings (Collins 1961; Fajvan and Wood 1996).

Mattson and Addy (1975) pointed out that plant-eating insects help maintain high primary productivity in forest communities by consuming less vigorous plants and opening the canopy for more vigorous plants. In addition to increased light in the understory, nutrient and moisture levels in the understory also increase because of removal of transpiring foliage in the overstory and additions of frass and debris to the forest floor (Eshleman et al. 1998). Several studies have documented increases in the abundance and growth of shrub species after insect attack (Ghent et al. 1957; Batzer and Popp 1985), but few have addressed changes in the herbaceous layer. Ehrenfeld (1980) found no significant differences in herb cover and no obvious changes in species composition between gap and control sites in a gypsy-moth-defoliated forest. She concluded, however, that the spatial pattern and relative densities of plants in the forest understory, in combination with the size and pattern of gaps, are the critical factors determining the pattern of forest understory response to canopy gaps created by insect defoliation.

Wind damage in forests has been the subject of considerable study (see review by Everham and Brokaw 1996). As with insect defoliation, the severity of wind disturbance can vary from small gaps created by broken branches or the fall of individual canopy trees (Moore and Vankat 1986) to catastrophic removal of the majority of the canopy (Dunn et al. 1983; Foster 1988a; Merrens and Peart 1992; Peterson and Pickett 1995; others cited in Everham and Brokaw 1996). The severity of damage is related to wind intensity in combination with a number of biotic and abiotic factors, including tree species, stem size, canopy evenness, previous weakening by pathogens, and topographic exposure (Everham and Brokaw 1996). Wind influences conditions for understory herbaceous vegetation by increasing light, altering substrate (snapped trees vs. uprooted), and depositing litter (Everham and Brokaw 1996). It is likely that the frequency of uprooted trees, and likewise the frequency of pits and mounds, is generally higher after wind disturbance than after insect defoliation.

Several studies (e.g., Beatty 1984; Peterson et al. 1990) have documented differences in species composition on pits and mounds and have demonstrated the importance of pit and mound microsites in maintaining species diversity of understory vegetation. Others have found subtle change in understory composition after creation of individual tree windthrow gaps in mature forests (Brewer 1980; Moore and Vankat 1986). Moore and Vankat (1986) studied the portions of the gaps where soil had not been disturbed by uprooting and found significantly higher solar radiation and soil moisture within 1 to 2-year-old gaps compared to older gaps and intact forest. They concluded that higher total herbaceous cover in the young gaps was related to increases in species that were present under the intact canopy before gap creation as opposed to invasion of new species. It is likely that invasion increases in importance with increasing soil disturbance and canopy removal.

Ice or glaze storms are common recurring disturbances in forests throughout eastern North America, with the exception of Florida (Oliver and Larson 1996; Rebertus et al. 1997; Lautenschlager and Nielsen 1999). Lemon (1961) showed that major ice storms occur more frequently in northeastern forests than do large wind storms or fires. Most studies of forest recovery after ice storm damage have focused on the canopy layer (e.g., Lemon 1961; Siccama et al. 1976; Bruederle and Stearns 1985; De Steven 1991b) and have concluded that the effects depend on storm intensity, landscape position, wind, and forest type and structure. Succession of forest tree species may be accelerated or retarded depending on the degree of damage and current successional stage; removal of early-successional canopy dominants may allow late-successional understory trees to fill canopy gaps, whereas extensive canopy damage might allow reproduction of early-successional species (Whitney and Johnson 1984).

Little information is available on the response of the herbaceous layer to ice storms. Given that storm damage typically occurs from breakage of limbs or whole trees (Runkle 1985; Oliver and Larson 1996), we would predict that (1) canopy gaps would be created, favoring shade-tolerant shrubs and herbs in smaller gaps and shade-intolerant species in larger gaps; (2) there would be relatively little uprooting and disruption of the forest floor with little mineral soil exposure, resulting in (3) growth stimulation of preexisting plants with relatively little invasion of new individuals (as compared to wind disturbance, for example); and (4) the input of coarse, woody debris to the forest floor can be substantial (Bruederle and Stearns 1985), eventually providing substrate for herbaceous plants.

The ice storm of 1998, which affected Ontario, Quebec and the Maritime Provinces of Canada, as well as portions of New England, in the United States (Irland 1998), was widespread. This storm was unprecedented in terms of the total area affected (603,654 ha in Ontario alone), as well as the duration and the amount of freezing rain that was deposited (Lautenschlager and Nielsen 1999; Van Dyke 1999). More than 80 hours of freezing rain and drizzle occurred between January 5 and January 10, in Ontario, producing 73–108 mm of precipitation. This was twice the duration and amount of previous ice storms (http://www.msc.ec.gc.ca/events/icestorm98/icestorm98_the_worst_e.html#top). The amount of woody litter produced by the 1998 ice storm in an old-growth forest at Mont St. Hilaire, Quebec (19.9 metric tons/ha), was the greatest of any ice storm on record and approached amounts produced by the most powerful hurricanes (Hooper et al. 2001). Such damage was found by Jones et al. (2001) to be predominantly a function of tree size for a mature, deciduous forest in eastern Ontario; that is, large stems in the canopy experienced much greater damage than did smaller stems in the subcanopy. Whether there are significant effects on the herbaceous layer after this and other ice storms will likely depend on stand structure and species composition, in addition to storm intensity (Van Dyke 1999). The magnitude of the response of herbs is likely to be greater where canopy openings are large and shade-tolerant tree seedlings, saplings, or shrubs are sparse.

Natural Disturbances That Affect Only the Forest Floor or Both Floor and Canopy

The natural fire return interval (before European settlement) in eastern forests varies from 2 to 1,000 years, depending on climate, site, species, stand growth patterns, and influences of other disturbances (Oliver and Larson 1996 and studies reviewed therein). The forest types exhibiting the shortest return intervals are the longleaf and loblolly pine forests of the southeastern U.S. coastal plain and jack pine forests in the Lake States. Upper-elevation conifer forests and mixed forests of northeastern Maine show the longest return intervals (Lorimer 1977; Wein and Moore 1977), whereas the eastern boreal forest, northern hardwoods of central New England, spruce/hemlock/pine types of eastern Canada, and the aspen/birch and birch/maple/hemlock types of the Lake States have intermediate return intervals of 50–350 years (Oliver and Larson 1996).

Various authors have emphasized the autecological characteristics of plant species in examining the response of plants to fire. We would expect plant species to have developed a wide range of adaptations, given the historical role of fire in eastern forests. Rowe (1983) cautioned, however, that there is considerable variability in the selection process resulting in only very broad, overlapping strategies among species for coping with fire. Correlations between the historical occurrence of fire and specific life-history adaptations are likely to be weakest within broad regions containing mostly wide-ranging species such as the boreal forest (Rowe 1983).

The effects of fires on vegetation are generally correlated with fire intensity and duration (Rowe 1983; Oliver and Larson 1996). The heat from fires remains principally above the soil surface, except where soils are extremely dry, fuel is very concentrated, or fires burn in underground roots (Oliver and Larson 1996). Depth of heat penetration can be expected to affect regeneration from buried propagules (Moore and Wein 1977). Martin (1955) found that all of the herbaceous and shrub species 2 years after a severe burn survived the fire either as underground stems or dormant seeds. The resprouting plants originated from underground stems buried 2.5 cm in the humus and occurred only in parts of the burn where the humus was not completely consumed. In response to the temporal and spatial variability of the fire regime, species have evolved a variety of strategies for survival or regeneration after fire. Skutch (1929) noted that *Pteridium aquilinum* (L.) Kuhn resprouted from deep-seated rhizomes even where the humus had been burned away on a severely burned site in Maine. He also observed sprouting of *Epilobium angustifolium* L. from the primary and secondary roots within the first year, a phenomenon that has also been documented for *Rubus ideaus* L. (Roberts and Dong 1993). The cryptogams *Marchantia polymorpha* L., *Polytrichum commune* Hedw. and *P. juniperinum* Hedw. were in high abundance on burned sites in Maine (Skutch 1929) and Nova Scotia (Martin 1955). The early establishment of these species on burned sites has been attributed to their widely dispersed spores and preference for high-light environments (Skutch 1929).

Ahlgren (1960) classified plants into three types: (1) species found only on unburned sites, consisting of shade-loving perennials reproducing mostly by shallow or surface rhizomes or bulbs, (2) species found only on burned-over sites, which includes mostly species that reproduce by seed, and (3) species occurring on both burned and unburned sites, including both seed and vegetatively reproduced species. Based on Noble and Slatyer's (1980) modes of persistence, Rowe (1983) proposed a classification of plant strategies in the context of fire. The first division in the classification is based on method of regeneration and reproduction, being either disseminule based or vegetative based. Disseminule-based species were subdivided into *invaders* (highly dispersive, pioneering fugitives with short-lived propagules), *evaders* (species with long-lived propagules stored in the soil or canopy), and *avoiders* (shade-tolerant, late successional species, often with symbiotic requirements). Vegetative-based species were subdivided into *resisters* (shade-intolerant species that can survive low-severity fires in the adult stage) and *endurers* (resprouting species with buried perennating buds). Lyon and Stickney (1976) also provided a classification of plants based on their autecological characteristics. They found that most plant species on site before intense fires in the northern Rocky Mountains survived or reestablished on the burns by virtue of on-site surviving parts and seeds or transport of seeds from adjacent, unburned communities. McLean (1969) found that fire resistance was related to rooting depth.

Another line of research has focused on changes in the physical environment caused by burning in explaining herbaceous-layer response. Ahlgren (1960) attributed differences in species composition on lightly and severely burned sites to seedbed characteristics. He also noted that some species requiring open conditions might be limited by competition with resprouting species on light burns. Martin (1955) noted higher species richness and more luxuriant growth in plots under shade cloth than in open plots 2 years after burning. He attributed the difference to higher moisture in the shaded plots. Gilliam and Christensen (1986) found higher species richness and total cover of the herbaceous layer after burning on infertile sites in the southeastern coastal plain of the United States. This effect was found only after winter burns and disappeared after 2 years. They attributed the result to decreased shading and increased nutrient availability immediately after burning. Indeed, the fertilization effect of burning has been well documented (e.g., Ahlgren 1960; Wilbur and Christensen 1983; Skre et al. 1998), although the possible counterbalancing losses of nutrients to volatilization, leaching, and overland flow must also be considered (MacLean et al. 1983). Gilliam (1991) suggested that some fire-prone ecosystems may depend on fire to maintain the availability of essential resources which would otherwise be growth limiting.

Herbivory by mammals is another understory disturbance that has been frequently documented in the forests of eastern North America (chapter 4, this volume). Because of dramatic increases in populations of white-tailed deer (*Odocoileus virginianus* Zimmermann) and moose (*Alces alces* L.) in specific regions throughout the 1900s, damage from these two animals is often cited. Crawley (1983) stresses that herbivores have both direct and indirect effects on plant species richness and community dynamics. Plants may be affected directly by being eaten to extinction or by being favored if they are unpalatable. Indirect effects operate through altering the relative competitive abilities of plants. Both categories of effects are clearly evident in studies that have documented browsing in eastern forests. As with other types of disturbances, most browsing studies have focused on the tree layer or tree regeneration; effects on the herbaceous layer have been seldom described.

Studies of moose browsing suggest that the effects on the herbaceous layer are mostly indirect. On Isle Royale, Michigan, moose browsing reduced the cover of canopy trees, particularly *Abies balsamea* and *Populus tremuloides* Michx., and promoted a well-developed understory of shrubs and herbaceous species (McInnes et al. 1992). The herbs, in general, were not browsed and became a greater proportion of total community biomass. Thompson et al. (1992) documented reduced densities of *Abies balsamea, Prunus pensylvanica* L.f., *Viburnum trilobum* Marsh., and *Amelanchier* spp., and increased densities of *Kalmia angustifolia* L. in response to moose browsing in Newfoundland. Although the herbaceous layer was not assessed in this study, one would expect the number and cover of herbaceous species to be low in areas where *Kalmia angustifolia* is dominant. Snyder and Janke (1976) found twice as much herbaceous-layer cover on moose-browsed sites as on unbrowsed sites, which they attributed to increased light.

Populations of white-tailed deer began to increase throughout much of eastern North America in the early twentieth century after enactment of numerous game laws (Marquis and Brenneman 1981). Heavy browsing of preferred species by deer has resulted in regeneration failures of many commercially important forest tree species, as well as noticeable changes in understory composition (Marquis and Brenneman 1981; Horsley and Marquis 1983). Preferred species of browse in northwestern Pennsylvania, included *Prunus serotina* (Ehrh.), *P. pensylvanica, Acer rubrum, A. saccharum* (Marsh.), *Fraxinus americana* (L.), *Betula* spp., and *Rubus* spp. (Horsley and Marquis 1983). Species favored in areas that are heavily browsed included *Fagus grandifolia* (Ehrh.), *Acer pensylvanicum* (L.), ferns, grasses, *Solidago* spp., and

Aster spp. (Marquis and Brenneman 1981). Horsley and Marquis (1983) documented an interaction between browsing and interference from ferns and grasses that caused regeneration failures of commercially important tree species in Pennsylvania. Deer browsing on *Rubus* spp. and tree seedlings promoted increases in cover of *Dennstaedtia punctilobula* (Michx.), *Thelypteris novaboracensis* (L.), and *Brachyelytrum erectum* (Schreb.), thereby reducing densities of preferred tree species. Any surviving tree seedlings that grew above the herb layer were browsed by deer. It is likely that richness and cover of other herbaceous species is low in areas with high abundance of the interfering ferns and grasses.

Heinen and Currey (2000) documented changes in tree species composition after browsing by introduced Rocky Mountain elk (*Cervus elaphus nelsonii* L.) and deer in Michigan. *Populus grandidentata* Michx. and *Acer rubrum* decreased in abundance, and *Populus tremuloides* increased in browsed areas. Approximately half of the browsed area was dominated by nontree vegetation, predominantly *Pteridium aquilinum* and *Rubus* spp.

Unlike moose, deer browse directly on herbaceous species. Deer forage on herbs during the spring and summer, then switch to woody browse in the winter months (Balgooyen and Waller 1995). Skinner and Telfer (1974) found that species of the lily family (Liliaceae), principally *Clintonia borealis* (Aiton) Raf. and *Maianthemum canadense* Desf., supplied 29% of the spring diet for white-tailed deer in New Brunswick, Canada. Other commonly browsed herbaceous species included *Gaultheria procumbens* L., *Erythronium americanum* Ker Gawler, and grasses (Poaceae) in the spring and *Cornus canadensis* L. in the fall (Skinner and Telfer 1974). Among woody species, *Acer spicatum* Lam., *Betula allegheniensis*, *Sorbus decora* (Sarg.) C. K. Schneider, and *Taxus canadensis* Marshall were reduced in abundance in areas with high deer densities in northern Wisconsin (Balgooyen and Waller 1995). Several herbaceous species, including *Aralia nudicaulis* L., *Maianthemum canadense*, and *Clintonia borealis*, also had decreased frequency and cover in areas with high densities of deer relative to areas with low densities. Other commonly grazed herbaceous species included *Aster macrophyllus* L., *Habenaria orbiculata* (Pursh) Torr., *Sanguinaria canadensis* L., *Smilacina racemosa* L. (Desf.), *Streptopus roseus* Michx., *Trillium cernuum* L., *T. grandiflorum* (Michx.) Salisb., and *Uvularia sessilifolia* L. (Balgooyen and Waller 1995). R. C. Anderson (1994) found that *Trillium grandiflorum*, *Erythronium americanum*, and *Claytonia virginica* L. were preferentially browsed in Illinois.

In some cases, herbaceous species have been used as indicators of deer browsing intensity. For example, percent cover, number of leaves/plant, scape height, and number of pedicels/umbel of *Clintonia borealis* were negatively correlated with deer densities in the study by Balgooyen and Waller (1995). R. C. Anderson (1994) found the height of *Trillium grandiflorum* to be a useful indicator of deer browsing intensity.

Deer browsing may also produce indirect effects on the herbaceous layer if the shrub, seedling, or sapling layers are heavily browsed. One example of the indirect effects of browsing is provided by Balgooyen and Waller (1995), who found greater herb diversity and cover in areas where *Taxus canadensis* was more heavily browsed.

Anthropogenic Disturbances: Agriculture and Forestry

The combined forces of clearing for agriculture and lumbering transformed eastern North America from a forested to a predominantly agricultural landscape in the two centuries following European settlement (Whitney 1994; chapter 9, this volume). Agriculture is the most severe form of these disturbances and has the greatest effect on the herbaceous layer because of the removal of vegetative propagules and seed banks associated with annual cultivation and the dramatic change in microenvironment (Runkle 1985). Grazing by pigs, sheep, and cattle can also dramatically affect woody and herbaceous vegetation composition in woodlands (Whitney 1994). The effects of forest harvesting depend on a number of factors, including type of harvesting system, equipment used, season of year, site conditions, soil type, subsequent treatments, and others.

Forest management approaches common in the eastern deciduous forest can be viewed as a gradient of disturbance intensity, varying from the least intense with single tree selection to the most intense with clearcutting (Gilliam and Roberts 1995; Hammond et al. 1998). Recent research has looked at the direct effects of harvesting alone, as well as in combination with other management techniques, on herb layer composition and diversity (Reader and Bricker 1992a, 1992b; Halpern and Spies 1995; Roberts and Gilliam 1995b; Elliott et al. 1997; Hammond et al. 1998; Thomas et al. 1999; He and Barclay 2000). Other work has examined herb layer recovery after more extreme treatments not associated with forest management, such as deforestation and regrowth suppression with repeated herbicide applications (Kochenderfer and Wendel 1983; Reiners 1992). Still others have provided valuable information on species dynamics of the herb layer by comparing patterns in second-growth to those in old-growth forest stands (Qian et al. 1997; Goebel et al. 1999; chapter 6, this volume). The lack of consistency in findings of these studies demonstrates the site-specific nature of herb layer responses to anthropogenic disturbances to forests (Roberts and Gilliam 1995b), precluding broad generalizations and underlining the importance of in-depth study of these responses for different forest types and harvest techniques.

Several studies have looked at postharvest responses of herb layer vegetation to clearcutting either alone or in combination with other silvicultural treatments. Halpern and Spies (1995) found that changes in herb layer diversity were short-lived after clearcutting and slash burning of Douglas-fir forests of western Oregon and Washington and that herb diversity returned to preharvest conditions before canopy closure (10–20 years). Roberts and Gilliam (1995b) found that responses of the herb layer to clearcutting were highly site-dependent in mesic and dry-mesic aspen stands. Harvested stands approximately 15 years old had higher diversity and richness than mature stands on mesic sites, whereas there were no significant differences on dry-mesic sites. De Grandpré and Bergeron (1997) found that site dependence of herb response to disturbance might be related to the age of the stand at the time of disturbance in southern boreal forest of Quebec (i.e., younger communities changed less after clearcutting than did more mature communities). Fredericksen et al. (1999) concluded that only the most intense levels of harvesting measurably affected herb layer richness, diversity, composition, and cover in northern hardwood and oak-hickory forests of Pennsylvania.

Several lines of evidence suggest that forest harvesting may have long-term detrimental effects on the composition and diversity of herbaceous plants. Based on direct comparisons of harvested and old-growth stands in southern Appalachian forests, Duffy and Meier (1992) argued that the harvested stands are impoverished even after 150 years. Because of methodological problems, the accuracy of the results of this study has been questioned (Elliott and Loftis 1993; Johnson et al. 1993). Replies to these criticisms (Duffy 1993a, 1993b) and further work (Meier et al. 1995) by these authors failed to resolve the problems (see chapter 6, this volume, for further discussion). Although similar findings have been reported for a variety of European and North American forest communities (e.g., Hill 1979b, Scanlan 1981, Peterken and Game 1984, Whitney and Foster 1988, Dzwonko and Loster 1989, and others reviewed in Matlack 1994a), all have compared second-growth stands or plantations on abandoned agricultural fields to natural forests. It is not clear from these studies what the effects of harvesting on the herbaceous layer would have been independent from the effects of agriculture. Thus, further research is needed to determine the influence of harvesting on herbaceous layer composition and diversity. Long-term studies along gradients of disturbance type and severity would be of particular value.

Evidence from early botanical descriptions and herbarium records suggests that many species have declined in abundance or have become extinct in North America as a result of the combined effects of European settlement. Indeed, the impact of historical land-use on forest floras has been frequently documented (Matlack 1994a); however, the extent to which these changes can be attributed to forest management activities, as distinct from other land uses, remains unknown.

Other studies have also found that the forest understory remains impoverished for long periods (perhaps centuries) after natural disturbances such as wildfire (Spies 1991; MacLean and Wein 1977). Thus, the effects of harvesting must be balanced against the natural patterns of succession (Roberts and Gilliam 1995a). It is not clear whether harvesting would cause more severe effects than natural disturbances, but any differences would likely be related to the intensity of the disturbance. Without doubt, more research is needed to identify patterns and mechanisms of herbaceous layer response to harvesting.

Published results of harvesting effects on herbaceous layer diversity of forest types throughout North America were reviewed by Battles et al. (2001). We have taken a similar approach with a narrower focus on studies in northeastern forests, including two new studies that were not reviewed in

Battles et al. (2001). We present the actual values as calculated in the various studies for species richness (*S*; usually the average number of species per plot) and the Shannon-Wiener index (*H'*; table 13.1). It was not possible to compare results among studies because of differences in methods; rather, we compared the treated condition to the reference condition within each study and noted the direction of change in *S* and *H'* for each case. Only treatments that involved clearcutting (with or without plantation management) were included because clearcutting constitutes the most severe management disturbance (Roberts and Zhu, 2002). Thus, if clearcutting shows no significant effect on species richness or diversity, it is unlikely that a less severe management treatment, such as selection harvest, would show a significant effect in that forest type.

The studies in eastern forests that we reviewed showed diverse patterns, including no differences in *S* or *H'* between clearcuts and controls (Goebel et al. 1999; Yorks and Dabydeen 1999; Gilliam 2002; Roberts, 2002), greater *S* or *H'* in clearcuts than in controls (Jenkins and Parker 1999; Roberts and Zhu 2002), and lower *S* or *H'* in clearcuts than in controls (Meier et al. 1995; Elliott et al. 1997; table 13.1). Roberts and Gilliam (1995b) found that the patterns depended on site type, with no difference on dry-mesic sites and greater *S* and *H'* in clearcuts compared to controls on mesic sites in northern lower Michigan (table 13.1). Neither type of control (second growth vs. old growth) nor age of treated stand appeared to explain any trends in *S* or *H'*. For example, Goebel et al. (1999) compared mature clearcuts (70–79 years) to old-growth reference stands, whereas Yorks and Dabydeen (1999), Gilliam (2002), and Roberts (2002) studied young clearcuts (2–26 years) in relation to mature second-growth reference stands, yet all found no significant differences between treatments and controls.

Of the studies that included a chronosequence of treated stands, conflicting patterns were reported. Yorks and Dabydeen (1999) found no statistically significant differences with stand age but noted that *S* and *H'* peaked in recent clearcuts (3–4 years) and controls (90 years) on the moister northwestern aspects but tended to decrease over time on drier southwestern aspects. Jenkins and Parker (1999) also found no significant differences with stand age, although there was a slight tendency for *H'* to decrease and *S* to increase with age (9–15 year age class vs. 16–24 year age class; table 13.1). In Elliott et al. (1997), two community types on moister sites (cove hardwoods and mixed-oak hardwoods) showed significantly lower *H'* in 1993 than in 1979, but the drier hardwood-pines site did not have significant differences with stand age (table 13.1). Certainly, changes in diversity over time are influenced by site conditions, as noted elsewhere (e.g., Roberts and Gilliam 1995b), along with a complex of additional factors, making generalizations difficult.

Without doubt, more research is needed to identify patterns and mechanisms of herbaceous layer response to harvesting. In the next section, we present a summary of theory that can be applied to this problem.

Table 13.1. Effects of clearcut harvesting on herbaceous-layer species diversity in selected forests of eastern North America

Study and forest type or site type	Type of comparison	Treated			Control			Effect on herbaceous species diversity
		H'	S	Age (years)	H'	S	Age (years)	
Yorks and Dabydeen (1999)	Clearcuts and mature second-growth reference stands							No effect on H'
Frostburg Watershed		3.14[a]		2	3.42		90	
		3.91		13–14				
		3.93		19				
Savage R. (NW aspect)		2.25		2	4.23		75	
		3.63		13				
		3.79		17				
		2.80		26				
Savage R. (SE aspect)		4.49		2	3.26		80	
		3.96		13				
		3.16		26				
Jenkins and Parker (1999)	Clearcuts and mature second-growth reference stands							S and H' significantly greater on mesic slopes
Dry-mesic slopes		2.20	28	9–15	2.40	27	>80	No differences in S or H' with clearcut age; S and H' not different from control
		2.10	30	16–24				
Mesic slopes		2.50	38	9–15	2.40	27	>80	No differences in S or H' with clearcut age; S greater in clearcuts than control but no difference in H'
		2.20	39	16–24				
Gilliam (2002)	Clearcuts and mature second-growth reference stands	1.64	14[b]	20	1.53	13[b]	>70	No differences in S or H'; variations in H' related more to changes in evenness than richness
Goebel et al. (1999)	Second growth compared to old growth	2.12[c]	14	70–79	2.54	19	>150	No significant differences in S or H'

(continued)

Table 13.1. *Continued*

Study and forest type or site type	Type of comparison	Treated			Control			Effect on herbaceous species diversity
		H'	S	Age (years)	H'	S	Age (years)	
Elliott et al. (1997)	Clearcut harvest vs. mature forest before harvest							Sustained decreases in S and H' after harvest compared to precut forest
Cove hardwoods		2.19	22	2	2.52[d]	27	>70[e]	
		0.82	20	16				
Mixed-oak hardwoods		2.04	18	2	3.14	49	>70[e]	
		1.32	16	16				
Hardwood-pines		2.28	25	2	2.40	45	>70[e]	
		1.90	27	16				
Roberts and Zhu (2002)	Clearcut harvest vs. mature forest before harvest							Significantly greater S and H' after harvest compared to precut forest
Clearcut only		0.80[f]	13[b]	2	0.60[f]	12[b]	>80	
Clearcut and planted		0.88[f]	19[b]	2	0.79[f]	16[b]	>80	
Roberts and Gilliam (1995b)	Clearcuts and mature second-growth reference stands							
Mesic sites		3.76	35	3–14	3.34	26	57–82	S and H' significantly greater in young clearcuts
Dry-mesic sites		3.04	24	3–12	2.76	23	55–80	No significant differences in S or H'
Meier et al. (1995)	Clearcut and old-growth reference stand		10[b]	5		14[b]	?	S significantly greater in control
Roberts (2002)	Plantations and mature reference stands	1.73	27	5–16	1.75	24	>90	No significant differences in S or H'

[a] Average of June and August sampling periods.
[b] Mean no. species/plot.
[c] Average of late spring and summer sampling periods.
[d] Preharvest inventory.
[e] Originated from cutting and burning, followed by selective logging.
[f] Mean H'/plot.

Community Patterns and Mechanisms of Successional Change in Understory Communities

Hypothesized Changes in Species Diversity with Succession

Peet and Christensen (1988) hypothesized that diversity would wax and wane during stand development according to the changing intensity of competition (chapter 9, this volume). For example, species richness is initially high after disturbance, then decreases to a minimum during the stem exclusion stage (sensu Oliver and Larson 1996), when competition among the overstory trees is at a maximum. Subsequently, richness should increase again during the transition to the old-growth stage. During the old-growth stage, richness should either decline slightly due to the loss of early successional species or reach a new peak as slowly dispersed climax-specialist species invade. Beta (between-habitat) diversity should increase during the periods of most intense competition because of a decrease in niche breadth induced by competition (Peet and Christensen 1988). Gilliam et al. (1995) suggested that the link between forest strata becomes tighter as competition increases after the establishment phase. We hypothesize that different strata show increasingly similar response to environmental gradients with successional time (chapter 8, this volume). Certainly, patterns of change in herbaceous layer diversity are related to the intensity of the disturbance as determined by the management regime.

Mechanisms of Change in Understory Communities During Succession: Initial Effects

Halpern and Spies (1995) have suggested that we separate disturbance effects into two broad classes: initial effects on existing plant populations and long-term effects on recovering plant populations. We follow this convention because different mechanisms are involved in each case (Roberts and Gilliam 1995a). The *initial effects* influence plant community response through destruction of vegetative stems and propagules, effects on propagule availability and species invasion, and modification of habitats, including seedbed conditions, light, temperature, and moisture. Initial effects could vary widely depending on the type and severity of disturbance. For example, management treatments such as clearcutting with mechanized site preparation influences early vegetation response quite differently from clearcutting alone (Roberts and Zhu, 2002). The type of disturbance in terms of its effects on the vegetation and environment is the important consideration here.

The *long-term effects* are manifested as changes in species composition, rate of stand development, and competitive interactions. Control of competing vegetation with herbicides is a typical example of a management treatment that has all of these long-term effects (e.g., Horsley 1994). Other long-term effects include composition and density of planting, timing and intensity of thinning, and length of harvest rotation (Halpern and Spies 1995). The analogs in natural disturbances are variations in initial stand density and stand structure, as well as the occurrence of minor disturbances during stand development.

One of the most important factors controlling the initial effects of disturbance on the herbaceous understory is the nature of the disturbance. As discussed above, characterizing the disturbance in terms of amount of forest canopy removed, amount of forest floor vegetation, forest floor and soil removed (severity), frequency, and size and shape (Oliver and Larson 1996) captures the processes involved in understory recovery. Classifying disturbances in this fashion eliminates the need to exhaustively assess each individual disturbance event. This approach allows us to accurately categorize each disturbance in terms of its effects on the physical environment and the forest community and to examine understory response to these factors. Many studies have documented the changes in light (e.g., Jackson 1959; Berry 1964), soil temperature and moisture (e.g., Minckler et al. 1973; Fowler 1974; Hungerford and Babbitt 1987; McInnis and Roberts 1995), and nutrient availability (Johnson and Schultz 1999) that occur after disturbance of varying severities. Others have related disturbance severity to differences in vegetation response (Halpern 1988; Roberts and Dong 1993).

Taken in conjunction with disturbance characteristics, the life-history characteristics (principally reproductive and survival strategies) provide the means of predicting initial vegetation response (Roberts and Gilliam 1995a). Various plant classifications based on functional (life-history) characteristics have been proposed, including life forms (Raunkiaer 1934), competitor/stress tolerator/ruderal strategies (Grime 1977, 1979), assembly/response rules (Keddy 1992), vital attributes (Noble and Slatyer 1980; Rowe 1983), functional types (reviewed in T. M. Smith et al. 1997), or subsets of specific life-history traits (Matlack 1994b; McIntyre et al. 1995). Numerous studies have demonstrated the importance of species' life histories in long-term successional community dynamics (e.g., Drury and Nisbet 1973; Sousa 1980; Roberts and Richardson 1985; Peet and Christensen 1988; Halpern 1989; others reviewed in McCook 1994). Different life-history traits have been found to be relevant depending on the nature of disturbance. Malanson (1984) pointed out the fundamental difference in regeneration mechanisms between light (survival in situ or vegetative regeneration) and severe disturbance (invasion). In the case of harvesting disturbance, we would expect to find correlations between species composition and a suite of life-history traits including rooting habit, shade tolerance, ability to reproduce vegetatively, seed storability, seed production, seed size, and dispersal mechanism.

Species in the herbaceous layer may reappear following disturbance via of one or more of four basic mechanisms:

1. Survival in situ. Depending on the disturbance severity, plants may survive in the vegetative form. Given the patchy nature of many types of disturbance and the low severity of others, considerable numbers of individual plants may survive the disturbance.

2. Vegetative regeneration. In situations where above-ground vegetation is damaged or killed, new individuals may reappear by vegetative means. Vegetative regeneration is the primary means of reproduction for many deciduous forest herbs (Bierzychudek 1982a). Rhizome growth rates may approach 1 m/year in some species (Sobey and Barkhouse 1977). Through a combination of these first two mechanisms, early vegetation composition often closely resembles the predisturbance composition (Lyon and Stickney 1976).
3. Regeneration from the seed bank. There is a potential for some species of the herbaceous layer to regenerate from the seed bank. Typically, however, buried seed reserves are modest in coniferous forests and are dominated by early successional species (Archibold 1989). Seed bank composition tends to greatly differ from the above-ground vegetation in most forest types (Pickett and McDonnell 1989).
4. Regeneration by dispersed propagules. Most temperate forest herbs flower and produce seeds regularly, and regeneration from seed may be the most important reproductive mechanism for many of these species (Bierzychudek 1982a; Cain et al. 1998; chapter 5, this volume). Seeds may be dispersed from individuals surviving in adjacent patches within the disturbed area or from nearby undisturbed communities.

Mechanisms of Change in Understory Communities During Succession: Long-Term Effects

Once beyond the initial effects of disturbance and the immediate response of the herbaceous layer, we must consider the long-term influences of the disturbance on the successional recovery of the herbaceous layer. Spies (1991) notes four factors that contribute to successional change:

1. The microclimate becomes more cool and humid during the dry season in old forests and soil moisture may be higher because of litter accumulation.
2. There is an increase in horizontal spatial heterogeneity of resources and environments, relating to the development of canopy gaps in a matrix of closed forest. Some species survive in deep shade but require gaps for long-term survival.
3. An increase in vertical environmental diversity occurs (e.g., increase in height and number of canopy layers). This may be more important for habitat for foliose lichens and epiphytes than for forest herbs.
4. Sensitivity to disturbance (e.g., fire) and slow rates of reestablishment and growth after the disturbance are important. In addition to affecting initial vegetation response, the type and intensity of the disturbance also has long-term consequences by influencing the initial mix of species, propagule availability, environmental conditions, competitive relationships and individual growth rates.

These factors help explain the changes in species richness with stand development hypothesized by Peet and Christensen (1988), Oliver and Larson (1996), and others. These patterns correspond to an invasion of shade-intolerant herbs and shrubs into the surviving community of shade-tolerant forest species soon after disturbance, followed by a decrease in diversity during the stem-exclusion stage of stand development, and then higher diversity in the old-growth stage in which the increases in horizontal and vertical heterogeneity and changes in microclimate are expressed.

A number of factors may modify this idealized pattern of successional change. Site conditions, such as soil moisture and fertility, may exert an overriding control on successional patterns of species diversity as noted by Auclair and Goff (1971), Roberts and Christensen (1988), and Roberts and Gilliam (1995b). Auclair and Goff (1971) hypothesized that a more nearly linear increase in species diversity with increasing stand age would occur on xeric, hydric or infertile sites as compared to mesic, fertile sites. Disturbance-related influences on resource availability may affect the rate and direction of succession, depending on the type and severity of disturbance. Indeed, several resource-mediated models of succession are based on the assumption that resource availability changes as a result of disturbance and continues to change with time (Grime 1977; Tilman 1985; Vitousek 1985). The competitive balance among species can vary as a consequence of differences in disturbance type and severity or chance factors (e.g., seed production, weather) controlling initial plant densities, resulting in different successional trajectories (Oliver and Larson 1996).

Conceptual Model of Herbaceous Layer Response to Disturbance

The factors discussed in the previous section have been incorporated into a general conceptual model that attempts to capture the essential processes controlling response of the herbaceous layer to variations in disturbance severity in both the short term and long term (fig. 13.1). Disturbance affects forest structure in terms of overstory and understory density and cover, presence and relative densities of different vegetation layers (vertical stratification), and abundance of standing snags and coarse, woody debris. These effects depend on disturbance severity (overstory and understory). In turn, forest structure affects the environment and substrate for the herbaceous layer through modifications in microclimate (light intensity, relative humidity, temperature, surface moisture) and substrate (abundance of coarse, woody-debris and stages of decay). Forest structure, of course, changes over successional time, with resulting changes in levels of competition between the herbaceous layer and overtopping vertical strata. Disturbance also affects the forest floor directly by creating or destroying pits and mounds, creating mineral soil substrates, and modifying abundance and condition of coarse, woody-debris substrates. Finally, disturbance affects the preexisting plant community by damaging or killing individual plants and changing propagule availability by

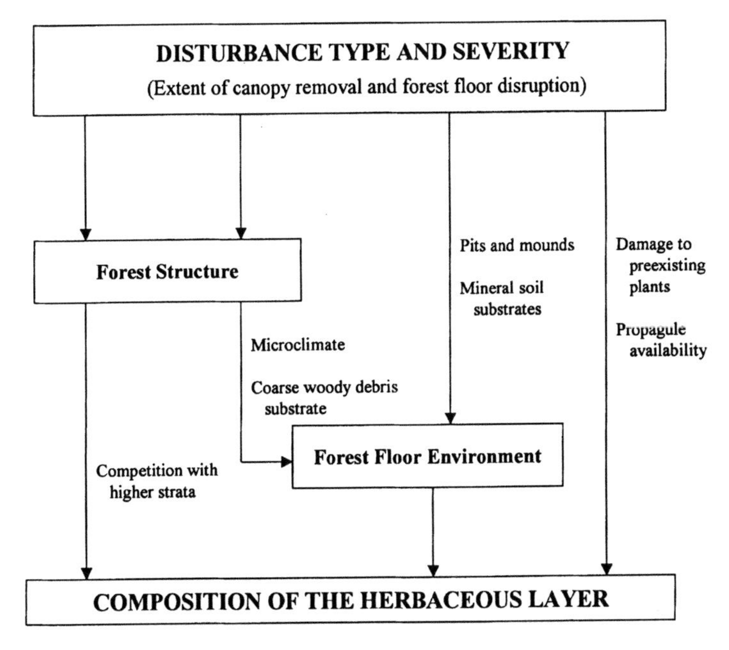

Figure. 13.1 Processes controlling short-term and long-term responses of the herbaceous layer to variations in disturbance severity. After Ramovs (2001).

modifying seed banks and seed rain. Using this framework, we are able to make improved predictions of the effects of disturbance on the herbaceous layer in eastern forests.

Conclusions and Recommendations

Many studies have documented effects of disturbance on eastern forests, but surprisingly few have looked at the herbaceous layer. In assessing the nature of disturbance and its effects on the herbaceous layer, it is useful to characterize disturbance severity in terms of the amount of destruction of both the overstory and understory. Insect defoliation, wind, and ice storms are typical natural disturbances in eastern forests that primarily affect the forest canopy. Their effects on the herbaceous layer and forest floor are expressed largely through changes in environment (increased light) and substrate (additions of litter, frass, and coarse, woody debris to the forest floor and creation of pits and mounds). In contrast to canopy disturbances, fire and herbivory by mammals exert their greatest effects on the understory layers. The herbaceous layer is affected by these disturbances by direct mortality and damage of plants, alteration of propagule availability (reduction or destruction of the seed bank and changes in composition of the seed rain), and changes in competitive relationships among plant species. Forest harvesting and other anthropogenic disturbances can vary greatly in disturbance severity and in the relative effects on canopy and understory, depending on the type of treatment and how it is applied.

Characterizing disturbance in terms of severity at both the canopy and understory levels allows us to examine the effects of disturbance based on its impact on the ecosystem independent of the causal agent. Thus, we are able to focus more clearly on processes that control ecosystem response. This is particularly useful in characterizing the nature of anthropogenic disturbances and evaluating their effects on the ecosystem. With our current emphasis on ecosystem management and the need to pattern forest management after natural disturbance regimes, it is important to be able to compare disturbances on a common basis.

In characterizing the effects of disturbance on the herbaceous layer, it is useful to recognize two general types of effects: direct effects, which include direct damage to preexisting plants and alterations in propagule availability, and indirect effects, comprising changes in microclimate and forest floor substrates, as well as modification of competitive relationships with other plants in all forest strata. In addition, there are initial effects and long-term effects, the former including direct and indirect effects of the disturbance on preexisting plants, propagule availability, microclimate and substrates, and the latter including the development of competitive relationships, propagule availability, microclimate, and substrates over succession time. The interaction of species' life-history characteristics with these disturbance factors determines the direction and outcome of forest succession.

IV

Synthesis

14

The Dynamic Nature of the Herbaceous Layer

Synthesis and Future Directions for Research

Frank S. Gilliam
Mark R. Roberts

Like many other plant ecologists of the past several generations, we have long been impressed and influenced by the work and writings of Henry Chandler Cowles, particularly his turn-of-the-twentieth-century work on the vegetation of the Indiana dunes. With his well-known, succinct description of succession as "a variable approaching a variable, not a constant" (Cowles 1901, p. 81), he articulated perhaps more clearly than did any other ecologist the dynamic nature of plant communities in general. We have adopted such a view in this book, attempting to highlight the spatially and temporally dynamic nature of the herbaceous layer of forests.

One of the reasons for conveying the story of the term *step-overs* as a collective synonym for *herbaceous layer* (chapter 1, this volume) was to emphasize the low regard forest managers commonly and traditionally have had for the lower vascular strata of forests. In this final chapter, we would like to bring back into focus several of the various conclusions brought out in previous chapters to demonstrate the misguided nature of such a view.

Certainly, one of the consequences of ignoring the ecological significance of herb layer species is their demise, particularly as a result of land-use practices. Although it is clear that, relative to shrub, hardwood, and conifer species, herbaceous species have the highest rates of natural extinction (Levin and Levin 2001; chapter 1, this volume), it is less clear what the specific effects of land-use practice are on herb layer biodiversity. The debate concerning the effects of forest harvesting practices on species diversity continues (Noble and Dirzo 1997; Reich et al. 2001) and has been presented in several forms throughout this book (chapters 5, 6, 10, 13, this volume). This is an area worthy of much more intensive research. In this concluding chapter we emphasize this need, along with other areas as future directions for herb layer research.

In addition to attempting to convey new knowledge regarding the ecology of the herbaceous layer of eastern North American forests, a major impetus behind this book was to bring together in a single volume, to the extent possible, what is known of herb layer ecology. Thus, we begin this chapter with a brief synthesis of the major points brought out in each chapter. Rather than simply providing summaries on a chapter-by-chapter basis, we use the parts of the book, namely, "The Environment of the Herbaceous Layer" (part I), "Population Dynamics of the Herbaceous Layer" (part II), "Community Dynamics of the Herbaceous Layer Across Spatial and Temporal Scales" (part III), and "Community Dynamics of the Herbaceous Layer and the Role of Disturbance" (part IV) as a basis for the synthesis.

Synthesis

Research on the forest herbaceous layer has increased dramatically in the past several decades. This increase has been as timely as it has been essential, given the naturally high diversity of the herb layer and the ongoing concern over loss of biodiversity. Indeed, this work has done much to increase our understanding of functional roles of herb species in forest ecosystems.

Plant ecologists with an interest in the ecology of the herb layer should expect to find a wide variety of synonyms for this vegetation stratum in the literature. Vegetation scientists of North America tend to use *herbaceous/herb layer* more than other terms, whereas those of Europe tend to use *ground vegetation* more often. Rather than calling for strict uniformity in use of terminology, we use this opportunity to advise researchers what to expect among published studies, particularly when carrying out literature searches.

Similarly, the literature contains numerous definitions of the herb layer, reflecting the considerable variation in vegetation structure and composition among forest types. Most definitions of the herbaceous layer focus on its physical aspects, especially height, rather than on growth form. Although we offered a commonly used definition of the herb layer as the forest stratum composed of all vascular species that are ≤ 1 m in height, the maximum height limit and exclusion/inclusion of nonvascular plant species varies substantially in the literature, with most height limits generally falling between 0.5 and 1.5 m.

Once again, we do not call for a uniform definition of the herb layer. Vegetation scientists should have the freedom of adapting their definitions in ways that are appropriate for the particular forest type being studied. Because the literature contains several studies that do not provide a clear definition of the herb layer, we ask that researchers state explicitly their working definition of the herb layer and base their definition on the biological and physical structure of the forest system.

We presented a simple conceptual framework for the forest herbaceous layer, comprising two functional groups: resident species and transient species. *Resident species* are those with life-history characteristics that confine them to maximum above-ground heights of no more than about 1.5 m. *Transient species* are those with the potential to develop and emerge into higher strata, and thus their existence in the herb layer is temporary, or transient. Juveniles of overstory species compete as transient species with resident species and either pass through this layer or die.

Thus, as a stratum of forest vegetation, the herb layer is the intimate spatial and temporal coincidence of resident and transient species—two otherwise disparate plant groups. In addition to the more obvious differences between them in growth habit and form, they differ in the factors that determine their distribution, patterns of reproduction, and respective mechanisms of seed dispersal. Whereas wind and vertebrate herbivores are predominant mechanisms for transient species, invertebrates are predominant dispersal vectors for resident species (especially myrmecochory, seed dispersal by ants). Such differences between resident and transient species in the herbaceous layer of forest ecosystems create a forest stratum with impressive spatial and temporal variability.

The Environment of the Herbaceous Layer

The microenvironment of the forest floor, the one that most closely influences plants of the herbaceous layer, provides stark contrast to that of an open field or that above the forest canopy. Forest overstory species alter the quality and/or quantity of virtually all aspects of the environment that are essential to the survivorship and growth of herb layer species. Here we focus primarily on nutrient and light availability.

Foliar concentrations of essential nutrients are generally much higher in herbaceous species than in woody overstory species, with spring ephemeral herbs having particularly high foliar concentrations of nitrogen, emphasizing yet another important contrast between resident and transient species of the herb layer (chapter 2, this volume). Surprisingly, foliar nutrient concentrations exhibit little variation within herb species occupying contrasting forested sites, possibly the result of shifting carbon sinks (i.e., whereas enhanced foliar growth would occur on nutrient-rich sites, increased mycorrhizal support would occur on nutrient-poor sites). There is evidence that herbaceous species are capable of rapid uptake and temporary storage of nutrients during periods of high nutrient availability; this is followed by retranslocation of these nutrients to support growth during periods of active biomass accumulation. Therefore, although herbaceous species do not exhibit luxury uptake of nutrients in the more traditional sense (i.e., as evidenced in intersite variation), they do exhibit seasonal patterns of luxury uptake supporting subsequent growth during periods of limited nutrient availability (chapter 2, this volume).

Clearly, the most spatially and temporally variable component of the environment of the forest floor is light availability. The light environment to which herb layer species are exposed varies at many levels of scale over space and through time. Accordingly, we refer to this environment as a *dynamic mosaic*, for light penetrates the forest canopy to reach the forest floor in a mosaic of discrete patches of varying size (i.e., sunflecks), the size and distribution of which vary at time scales from the diurnal to the seasonal. Time-lapse photography would reveal the constant dance of these sunflecks across the forest floor.

The presence of herbs in the understory of temperate deciduous forests depends greatly on their ability to grow in this dynamic mosaic light environment. In turn, variations in irradiance can influence other microclimatic factors, including temperature, relative humidity, and water availability. Adaptations to the dynamic nature of this environment are expressed physiologically and morphologically at multiple scales—from subcellular to leaf to whole plant (chapter 3, this volume).

There is a selective advantage for herb species that are physiologically active during the warmer portions of the year when temperatures are more favorable for photosynthesis and nutrient uptake. Other selective advantages among forest herbs include adaptations to increase light absorption in the forest understory, such as a well-developed spongy mesophyll to scatter light, reflective lower leaf surface to direct light back into the leaf, and adaxial surface cells that are concavely shaped to direct light toward cells containing chloroplasts (chapter 3, this volume).

Despite the fact that the importance of nutrients in influencing plant growth and survivorship often is considered independent of that of light, the two factors exert their influences simultaneously and synergistically. Thus, although the nutrient environment of the herbaceous layer of deciduous forests may be relatively rich, the light environment may affect herbaceous species' ability to exploit available nutrients (chapter 4, this volume). Severe light limitation may decrease demand for nutrients because nutrient uptake may be limited by accumulation of biomass by individual plants and their specific nutrient saturation points. Efficiency of nutrient resorption may also decrease in herb layer plants in an environment of nutrient availability that is high relative to demand.

Nutrient availability may mediate interactions between herbaceous species and organisms outside the plant kingdom. Mycorrhizal associations are more common when nutrients are limited and may indirectly mediate water stress for some species, although such associations have a high carbon cost to the host plant. Nutrients may also increase susceptibility of herbaceous species to invertebrate or vertebrate herbivores. Finally, natural or anthropogenically altered nutrient environments may shift physiological responses and community patterns in such a way that nutrients reside in the herbaceous layer for shorter periods of time and possibly become more susceptible to being leached from forest ecosystems (chapter 4, this volume).

Population Dynamics of the Herbaceous Layer

An appreciation of the highly variable nature of the environment in which species of the herbaceous layer have evolved is essential to understand the complexity of life-history strategies that governs population dynamics of herb layer species. This is particularly pronounced for forest herbs and their light environment. The distinct seasonality of the light environment has resulted in the evolution of diverse phenological patterns among herb species—spring ephemerals, summer greens, wintergreens, evergreens, heteroptics, and parasitic and saprophytic plants. In addition to their obvious contrasts in temporal variation in growth characteristics, these phenological groups exhibit contrasting reproductive modes.

Because many forest herbs are cryptophytes, vegetative reproduction is commonly considered the predominant reproduction mode for forest herbs. Sexual reproduction, however, plays a major role in the persistence of these species, many, if not most, of which are relatively long lived (e.g., 15–25 years). Still, it is difficult to generalize about breeding biology and mating systems of forest herbs because current estimates of breeding systems are often inaccurate. Pollination is often insect mediated. Population spread can be limited by seed dispersal, with rates of spread often < 1 m per year and rarely > 10 m per year. Vegetative spread may equal or exceed spread by seed dispersal. The role of the seed bank for most taxa is unknown. Most species exhibit some type of seed dormancy (largely physiological or morphological) at dispersal, which can reduce the risk of extinction (chapter 5, this volume).

Numerous vascular taxa of the forests of eastern North America have been identified as endangered and threatened. Demographic studies have been conducted on only a limited number of species. In spite of the usefulness of matrix projection models, few have been used for herbs of forested eastern North America, especially rare taxa. Newer methods, including elasticity and sensitivity analysis and variance decomposition, are potentially useful for predicting population changes through time. The metapopulation approach that uses measures of site occupancy, recruitment, and extinction is potentially useful for community-wide surveys of species in declining habitats, but unfortunately it has seldom been used. Furthermore, adequate demographic data for transition matrices or metapopulation analyses are not widely available for many species.

Despite their importance, quantitative modeling of population sizes using metapopulation dynamics and population viability analysis (PVA) has been underutilized for forest herbs of eastern North America. Indeed, only nine herbaceous species that occur in forests of eastern North America have been used in notable PVA studies (chapter 5, this volume).

After 20 years, the questions originally posed in the seminal paper by Bierzychudek (1982a) are still relevant today and warrant further attention in the future (see section on "Future Research," this chapter): what factors regulate population sizes of forest herbs? How stable are population sizes of forest herbs? How much site-to-site variation occurs in population behavior?

Community Dynamics of the Herbaceous Layer Across Spatial and Temporal Scales

Land-use practices and natural disturbances have created a complex patchwork within the landscape of eastern North America. Beginning at the time of the first European settlers, agriculture and logging throughout this region, once essentially covered with pristine forests, eliminated much of the primeval forest, resulting in limited coverage of what is often called *old-growth forest*. Although this term has been the subject of much debate, consensus is growing as to how to define old-growth forest. Unfortunately, the herbaceous layer is generally not included in these definitions, a serious omission considering the high species diversity of this stratum. Preservation or conservation efforts using functional groups, rather than individual species, within the herb layer should allow better standards to emerge for assessing old-growth status, ultimately assisting with land management decisions (chapter 6, this volume).

Likely because of the logistical demands of field sampling, studies based on one-time samples of the herb layer are prevalent in the literature. Unfortunately, this approach does not lead to an appreciation of the great temporal variability of forest understory communities. A large emphasis on the spring vernal herbs often occurs at the expense of studies of the flora throughout the rest of the growing season. Certainly, we must expand our observations beyond the level of the stand and single-year study to fully understand spatial and temporal patterns in the understory.

Current emphases on sustainable use for forest ecosystems have focused on management questions. In spite of this, old-growth forests will remain an important component in our understanding of the structure and function of forest ecosystems. These will serve as benchmark ecosystems in the heavily disturbed landscape of forests of eastern North America (chapter 6, this volume).

One characteristic of older forest stands, whether true old growth or mature secondary growth, is the high degree of spatial heterogeneity in microtopography (also called *microrelief*) that naturally results when overmature trees die and tip over, creating paired, contrasting conditions of mounds (the vertically displaced root system and associated soil) and pits (the space formerly occupied by the root system). This spatial heterogeneity in microrelief provides a means by which many species are maintained in the community (chapter 7, this volume). Those species with spatial distributions across multiple microsites have the advantage of being buffered against moderate to severe environmental fluctuations. Those species restricted to one microsite, however, suffer a greater variability in population size with environmental fluctuation and have a higher risk of local extinction. For these species, microtopographic heterogeneity may limit available space, placing such species at a disadvantage.

Given the role of spatial heterogeneity and long-term, on-going changes in land use in forest communities of eastern North America, the interaction of these environmental factors should be taken into account in any study of herb layer communities. It has been proposed that an intermediate level of heterogeneity will promote greater species richness in a community (chapter 7, this volume). Extreme microsites may shift the balance to restrict species composition, whereas lack of microsites may increase overall competition with greater species overlap, eliminating safe sites necessary for some species. It has also been suggested that the current level of heterogeneity in a given forest community is likely the result of past events that either minimize or maximize microsites (chapter 7, this volume). For example, plowing and intense compaction from overgrazing eliminate mounds and pits, and the resulting second-growth stand will have a much lower spatial heterogeneity for the understory. In contrast, mature second-growth stands on unplowed land retain microtopography from the original stand and experience new treefall mounds and pits formation with age, maintaining higher spatial heterogeneity. Because these confounding factors can influence community richness, stand age alone is likely a poor predictor of patterns of species richness through succession.

Mound and pit microtopography represents only one of several ways in which the forest overstory can directly and indirectly influence the herbaceous layer. Through competitive interactions, however, the herb layer can, in turn, influence the composition of the overstory. In addition, species of both strata respond to spatial and temporal shifts in a suite of environmental factors. These reciprocating effects and responses to environmental gradients can lead to a measurable spatial correlation of the occurrence of plant species between the overstory and the herb layer. When this condition develops, the strata are said to be *linked*. It has been suggested that such linkage arises from similarities among forest strata in the responses of their respective species to environmental gradients (chapter 8, this volume). These responses may change through secondary forest succession and thus may be a function of stand age.

Although the concept of linkage among vegetation strata of forest communities has been the subject of considerable debate among vegetation scientists in the past, it is gaining wider acceptance and exhibits great potential as a concept with a high degree of ecological importance and practical application. That is, it furthers our understanding of and appreciation for the complexities underlying the structure and function of forest ecosystems, e.g., responses to disturbance and mechanisms of secondary succession. In addition, it may be applied toward landscape-level investigations of forest cover types and remote sensing.

A particularly intense disturbance that severely disrupts overstory–herbaceous layer interactions is that involving clearing of forests for agricultural practices, the agricultural practices themselves, and abandonment of agricultural fields. The response of vegetation following this abandonment is called *old-field succession*. In the Piedmont of North Carolina, where old-field succession has been studied in greatest detail, the sequence begins with a complex assemblage of herbaceous species and ends with pine and, finally, hardwood dominance. Patterns associated with this response are best understood as a consequence of individualistic responses of species related to their ability and opportunities to disperse to and compete at particular sites (chapter 9, this volume). Although a few of the individuals of woody plant species that will dominate late in succession may arrive early and simply outlive pioneers, virtually none of the herb species common in pine and hardwood stands is found in old fields initially after abandonment.

A significant proportion of the variability in herbaceous species composition of old fields is correlated with soil variables, regardless of successional stage. At the landscape scale, soil moisture conditions, a function of topography and proximity to streams, accounts for much of the variation. Soil chemistry is also highly correlated with herb species distributions and overall species richness. The correlation between soil site variables and species composition diminishes sharply among late-stage pine stands, a decline that has been suggested to be a consequence of changes in stand structure (e.g., creation of canopy gaps from increased pine mortality) that increase variability of the light environment (chapter 9, this volume).

Temporal change in species richness during old-field succession is dependent on spatial scale. At the spatial scale of a 0.1-ha sample, mean herbaceous species richness is remarkably similar (i.e., 50 species) among successional ages. However, when comparing among ages the total list of species encountered across the full range of environments, species richness appears to increase over time. The key difference among these ages is in the numbers of species occurring in only a few stands at relatively low abundance, coinciding with the prediction that low-abundance, stress-tolerant species should become more abundant in later stages of succession.

Although it is clear that changes in the relative availability of resources are important in explaining successional patterns, much effort to explain variability in the distribution of herbs on successional landscapes has focused on the importance of competitive interactions in the context of changes in resource availability. If such interactions were the only factors shaping temporal and spatial variability of herb species, we should expect high correlations between compositional variations and patterns of environmental variation. Although such correlations do explain significant amounts of variation in composition, often the majority of such variation remains unexplained. Other mechanisms are likely equally important, precluding the creation of a unified theory of change (chapter 9, this volume).

The combination of forest clearing and agricultural practices that ultimately leads to old-field succession represents an extremely intense form of disturbance. Forests of the boreal region (often called taiga) also experience intense, often frequent, disturbances of a very different nature—fire and forest management practices. It is not surprising that variability in disturbance regime influences the distribution of herbaceous layer species of boreal forests (chapter 10, this volume).

Whereas high fire frequencies favor the presence of ericaceous species and terricolous lichens, lower frequencies are associated with higher richness of herbs and shrubs species. Because boreal species have evolved to persist in a context of fire, their spatial distribution is closely linked to fire frequency. Thus, the apparent co-occurrence of overstory cover types and herb layer species may only reflect similar responses to a particular disturbance regime. Although the canopy cover can modify the abiotic conditions of a stand, thus affecting the composition of the understory, many understory species are not restricted to a specific canopy, but rather to specific abiotic conditions (chapter 10, this volume). Herbaceous layer diversity at the boreal landscape level is generally highest with intermediate frequency of fire, independent of cover and site type, supporting the intermediate disturbance hypothesis.

It is clear that herb layer species of boreal forests have developed different reproductive strategies to persist under particular fire regimes. It is not surprising, then, that patterns of herb layer richness and abundance will be directly affected by disturbance characteristics. Accordingly, when the predominant disturbance regime shifts from fire to forest harvesting, herb species are exposed to a very different disturbance regime for which they lack particular adaptive traits. Clearcutting does not necessarily have a direct negative impact on the diversity of vascular herb species; however, it does change the relationships among species in boreal communities and contributes to change in community composition. Clearcutting has been observed to greatly alter the successional processes under certain abiotic conditions. Herbaceous layer communities are not only part of the rich biodiversity of boreal forests, they can also play important roles in boreal forest dynamics (chapter 10, this volume).

Community Dynamics of the Herbaceous Layer and the Role of Disturbance

Although the study of community dynamics and response of the herbaceous layer to disturbance have long been popular and important avenues of inquiry, the appreciation of the importance of *novel species* (nonindigenous species) in the forests of eastern North America is more recent (chapter 12, this volume). The presence of novel species in eastern forests is increasingly being noted, and we have come to realize that temperate forests may be more susceptible to invasions of novel species than other types of communities. Indeed, different regions of eastern forests support different novel species, with species often emerging as invasive only in subsections of their range.

The impacts of novel species can be better understood when we categorize them by their degree of invasiveness rather than painting them all with a broad brush as invaders (e.g., capable of high impacts). Novel species may not be fundamentally different from indigenous species in terms of their ability to colonize. Although invaders of forest communities share traits of colonizers of successional communities, they must also be adapted to low light. In addition, some are either less susceptible or more resilient to herbivory, or both, compared to native species. The ability to invade may be linked more closely with phenotypic plasticity than with a single trait.

Invaders are most likely to appear in forest habitats that have higher light availability, such as light gaps or along forest edges, and that are proximate to a seed source. Invasion susceptibility should increase in areas with higher degrees of anthropogenic disturbance, especially those close to populated areas. Using our categorization of disturbance severity as the degree of disruption of the overstory and understory/forest floor (chapter 13, this volume), we would anticipate greater invasions in environments where both understory and overstory disturbance is severe.

Novel species may be competitively superior to indigenous species and better able to respond to disturbance. In the long term, novel species can change the composition of the bank of potential colonizers, thereby reducing the ability of indigenous species to respond. The implication for management is that removal of novel species may lead to further invasions. Thus, novel species add another dimension to the process of community recovery following disturbance, as depicted in figure 13.1. The influence of novel species may result in a very different postdisturbance composition of the herbaceous layer. As a result, invasion by novel species may be not only an immediate response to disturbance, but also a chronic disturbance itself.

Past management of novel species has been based on the assumptions that novel species are undesirable and that novel species must have negative impacts on preserved forests. In reality, gains and losses of species over time in preserves is common. Management actions may be justified only where management goals are explicit, impacts of novel species clearly compromise management goals, and the impacts of management activities are known.

Identifying the nature of understory—overstory interactions is one of the keys to understanding herbaceous layer dynamics; indeed, it has been one of the main themes of this book. Again, using the terminology set forth in the opening chapter and summarized at the beginning of this chapter, transient species influence the microclimate and competitive relationships in the understory and control the composition and spatial distribution of understory plants. There may also be an element of passive linkage among understory and overstory canopy strata which results from similar responses of different strata to the physical environment (Chapter 8, this volume). Another important form of interaction is that resident species can exert control over the germination, survival, and growth of transient species. Indeed, the influence of herbaceous plants on tree seedlings can be highly selective (i.e., species-specific), with long-term consequences for the composition of tree species (chapter 11, this volume).

The microenvironment under the herbaceous layer is characterized by lower light, more litter (under some ferns and shrub species), and in some cases increased seed predation. These conditions decrease seed germination for some tree species, as well as decreasing seedling survival and reducing overall seedling density. Shifts in tree species composition occur under some herbaceous layer species, such as ferns and shrubs. Species-selective reductions in seedling growth changes the competitive relationships among the trees.

Different understory species alter the microenvironment below their canopies to varying degrees, depending on leaf morphology and area, clonal density, phenology, stature (height), litter quantity and quality, and belowground resource capture. The clonal growth habit of many understory species and the segregation of understory species in different microsites leads to spatial heterogeneity in the intensity of the selective filtering influence on tree seedlings. The understory mosaic, then, can result in the aggregation of seedlings of different tree species into patches. Once these distinct patches of transient species emerge from the herbaceous layer and become the overtopping canopy, they can be expected to have their own reciprocating influence on the herbaceous layer through their effects on forest structure and the forest floor microenvironment (fig. 13.1). Thus, extending the concept of the selective filter (chapter 11, this volume) to multiple canopy layers, we would expect the resident species of the herbaceous layer and the transient species to exert selective filtering effects on each other in turn as dominance shifts from one group of species to the other.

After canopy disturbance, understory plants may increase in cover, compensating at least partially for canopy opening below them. Other patterns of change in understory canopy density are also possible, including no change or a decrease in density, depending on the nature of the disturbance and other factors. For example, the major changes after gap formation are often in species abundance rather than species presence. Disturbance to the litter layer, however, can promote the establishment of new species. In the final analysis, the extent of the change in the filtering activity of the understory layer depends on the degree to which understory plants preempt, neutralize, and/or modify available resources.

Herbaceous layer dynamics in response to disturbance has been another main theme of this book. Novel species and the influence of the understory layer on tree seedlings are intimately tied to the process of disturbance dynamics. Disturbance severity, in terms of the degree of destruction of both the overstory and understory layers (including the litter and surface soil layers) exerts control over the response of the herbaceous layer, including the invasion of novel species and the selective filtering effect of the understory layer on tree seedlings (chapters 11 and 12, this volume). We have emphasized the importance of evaluating disturbance in terms of these two axes of severity in addition to disturbance frequency, size, and shape.

It is important to consider the reciprocal effects of the overstory and understory layers on each other, as noted above, in assessing herbaceous layer dynamics after disturbance. The predisturbance canopy structure and presence of a subcanopy of shade-tolerant tree or shrub species have important effects on herbaceous-layer response. At the same time, the selective filtering effect of the understory on tree seedlings leaves its mark over time on the composition and structure of the canopy above the herbaceous layer.

The life-history characteristics of species interact with disturbance characteristics to determine herbaceous layer response. For example, whether the disturbance primarily affects the overstory canopy or the understory, along with the severity of those effects in each canopy layer, will control in situ survival, vegetative regeneration, regeneration from the seed bank, or regeneration from dispersed propagules. In addition to the direct effects of disturbance on the herbaceous layer, there are also indirect effects, such as herbivory on trees and shrubs, that influence the microenvironment for the understory. Most forms of agriculture constitute the extreme in disturbance severity for the herbaceous layer because of the outright destruction of preexisting plants and the removal of propagules of forest species. Forestry practices typically engender less dramatic changes in the herbaceous layer than do agricultural practices, although treatments that severely disturb both the canopy and understory (e.g., whole-tree harvesting with heavy mechanical site preparation) can greatly modify the herbaceous layer. Our knowledge of the long-term effects of forestry practices on the herbaceous layer is limited by the lack of long-term studies addressing forestry practices in isolation from other anthropogenic disturbances such as agriculture.

Although we tend to think of disturbances in terms of their initial effects only, it is important to consider the long-term effects of the disturbance on herbaceous-layer recovery. Our ultimate goal is to predict herbaceous-layer response over the long term to disturbances of both natural and anthropogenic sources and to apply this knowledge to the wise conservation and management of the forests of eastern North America.

Future Research

Before putting together this final synthesis, we asked chapter authors the following question: what do you see as important areas of future research for better understanding of the ecology of the herbaceous layer? As with the synthesis, we have organized these responses by the major sections of the book instead of organizing them by individual chapters. In contrast to the synthesis, however, we are presenting these areas for future research without additional narrative. Rather, we present them as separate, individual topics within the general sections of the book.

The Environment of the Herbaceous Layer

- Role of high potassium concentrations in physiology of herbaceous leaves, especially water relations.
- Phenological (seasonal) patterns of nutrient uptake in relation to external availability and internal demand.
- Role of herbaceous layer in altering throughfall chemistry via either additions of leached nutrients or removal of absorbed nutrients.
- Detailed analysis of the phenological patterns of nutrient absorption and

release by vernal herbs in comparison to the same patterns in the soil microbial community.
- Potential role of spring ephemerals in promoting nutrient loss by stimulating decomposition processes.
- Morphological, biochemical, and physiological processes involved in the response of herbs to the dynamic light environment of the deciduous forest understory.
- Water relations of understory herbs across the phenological strategies.
- Contribution of understory herbs to forest community-level carbon gain, production, and CO_2/H_2O exchange with the atmosphere.
- Influence of herbivory on light—nutrient interactions in herbaceous layer species.

Population Dynamics of the Herbaceous Layer

- Better understanding of the basic life-history characteristics of herbaceous layer species.
- Long-term demographic monitoring and quantifying difficult phases of the life cycle (e.g., dispersal and the seed bank) of rare taxa of concern, notably endangered and threatened species.
- Collection of data for analysis with and empirical tests of demographic models, including transition matrices and population projections, metapopulation dynamics, elasticity and loop analysis, and PVA.
- Analysis of the spatial structure of populations to allow scaling up from individual- and population-level foci to landscape-level inferences.
- Synthetic approaches to population biology that combine the development of theoretical, empirical, and experimental tests of models.
- Development of user-friendly computer software to apply such models.
- Metapopulation studies of herb layer populations of ecosystems controlled by large disturbances and submitted to fragmentation.
- Use of new genetic markers to more fully understand the genetic structure of herb layer populations.

Community Dynamics of the Herbaceous Layer Across Spatial and Temporal Scales

- Assessment of patterns of herbaceous layer diversity along chronosequences that include old growth.
- Determination of sampling methods and sample sizes required to encounter a greater percentage of the complete flora of the herbaceous layer, including observations beyond the level of the stand and single-year study, as well as long-term permanent plot studies arranged in a stratified fashion throughout the landscape.
- Improving understanding of the relationships between diversity and stability of natural ecosystems in the context of their invasibility by non-indigenous plant species.

- Testing the intermediate heterogeneity hypothesis, which suggests that an intermediate level of spatial heterogeneity in microtopography will promote greater species richness in an herb layer community.
- Testing the heterogeneity cycle hypothesis, which suggests that the current level of heterogeneity in a community is likely the result of past events that either create/enhance the number of microsites or destroy/obliterate microsites.
- Testing the linkage hypothesis, which states that linkage among forest strata arises from parallel responses of strata to similar environmental gradients, using more North American forests, including varying stand types, stand ages, land-use histories, and levels of β diversity.
- Assessment of the relative importance of historical effects versus interspecific competition among mature plants in explaining secondary succession.

Community Dynamics of the Herbaceous Layer and the Role of Disturbance

- Determination of effects of the understory on future forest composition and structure in the context of both predisturbance and postdisturbance conditions.
- Determination of functional guilds in the herb layer and how are these guilds modified by landscape configuration and historical disturbances.
- Elucidation of the mechanisms by which invaders of the herb layer overcome light limitation.
- Further study of herbaceous layer response to the principal types of natural and anthropogenic disturbances in eastern forests.
- Development of new methods of characterizing disturbance that can be used across all disturbance types.
- Long-term studies that address changes in herbaceous layer composition and structure in response to anthropogenic disturbances and along gradients of natural disturbance, including disturbance type and severity.
- Development of mechanistic models of herbaceous layer response to disturbance based on interaction of disturbance type/severity and life-history characteristics of the species.

Summary

It should be clear from this chapter that, to rephrase a well-known line in Alfred Lord Tennyson's *Ulysses* ("Tho' much is taken, much abides"), although much has been learned of the ecological dynamics of the herbaceous layer of forests of eastern North America, much remains to be learned. Our awareness of what lies ahead for future research in herb layer ecology remains in spite of the increasing amount of work that is being done. Some of this awareness, however, is also because of the work being done. That is, just like in any field of scientific endeavor, new knowledge begets new questions.

A great deal of the impetus behind assembling this book was that we believed that the time had come to synthesize our understanding of the basic ecology of the herb layer. One of the implied themes of this book, however, has been the conservation ecology of the herb layer. In this regard, the book clearly enters the realm of applied ecology. Indeed, both basic and applied approaches are relevant to studying and understanding the spatial and temporal dynamics of the herbaceous layer, particularly as they relate to forests of eastern North America.

Conservation ecology is emerging as a prominent ecological subdiscipline. Although it has been broadly defined, it generally focuses on the nature and extent of deviations of anthropogenically altered ecosystems from minimally altered states. As we learned in chapter 6, examples of such minimally altered states (i.e., old-growth stands) are all too infrequent in eastern North America. We see several lessons emerging from this observation.

First, existing old-growth stands of the region must be preserved, both as ecological legacies and as benchmarks for conservation ecologists, especially for studies of the herbaceous layer. Accordingly, we call for the preservation of such areas. Second, the remainder of forest stands of eastern North America represents a variety of responses to a variety of intensities and types of disturbances over many spatial and temporal scales. In short, if there were ever a case of a variable approaching a variable, this would be it. Thus, we remind researchers, particularly those working on the landscape scale, to bear this in mind when conducting their studies.

The highly disturbed nature of eastern North American forests will not change in the future. Rather, projected increases in human populations in the region will place an even greater demand on forested areas. The authors who have contributed to this book have provided ample evidence that responsible use of our forest resources is not necessarily inconsistent with protection of the herbaceous layer. We embrace the concept of sustainable use of natural resources, such as forests, that, by definition, allows continued use of those resources by future generations. Among those uses would be an appreciation for those forests' diminutive vegetation stratum.

References

Abrahamson, W.G. 1980. Demography and vegetative reproduction. Pages 89–106 in O.T. Solbrig, ed., Demography and evolution in plant populations. Berkley: University of California Press.

Abrams, M.D., and D.I. Dickmann. 1982. Early revegetation of clear-cut and burned jack pine sites in northern lower Michigan. Canadian Journal of Botany 60:946–954.

Adams, M.S. 1970. Adaptations of *Aplectrum hyemale* to the environment: effects of preconditioning temperature on net photosynthesis. Bulletin of the Torrey Botanical Club 97:219–224.

Aerts, R. 1996. Nutrient resorption from senescing leaves of perennials: are there general patterns? Journal of Ecology 84:597–608.

Ågren, J., and M.F. Willson. 1992. Determinants of seed production in *Geranium maculatum*. Oecologia 92:177–182.

Ahlgren, C.E. 1960. Some effects of fire on reproduction and growth of vegetation in northeastern Minnesota. Ecology 41:431–445.

Ahlgren, I.F., and C.E. Ahlgren. 1960. Ecological effects of forest fires. The Botanical Review 26:483–533.

Albert, D.A., S.A. Denton, and B.V. Barnes. 1986. Regional landscape ecosystems of Michigan. Ann Arbor: School of Natural Resources, University of Michigan.

Allen, T.R., and S.J. Walsh. 1996. Spatial and compositional pattern of alpine treeline, Glacier National Park, Montana. Photogrammetric Engineering and Remote Sensing 62:1261–1268.

Al-Mufti, M.M., C.L. Sydes, S.B. Farness, J.P. Grime, and S.B. Band. 1977. A quantitative analysis of shoot phenology and dominance in herbaceous vegetation. Journal of Ecology 65:759–791.

Alpert, P., and H.A. Mooney. 1986. Resource sharing among ramets in the clonal herb, *Fragaria chiloensis*. Oecologia 70:227–233.

Alvarez-Buylla, E. R., and M. Slatkin. 1994. Finding confidence limits on population growth rates: three real examples revisited. Ecology 75:255–260.

Alverson, W.S., D.M. Waller, and S.L. Solheim. 1988. Forests too deer: edge effects in northern Wisconsin. Conservation Biology 2:348–358.

Anderson, J.P. Jr., and F.E. Egler. 1988. Patch studies in the stability of non-diversity: *Dennstaedtia, Solidago, Spiraea, Kalmia*. Phytologia 64:349–364.

Anderson, M.C. 1964. Studies of the woodland light climate. I. The photographic computation of light conditions. Journal of Ecology 52:27–41.

Anderson, R.C. 1994. Height of white-flowered Trillium (*Trillium grandiflorum*) as an index of deer browsing intensity. Ecological Applications 4:104–109.

Anderson, R.C., J.S. Fralish, J.E. Armstrong, and P.K. Benjamin. 1993. The ecology and biology of *Panax quinquefolium* L. (Araliaceae) in Illinois. American Midland Naturalist 129:357–372.

Anderson, R.C., S.S. Khillion, and T.M. Kelley. 1996. Aspects of the ecology of an invasive plant, garlic mustard (*Alliaria petiolata*), in central Illinois. Restoration Ecology 4:181–191.

Anderson, R.C., and O. Loucks. 1973. Aspects of the biology of *Tridentalis borealis*. Ecology 54:798–808.

Anderson, R.C., O.L. Loucks, and A.M. Swain. 1969. Herbaceous response to canopy cover, light intensity and through fall precipitation in coniferous forests. Ecology 50:235–263.

Anderson, W.B. 1998. The role of spring ephemeral herbs in deciduous forest nutrient cycling, with special reference to *Claytonia virginica* L. (Portulacaceae). Ph.D. dissertation. Nashville, TN: Vanderbilt University.

Anderson, W.B., and W.G. Eickmeier. 1998. Physiological and morphological responses to shade and nutrient additions of *Claytonia virginica* (Portulacaceae): implications for the "vernal dam" hypothesis. Canadian Journal of Botany 76:1340–1349.

Anderson, W.B., and W.G. Eickmeier. 2000. Nutrient resorption in *Claytonia virginica* L.: implications for deciduous forest nutrient cycling. Canadian Journal of Botany 78:832–839.

Andersson, T. 1992. Significance of foliar nutrient absorption in nutrient-rich low-light environments—as indicated by *Mercurialis perennis*. Flora 187:429–433.

Andersson, T. 1997. Seasonal dynamics of biomass and nutrients in *Hepatica nobilis*. Flora 192:185–195.

Angevine, M.W., and S.N. Handel. 1986. Invasion of forest floor space, clonal architecture, and population-growth in the perennial herb *Clintonia borealis*. Journal of Ecology 74:547–560.

Antonovics, J., and R.B. Primack. 1982. Experimental ecological genetics in Plantago. VI. The demography of seedling transplants of *P. lanceolata*. Journal of Ecology 70:55–75.

Antos, J.A., and D.B. Zobel. 1984. Ecological implications of belowground morphology of 9 coniferous forest herbs. Botanical Gazette 145:508–517.

Archibold, O.W. 1979. Buried viable propagules as a factor in postfire regeneration in northern Saskatchewan. Canadian Journal of Botany 57:54–58.

Archibold, O.W. 1989. Seed banks and vegetation processes in coniferous forests. Pages 107–122 in M.A. Leck, V.T. Parker, and R.L. Simpson, ed., Ecology of soil seed banks. New York: Academic Press.

Ashmun, J. W, and L. F. Pitelka. 1984. Light-induced variation in the growth and dynamics of transplanted ramets of the understory herb, *Aster acuminatus*. Oecologia 64:255–262.

Ashmun, J.W., R.J. Thomas, and L.F. Pitelka. 1982. Translocation of photoassimilates between sister ramets in two rhizomatous forest herbs. Annals of Botany 49: 403–416.

Auclair, A.N., and F.G. Goff. 1971. Diversity relations of upland forests in the western Great Lakes area. American Naturalist 105:499–528.

Auten, J. 1941. Notes on old-growth in Ohio, Indiana, and Illinois. Technical Note 49. Columbus, OH: U.S.D.A. Forest Service.

Axelrod, D.I. 1966. Origin of deciduous and evergreen habits in temperate forests. Evolution 20:1–15.

Baker, T.T., and D.H. van Lear. 1998. Relations between density of rhododendron thickets and diversity of riparian forests. Forest Ecology and Management 109: 21–32.

Baldocchi, D.D., and S. Collineau. 1994. The physical nature of solar radiation in heterogeneous canopies: spatial and temporal attributes. Pages 21–72 in M.M. Caldwell, and R.W. Pearcy, eds., Exploitation of environmental heterogeneity by plants: ecophysiological processes above and belowground. San Diego, CA: Academic Press.

Baldocchi, D.D., B. Hutchison, D. Matt, and R. McMillen. 1984. Seasonal variations in the radiation regime within an oak-hickory forest. Agricultural and Forest Meteorology 33:177–191.

Baldocchi, D.D., S.B. Verma, and D.R. Matt. 1986. Eddy correlation measurements of CO_2 efflux from the floor of a deciduous forest. Journal of Applied Ecology 23: 967–976.

Balgooyen, C.P., and D.M. Waller. 1995. The use of *Clintonia borealis* and other indicators to gauge impacts of white-tailed deer on plant communities in northern Wisconsin, USA. Natural Areas Journal 15:308–318.

Banks, J.O. 1980. The reproductive biology of *Erythronium propullans* Gray and sympatric populations of *E. albidum* Nutt. (Liliaceae). Bulletin of the Torrey Botanical Club 107:181–188.

Banner, I.C. 1998. Leaf folding and photoprotective responses in *Oxalis acetosella* (L.). Ph.D. thesis. Newcastle Upon Tyne: University of Newcastle upon Tyne.

Bao, Y., and E.T. Nilsen. 1988. The ecophysiological significance of leaf movements in *Rhododendron maximum*. Ecology 69:1578–1587.

Barber, S.A. 1995. Soil nutrient bioavailability: a mechanistic approach, 2nd ed. New York: John Wiley & Sons.

Barbour, M.G., J.H. Burk, W.D. Pitts, F.S. Gilliam, and M.W. Schwartz. 1999. Terrestrial plant ecology, 3rd ed. Menlo Park, CA: Benjamin/Cummings.

Bard, G.E. 1945. The mineral nutrient content of the foliage of forest trees on three soil types of varying limestone content. Proceedings of the Soil Science Society of America 10:419–422.

Bard, G.E. 1949. The mineral nutrient content of the annual parts of herbaceous species growing on three New York soil types varying in limestone content. Ecology 30:384–389.

Basinger, M.A., J.S. Huston, R.J. Gates, and P.S. Robertson. 1997. Vascular flora of Horseshoe Lake Conservation Area, Alexander County, Illinois. Castanea 62:82–99.

Baskin, C.C., and J.M. Baskin. 1998. Seeds: ecology, biogeography, and evolution of dormancy and germination. New York: Academic Press.

Baskin, J.M., and C.C. Baskin. 1972. Influence of germination date on survival and seed production in a natural population of *Leavenworthia stylosa*. American Midland Naturalist 88:318–323.

Baskin, J.M., and C.C. Baskin. 1984. The ecological life cycle of *Campanula americana* in northcentral Kentucky. Bulletin of the Torrey Botanical Club 111:329–337.

Baskin, J.M., and C.C. Baskin. 1985. Role of dispersal date and changes in physiological responses in controlling timing germination in achenes of *Geum canadense*. Canadian Journal of Botany 63:1654–1658.

Baskin, J.M., and C.C. Baskin. 1994. Nondeep simple morphophysiological dormancy in seeds of the mesic woodland winter annual *Corydalis flavula* (Fumariaceae). Bulletin of the Torrey Botanical Club 121:40–46.

Baskin, J.M., and C.C. Baskin. 1997. Methods of breaking seed dormancy in the endangered species *Iliamna corei* (Sherff) Sherff (Malvaceae), with special attention to heating. Natural Areas Journal 17:313–323.

Baskin, J. M, K. M. Snyder, J. L. Walck, and C. C. Baskin. 1997. The comparative autecology of endemic, globally-rare, and geographically-widespread, common species: three case studies. The Southwestern Naturalist 42:384–399.

Bastrenta, B., J. D. Lebreton, and J. D. Thompson. 1995. Predicting demographic-change in response to herbivory—a model of the effects of grazing and annual variation on the population dynamics of *Anthyllis-vulneraria*. Journal of Ecology 83:603–611.

Battles, J.J., A.J. Shlisky, R.H. Barrett, R.C. Heald, and B.H. Allen-Diaz. 2001. The effects of forest management on plant species diversity in a Sierran conifer forest. Forest Ecology and Management 146:211–222.

Batzer, H.O., and M.P. Popp. 1985. Forest succession following a spruce budworm outbreak in Minnesota. Forestry Chronicle 61:75–80.

Bauhus, J., and C. Messier. 1999. Soil exploitation strategies of fine roots of different tree species of the southern boreal forest of eastern Canada. Canadian Journal of Forest Research 29:260–273.

Bazzaz, F.A. 1986. Life history of colonizing plants: some demographic, genetic, and physiological features. Pages 96–110 in H.A. Mooney and J.A. Drake, ed., Ecology of biological invasions of North America and Hawaii. New York: Springer-Verlag.

Bazzaz, F.A. 1996. Plants in changing environments. Linking physiological, population, and community ecology. Cambridge: Cambridge University Press.

Bazzaz, F.A., and L.C. Bliss. 1971. Net primary production of herbs in a central Illinois deciduous forest Bulletin of the Torrey Botanical Club 98:90–94.

Bazzaz, F.A., and W.E. Williams. 1991. Atmospheric CO_2 concentrations within a mixed forest: implications for seedling growth. Ecology 72:12–26.

Beattie, A.J., and D.C. Culver. 1981. The guild of myrmecochores in the herbaceous flora of West Virginia forests. Ecology 62:107–115.

Beatty, S.W. 1984. Influence of microtopography and canopy species on spatial patterns of forest understory plants. Ecology 65:1406–1419.

Beatty, S.W. 1987a. Spatial distributions of *Adenostoma* species in southern California chaparral: an analysis of niche separation. Annals of the Association of American Geographers 77:255–264.

Beatty, S.W. 1987b. Origin and role of soil variability in southern California chaparral. Physical Geography 8:1–17.

Beatty, S.W. 1989. Fire effects on soil heterogeneity beneath chamise and redshanks chaparral. Physical Geography 10:44–52.

Beatty, S.W. 1991. Colonization dynamics in a mosaic landscape: the buried seed pool. Journal of Biogeography 18:553–563.

Beatty, S.W., and D.L. Licari. 1992. Invasion of fennel (*Foeniculum vulgare* Mill.) into shrub communities on Santa Cruz Island, California. Madroño 39:54–66.

Beatty, S.W., and O.D.V. Sholes. 1988. Leaf litter effect on plant species composition of deciduous forest treefall pits. Canadian Journal of Forest Research 18:553–559.

Beatty, S.W., and E.L. Stone. 1986. The variety of soil microsites created by treefalls. Canadian Journal of Forest Research 16:539–548.

Beckage, B., J.S. Clark, B.D. Clinton, and B.L. Haines. 2000. A long-term study of tree seedling recruitment in southern Appalachian forests: the effects of canopy gaps and shrub understories. Canadian Journal of Forest Research 30:1617–1631.

Becker, P., and A.P. Smith. 1990. Spatial autocorrelation of solar radiation in a tropical moist forest understory. Agricultural and Forest Meteorology 52:373–379.

Bell, A.D., and P.B. Tomlinson. 1980. Adaptive architecture in rhizomatous plants. Botanical Journal of the Linnean Society 80:125–160.

Bendix, J. 1997. Flood disturbance and the distribution of riparian species diversity. Geographical Review 87:468–483.

Bennington, C.C., and J.B. McGraw. 1995. Natural-selection and ecotypic differentiation in *Impatiens pallida*. Ecological Monographs 65:303–323.

Benton, T.G., and A. Grant. 1999. Elasticity analysis as an important tool in evolutionary and population ecology. Trends in Ecology and Evolution 14:467–471.

Berg, H. 2000. Differential seed dispersal in *Oxalis acetosella*, a cleistogamous perennial herb. Acta Oecologica-International Journal of Ecology 21:109–118.

Berger, A., and K.J. Puettmann. 2000. Overstory composition and stand structure influence herbaceous plant diversity in the mixed aspen forest of northern Minnesota. American Midland Naturalist 143:111–125.

Bergeron, J.-F., J.-P. Saucier, A. Robitaille, and D. Robert. 1992. Québec forest ecological classification program. Forestry Chronicle 68:53–63.

Bergeron, Y. 2000. Species and stand dynamics in the mixed woods of Quebec's southern boreal forest. Ecology 81:1500–1516.

Bergeron, Y., and A. Bouchard. 1984. Use of ecological groups in analysis and classification of plant communities in a section of western Québec. Vegetatio 56:45–63.

Bergeron, Y., and D. Charron. 1994. Postfire stand dynamics in a southern boreal forest (Québec): A dendroecological approach. Écoscience 1:173–184.

Bergeron, Y., and M. Dubuc. 1989. Succession in the southern part of the Canadian boreal forest. Vegetatio 79:51–63.

Bergeron, Y., O. Engelmark, B. Harvey, H. Morin, and L. Sirois. 1998. Key issues in disturbance dynamics in boreal forests: introduction. Journal of Vegetation Science 9:464–468.

Bergeron, Y., and A. Leduc. 1998. Relationships between change in fire frequency and mortality due to spruce budworm outbreak in the southeastern Canadian boreal forest. Journal of Vegetation Science 9:492–500.

Bergeron, Y., and B. Harvey. 1997. Basing silviculture on natural ecosystem dynamics: an approach applied to the southern boreal mixedwood forest of Quebec. Forest Ecology and Management 92:235–242.

Bergeron, Y., B. Harvey, A. Leduc, and S. Gauthier. 1999. Forest management guidelines based on natural disturbance dynamics: stand- and forest-level considerations. Forestry Chronicle 75:49–54.

Berkowitz, A.R., C.D. Canham, and V.R. Kelly. 1995. Competition vs. facilitation of tree seedling growth and survival in early successional communities. Ecology 76:1156–1168.

Berry, A.B. 1964. Effect of strip width on proportion of daily light reaching the ground. Forestry Chronicle 40:130–131.

Bever, J.D. 1994. Feedback between plants and their soil communities in an old field community. Ecology 75:1965–1977.

Bever, J.D., K.M. Westover, and J. Antonovics. 1997. Incorporating the soil community into plant population dynamics: the utility of the feedback approach. Journal of Ecology 85:561–573.

Bevill, R.L., and S.M. Louda. 1999. Comparisons of related rare and common species in the study of plant rarity. Conservation Biology 13:493–498.

Bicknell, S.H. 1979. Pattern and process of plant succession in a revegetating northern hardwood ecosystem. Ph.D. dissertation. New Haven, CT: Yale University.

Bierzychudek, P. 1982a. Life histories and demography of shade-tolerant temperate forest herbs: a review. New Phytologist 90:757–776.

Bierzychudek, P. 1982b. The demography of jack-in-the pulpit, a forest perennial that changes sex. Ecological Monographs 62:335–351.

Bierzychudek, P. 1999. Looking backwards: assessing the projections of a transition matrix model. Ecological Applications 9:1278–1287.

Billings, W.D. 1938. The structure and development of old field short-leaf pine stands and certain associated physical properties of the soil. Ecological Monographs 8: 437–499.

Birch, C.P.D., and M.J. Hutchings. 1994. Exploitation of patchily distributed soil resources by the clonal herb *Glechoma hederacea*. Journal of Ecology 82:653–664.

Bisbee, K.E., S.T. Gower, J.M. Norman, and E.V. Nordheim. 2001. Environmental controls on ground cover species composition and productivity in a boreal black spruce forest. Oecologia 129:261–270.

Björkman, O. 1981. Responses to different quantum flux densities. Pages 57–107 in O.L. Lange, P.S. Nobel, C.B. Osmond, and H. Ziegler, eds., Physiological plant ecology I. Encyclopedia of plant physiology, ns, vol. 12A. Berlin: Springer-Verlag.

Björkman, O., and B. Demmig-Adams. 1995. Regulation of photosynthetic light energy capture, conversion, and dissipation in leaves of higher plants. Pages 17–47, in E.D. Schulze and M.M. Caldwell, eds., Ecophysiology of photosynthesis. Berlin: Springer-Verlag.

Björkman, O., and S.B. Powles. 1981. Leaf movement in the shade species *Oxalis oregana*. I. Response to light level and light quality. Carnegie Institute of Washington Yearbook 80:59–62.

Black, R.A., and L.C. Bliss. 1978. Recovery sequence of *Picea mariana—Vaccinium ulginosum* forests after burning near Inuvik, Northwest Territories, Canada. Canadian Journal of Botany 56:2020–2030.

Blackman, G.E., and A.J. Rutter. 1946. Physiological and ecological studies in the analysis of plant environment. I. The light factor and the distribution of the bluebell (*Scilla non-scripta*) in woodland communities. Annals of Botany 10:361–390.

Blais, J.R. 1983. Trends in the frequency, extent, and severity of spruce budworm outbreaks in eastern Canada. Canadian Journal of Forest Research 13:539–547.

Blank, J.L., R.K. Olson, and P.M. Vitousek. 1980. Nutrient uptake by a diverse spring ephemeral community. Oecologia 47:96–98.

Bloom, A.J., F.S. Chapin, and H.A. Mooney. 1985. Resource limitation in plants—an economic analogy. Annual Review of Ecology and Systematics 16:363–392.

Boardman, N.K. 1977. Comparative photosynthesis of sun and shade plants. Annual Review of Plant Physiology 28:355–377.

Boerner, R.E.J. 1986. Seasonal nutrient dynamics, nutrient resorption, and mycorrhizal infection intensity of two perennial forest herbs. American Journal of Botany 73: 1249–1357.

Bormann, F.H., and G.E. Likens. 1979. Pattern and process in a forested ecosystem. New York: Springer-Verlag.

Bossuyt, B., M. Hermy, and J. Deckers. 1999. Migration of herbaceous plant species across ancient-recent forest ecotones in central Belgium. Journal of Ecology 87: 628–638.

Boufford, D.E., and S.A. Spongberg. 1983. Eastern Asia-Eastern North American phytogeographical relationships—A history from the time of Linnaeus to the twentieth century. Annals of the Missouri Botanical Garden 70:423–439.

Bowersox, T.W., and L.H. McCormick. 1987. Herbaceous communities reduce the juvenile growth of northern red oak, white ash, yellow poplar, but not white pine. Pages 39–43 in R.L. Hay, F.W. Woods, and H. De Selm, eds., Proceedings of the Sixth Central Hardwood Forestry Conference, 24–26 February 1987. Knoxville, TN: University of Tennessee.

Bowles, M.L., and J.L. Mcbride. 1998. Vegetation composition, structure, and chronological change in a decadent midwestern North American savanna remnant. Natural Areas Journal 18:14–27.

Brach, A.R., and D.J. Raynal. 1992. Effects of liming on *Oxalis acetosella* and *Lycopodium lucidulum* in a northern hardwood forest. Journal of Applied Ecology 29:492–500.

Bratton, S.P. 1975. A comparison of the beta diversity functions of the overstory and herbaceous understory of a deciduous forest. Bulletin of the Torrey Botanical Club 102:55–60.

Bratton, S.P. 1976. Resource division in an understory herb community: responses to temporal and microtopographic gradients. American Naturalist 110:679–693.

Bratton, S.P., J.R. Hapeman, and A.R. Mast. 1994. The lower Susquehanna River gorge and floodplain (U.S.A.) as a riparian refugium for vernal, forest-floor herbs. Conservation Biology 8:1069–1077.

Braun, E.L. 1950. Deciduous forests of eastern North America. Philadelphia: Blakiston.

Brewer, R. 1980. A half-century of changes in the herb layer of a climax deciduous forest in Michigan. Journal of Ecology 68:823–832.

Bridgham, S.D., J. Pastor, C.A. McClaugherty, and C.J. Richardson. 1995. Nutrient-use efficiency: A litterfall index, a model, and a test along a nutrient-availability gradient in North Carolina peatlands. American Naturalist 145:1–21.

Brokaw, N.V.L., and S.M. Scheiner. 1989. Species composition in gaps and structure of a tropical forest. Ecology 70:538–541.

Brooks, J.R., L.B. Flanagan, N. Buchman, and J.R. Ehleringer. 1997. Carbon isotope composition of boreal plants: functional grouping of life forms. Oecologia 110: 301–311.

Brothers, T.S., and A. Spingarn. 1992. Forest fragmentation and alien plant invasion of central Indiana old-growth forests. Conservation Biology 6:91–100.

Brown, A.H.F. 1974. Nutrient cycles in oakwood ecosystems in NW England. Pages 141–161 in M.C. Morris and F.H. Perring, eds., The British oak. Farrington, UK: E. W. Classey.

Brown, M.J., and G.G. Parker. 1994. Canopy light transmittance in a chronosequence of mixed-species deciduous forests. Canadian Journal of Forest Research 24:1694–1703.

Brown, M.J., G.G. Parker, and N.E. Posner. 1994. A survey of ultraviolet-B radiation in forests. Journal of Ecology 82:843–854.

Brown, S.E., and G.R. Parker. 1997. Impact of white-tailed deer on forest communities within Brown County State Park, Indiana. Proceedings of the Indiana Academy of Science 106:39–51.

Bruederle, L.P., and F.W. Stearns. 1985. Ice storm damage to a southern Wisconsin mesic forest. Bulletin of the Torrey Botanical Club 112:167–175.

Brumelis, G., and T.J. Carleton. 1989. The vegetation of post-logged black spruce lowlands in central Canada. II. Understorey vegetation. Journal of Applied Ecology 26:321–339.

Brundrett, M., and B. Kendrick. 1990. The roots and mycorrhizas of herbaceous woodland plants. I. Quantatitve aspects of morphology. New Phytologist 114: 457–468.

Buell, M.F., and R.E. Wilbur. 1948. Life form spectra of the hardwood forests of the Itasca Park region, Minnesota. Ecology 29:352–359.

Bullock, J. M., B. C. Hill, and J. Silvertown. 1994. Demography of *Cirsium vulgare* in a grazing experiment. Journal of Ecology 82:101–111.

Burton, P.J., A.C. Balisky, L.P. Coward, S.G. Cumming, and D.D. Kneeshaw. 1992. The value of managing for biodiversity. Forestry Chronicle 68:225–237.

Burton, P.J., and F.A. Bazzaz. 1991. Tree seedling emergence on interactive temperature and moisture gradients in patches of old-field vegetation. American Journal of Botany 78:131–149.

Burton, P.J., and F.A. Bazzaz. 1995. Ecophysiological responses of tree seedlings invading different patches of old-field vegetation. Journal of Ecology 83:99–112.

Byers, D.L., and J.A. Quinn. 1998. Demographic variation in *Alliaria petiolata* (Brassicaceae) in four contrasting habitats. Journal of the Torrey Botanical Society 125:138–149.

Cabin, R.J. 1996. Genetic comparisons of seed bank and seedling populations of a perennial desert mustard, *Lesquerella fendleri*. Evolution 50:1830–1841.

Cain, M.L., and H. Damman. 1997. Clonal growth and ramet performance in the woodland herb, *Asarum canadense*. Journal of Ecology 85:883–887.

Cain, M.L., H. Damman, and A. Muir. 1998. Seed dispersal and the Holocene migration of woodland herbs. Ecological Monographs 68:325–347.

Cain, M.L., D.A. Dudle, and J.P. Evans. 1996. Spatial models of foraging in clonal plant species. American Journal of Botany 83:76–85.

Cain, M.L., B.G. Milligan, and A.E. Strand. 2000. Long-distance seed dispersal in plant populations. American Journal of Botany 87:1217–1227.

Cain, S.A. 1936. Synusiae as a basis for plant sociological fieldwork. American Midland Naturalist 17:665–725.

Cain, S.A. 1950. Life forms and phytoclimate. Botanical Review 16:1–32.

Cajander, A.K., and Y. Ilvessalo. 1921. Uber Waldtypen II. Acta Forestalia Fennica 20:1–77.

Cajander, A.K. 1926. The theory of forest types. Acta Forestalia Fennica 29:1–108.

Camacho-B, S.E., A.E. Hall, and M.R. Kaufmann. 1974. Efficiency and regulation of water transport in some woody and herbaceous species. Plant Physiology 54: 169–172.

Campbell, R.W., and R.J. Sloan. 1977. Forest stand responses to defoliation by the gypsy moth. Forest Science Monograph 19:1–34.

Cane, J.H., and V.J. Tepedino. 2001. Causes and extent of declines among native

North American invertebrate pollinators: detection, evidence, and consequences. Conservation Ecology 5(1):1.

Canham, C.D. 1984. Canopy recruitment in shade tolerant trees: The response of *Acer saccharum* and *Fagus grandifolia* to canopy openings. Ph.D. dissertation. Ithaca, NY: Cornell University.

Canham, C.D. 1985. Suppression and release during canopy recruitment in *Acer saccharum*. Bulletin of the Torrey Botanical Club 112:134–145.

Canham, C.D. 1988. Growth and canopy architecture of shade-tolerant trees: response to canopy gaps. Ecology 69:786–795.

Canham, C.D., J.S. Denslow, W.J. Platt, J.R. Runkle, T.A. Spies, and P.S. White. 1990. Light regimes beneath closed canopies and tree-fall gaps in temperate and tropical forests. Canadian Journal of Forest Research 20:620–631.

Canham, C.D., A.C. Finzi, S.W. Pacala, and D.H. Burbank. 1994. Causes and consequences of resource heterogeneity in forests: interspecific variation in light transmission by canopy trees. Canadian Journal of Forest Research 24:337–349.

Cantlon, J.E. 1953. Vegetation and microclimate of north and south slopes of Cushetunk Mountain, New Jersey. Ecological Monographs 23:241–270.

Carleton, T.J., R.K. Jones, and G.F. Pierpoint. 1985. The prediction of understory vegetation by environmental factors for the purpose of site classification in forestry: an example from northern Ontario using residual ordination analysis. Canadian Journal of Forest Research 15:1099–1108.

Carleton, T.J., and P. MacLellan. 1994. Woody vegetation responses to fire versus clear-cutting logging: a comparative survey in the central Canadian boreal forest. Écoscience 1:141–152.

Carleton, T.J., and P.F. Maycock. 1978. Dynamics of the boreal forest south of James Bay. Canadian Journal of Botany 56: 1157–1173.

Carleton, T.J., and P.J. Maycock. 1980. Vegetation of the boreal forests south of James Bay: non-centered component analysis of the vascular flora. Ecology 61:1199–1212.

Carleton T.J., and P.F. Maycock. 1981. Understorey-canopy affinities in boreal forest vegetation. Canadian Journal of Botany 59:1709–1716.

Carleton, T.J., and S.J. Taylor. 1983. The structure and composition of a wooded urban ravine system. Canadian Journal of Botany 61:1392–1401.

Carlisle, A., A.H.F. Brown, and E.J. White. 1967. The nutrient content of tree stem flow and ground flora litter and leachates in a sessile oak (*Quercus petraea*) woodland. Journal of Ecology 55:615–627.

Carlton, G.C., and F.A. Bazzaz. 1998. Resource congruence and forest regeneration following an experimental hurricane blowdown. Ecology 79:1305–1319.

Carpenter, J.S., and E.W. Chester. 1987. Vascular flora of the Bear Creek Natural Area, Stewart County, Tennessee. Castanea 52:112–128.

Carroll, G. 1995. Forest endophytes—pattern and process. Canadian Journal of Botany 73:1316–1324.

Casper, B.B. 1987. Spatial patterns of seed dispersal and post-dispersal seed predation of *Cryptantha flava* (Boraginaceae). American Journal of Botany 74:1646–1655.

Castelli, J.P., B.B. Casper, J.J. Sullivan, and R.E. Latham. 1999. Early understory succession following catastrophic wind damage in a deciduous forest. Canadian Journal of Forest Research 29:1997–2002.

Caswell, H. 1978. A general formula for the sensitivity of population growth rate to changes in life history parameters. Theoretical Population Biology 14:215–230.

Caswell, H. 2000. Prospective and retrospective perturbation analyses: their roles in conservation biology. Ecology 81:619–627.

Caswell, H., and P.A. Werner. 1978. Transient behavior and life-history analysis of teasel (*Dipsacus sylvestris* Huds). Ecology 59:53–66.

Chabot, B.F., and D.J. Hicks. 1982. The ecology of leaf life spans. Annual Review of Ecology and Systematics 13:229–259.

Chalker-Scott, L. 1999. Environmental significance of anthocyanins in plant stress responses. Photochemistry and Photobiology 70:1–9.

Chapin, F.S. III. 1980. The mineral nutrition of wild plants. Annual Review of Ecology and Systematics 11:233–260.

Chapin, F.S. III, K. Autumn, and F. Pugnaire. 1993. Evolution of suites of traits in response to environmental stress. American Naturalist 142:S78–S92.

Chapin, F.S. III, and K. Van Cleve. 1981. Plant nutrient absorption and retention under differing fire regimes. Pages 301–321 in Fire regimes and ecosystem properties, General Technical Report WO-26. U.S. Department of Agriculture, Forest Service.

Charron, D., and D. Gagnon. 1991. The demography of northern populations of *Panax quinquefolium* (American ginseng). Journal of Ecology 79:431–445.

Chazdon, R.L. 1988. Sunflecks and their importance to forest understory plants. Advances in Ecological Research 18:1–63.

Chazdon, R.L., R.K. Colwell, J.S. Denslow, and M.R. Guariguata. 1998. Statistical methods for estimating species richness of woody regeneration in primary and secondary rain forests of northeastern Costa Rica. Pages 285–309 in F. Dallmeier and J.A. Comiskey, eds., Forest biodiversity research, monitoring and modeling. Washington, DC: Parthenon.

Chazdon, R.L., and N. Fetcher. 1984. Photosynthetic light environments in a lowland tropical rainforest in Costa Rica. Journal of Ecology 72:553–564.

Chazdon, R.L., and R.W. Pearcy. 1986a. Photosynthetic responses to light variation in rainforest species. I. Induction under constant and fluctuating light conditions. Oecologia 69:517–523.

Chazdon, R.L., and R.W. Pearcy. 1986b. Photosynthetic responses to light variation in rainforest species. II. Carbon gain and photosynthetic efficiency during lightflecks. Oecologia 69:524–531.

Chazdon, R.L., and R.W. Pearcy. 1991. The importance of sunflecks for forest understory plants. Bioscience 41:760–766.

Cheng, Z. 1983. A comparative study of the vegetation in Hubei Province, China, and in the Carolinas of the United States. Annals of the Missouri Botanical Garden 70:571–575.

Cheplick, G. P. 1998. Seed dispersal and seedling establishment in grass populations. Pages 84–105 in G. P. Cheplick, ed., Population ecology of grasses. Cambridge: Cambridge University Press.

Christensen, N.L. 1989. Landscape history and ecological succession on the Piedmont of North Carolina. Journal of Forest History 33:116–124.

Christensen, N.L., and R.K. Peet. 1984. Convergence during secondary forest succession. Journal of Ecology 72:25–36.

Cipollini, M.L., D.A. Wallacesenft, and D.F. Whigham. 1994. A model of patch dynamics, seed dispersal, and sex-ratio in the dioecious shrub *Lindera benzoin* (Lauraceae). Journal of Ecology 82:621–633.

Cipollini, M L., D.F. Whigham, and J. O'Neill. 1993. Population growth, structure, and seed dispersal in the understory herb *Cynoglossum virginianum*: a population and patch dynamics model. Plant Species Biology 8:117–129.

Clark, D.B., D.A. Clark, P.M. Rich, S. Weiss, and S.F. Oberbauer. 1996. Landscape-scale evaluation of understory light and canopy structure: methods and application in a neotropical lowland rain forest. Canadian Journal of Forest Research 26:747–757.

Clark, D.B., M.W. Palmer, and D.A. Clark. 1999. Edaphic factors and the landscape-scale distributions of tropical rain forest trees. Ecology 80:2662–2675.

Clark, J.S. 1991. Disturbance and population structure on the shifting mosaic landscape. Ecology 72:1119–1137.

Clark, J.S. 1998. Why trees migrate so fast: confronting theory with dispersal biology and paleorecord. American Naturalist 152:204–224.

Clark, J.S., M. Lewis, and L. Horvath. 2001. Invasion by extremes: population spread with variation in dispersal and reproduction. American Naturalist 157:537–554.

Clarke, R.T., J.A. Thomas, G.W. Elmes, and M.E. Hochberg. 1997. The effects of spatial patterns in habitat quality on community dynamics within a site. Proceedings of the Royal Society of London, Series B-Biological Sciences 264:347–354.

Clay, K. 1984. The effect of the fungus *Atkinsonella hypoxylon* (Clavicipitaceae) on the reproductive system and demography of the grass *Danthonia spicata*. New Phytologist 98:165–175.

Clebsch, E.C., and R.T. Busing. 1989. Secondary succession, gap dynamics, and community structure in southern Appalachian cove forest. Ecology 70:728–735.

Clements, F.E. 1916. Plant succession: an analysis of the development of vegetation. Carnegie Institute of Washington Publication 242.

Clements, F.E. 1936. Nature and structure of the climax. Journal of Ecology 24:252–284.

Clinton, B.D. 1995. Temporal variation in photosynthetically active radiation (PAR) in mesic Southern Appalachian hardwood forests with and without rhododendron understories. Pages 534–540 in Gottschalk, K.W., and S.L.C. Fosbroke, eds., Proceedings of the Tenth Central Hardwood Forest Conference, Morgantown, WV, General Technical Report NE-197. Radnor, Pa: Department of Agriculture, Forest Service, Northeastern Forest Experiment Station.

Clinton, B.D., L.R. Boring, and W.T. Swank. 1994. Regeneration patterns in canopy gaps of mixed-oak forests of the Southern Appalachians: influences of topographic position and evergreen understory. American Midland Naturalist 132:308–319.

Clinton, B.D., and J.M. Vose. 1996. Effects of *Rhododendron maximum* L. on *Acer rubrum* L. seedling establishment. Castanea 6:38–45.

Cochran, M. E., and S. Ellner. 1992. Simple methods for calculating age-based life history parameters for stage-structured populations. Ecological Monographs 62: 345–365.

Cogbill, C.V. 1985. Dynamics of the boreal forests of the Laurentian highlands, Canada. Canadian Journal of Forest Research 15:252–261.

Collins, B.S., K.P. Dunne, and S.T.A. Pickett. 1985. Response of forest herbs to canopy gaps. Pages 218–234 in S.T.A. Pickett and P.S. White, eds., The ecology of natural disturbance and patch dynamics. Orlando, FL: Academic Press.

Collins, B.S., and S.T.A. Pickett. 1987. Influence of canopy opening on the environment and herb layer in a northern hardwoods forest. Vegetatio 70:3–10.

Collins, B.S., and S.T.A. Pickett. 1988. Response of herb layer cover to experimental canopy gaps. American Midland Naturalist 119:282–290.

Collins, S. 1961. Benefits to understory from canopy defoliation by gypsy moth larvae. Ecology 42:836–838.
Colwell, R.K., and J.A. Coddington. 1994. Estimating terrestrial biodiversity through extrapolation. Philosophical Transactions of the Royal Society 345:101–118.
Connell, J.H. 1978. Diversity in tropical rainforests and coral reefs. Science 199:1302–1310.
Connell, J.H. 1980. Diversity and coevolution of competitors, or the ghost of competition past. Oikos 35:131–138.
Connell, J.H. 1989. Some processes affecting the species composition in gaps. Ecology 70:560–562.
Connell, J.H., and R.O. Slatyer. 1977. Mechanisms of succession in natural communities and their role in community stability and organization. American Naturalist 111:1119–1144.
Connell, J.H., and W.P. Sousa. 1983. On the evidence needed to judge ecological stability or persistence. American Naturalist 121:789–824.
Constabel, A.J., and V.J. Lieffers. 1996. Seasonal patterns of light transmission through boreal mixedwood canopies. Canadian Journal of Forest Research 26:1008–1014.
Coomes, D.A., and P.J. Grubb. 2000. Impacts of root competition in forests and woodlands: a theoretical framework and review of experiments. Ecological Monographs 70:171–207.
Cooper-Ellis, S., D.R. Foster, G. Carlton, and A. Lezberg. 1999. Forest response to catastrophic wind: results from an experimental hurricane. Ecology 80:2683–2696.
Coulson, T., G. M. Mace, E. Hudson, and H. Possingham. 2001. The use and abuse of population viability analysis. Trends in Ecology and Evolution 16:219–221.
Cowles, H.C. 1899. The ecological relations of the vegetation on the sand dunes of Lake Michigan. Botanical Gazette 27:95–117, 167–202, 281–308, 361–391.
Cowles, H.C. 1901. The physiographic ecology of Chicago and vicinity: a study of the origin, development, and classification of plant societies. Botanical Gazette 31: 73–108, 145–182.
Cowles, H.C. 1911. The causes of vegetation cycles. Botanical Gazette 51:161–183.
Crafton, WM., and B.W. Wells. 1934. The old field prisere: an ecological study. Journal of the Elisha Mitchell Science Society 49:225–246.
Crawford, R.M.M. 1989. Survival on the forest floor. Pages 159–176 in Studies in plant survival. Oxford: Blackwell Scientific.
Crawley, M. J. 1989. Insect herbivores and plant population dynamics. Annual Review of Entomology 34:531–564.
Crawley, M.J. 1983. Herbivory: the dynamics of animal-plant interactions. Berkeley: University of California Press.
Crick, J.C., and J.P. Grime. 1987. Morphological plasticity and mineral nutrient capture in two herbaceous species of contrasted ecology. New Phytologist 107:403–414.
Critchley, C. 1998. Photoinhibition. Pages 264–272 in A. S. Raghavendra, ed., Photosynthesis: a comprehensive treatise. Cambridge: Cambridge University Press.
Cromack, K., and C.D. Monk. 1975. Litter production, decomposition, and nutrient cycling in a mixed hardwood watershed and a white pine watershed. Pages 609–624 in F.G. Howell, J.B. Gentry, and M.H. Smith, eds., Mineral cycling in southeastern ecosystems, ERDA Symposium Series (CONF-740513).
Crouch, V.E., and M.S. Golden. 1997. Floristics of a bottomland forest and adjacent uplands near the Tombigbee River, Choctaw County, Alabama. Castanea 62:219–238.

Crozier, C.R., and R.E.J. Boerner. 1984. Correlations of understory herb distribution patterns with microhabitats under different tree species in a mixed mesophytic forest. Oecologia 62:337–343.

Cubbedge, A.W., and C.H. Nilon. 1993. Adjacent land use effects on the flood-plain forest bird community of Minnesota valley national wildlife refuge. Natural Areas Journal 13:220–230.

Cui, M., and M.M. Caldwell. 1997. Shading reduces exploitation of soil nitrate and phosphate by *Agropyron desertorum* and *Artemisia tridentata* from soils with patchy and uniform nutrient distributions. Oecologia 109:177–183.

Culver, D.C., and A.J. Beattie. 1980. The fate of *Viola* seeds dispersed by ants. American Journal of Botany 67:710–714.

Currie, W.S., and J.D. Aber. 1997. Modeling leaching as a decomposition process in humid montane forests. Ecology 78:1844–1860.

Curtis, W.F., and D.T. Kincaid. 1984. Leaf conductance responses of *Viola* species from sun and shade habitats. Canadian Journal of Botany 62:1268–1272.

Czaran, T., and S. Bartha. 1992. Spatiotemporal dynamic models of plant populations and communities. Trends in Ecology and Evolution 7:38–42.

Daehler, C.C., and D.R. Strong. 1993. Prediction and biological invasions. Trends in Ecology and Evolution 8:380.

Dalling, J.W., S.P. Hubbell, and K. Silvera. 1998. Seed dispersal, seedling establishment and gap partitioning among tropical pioneer trees. Journal of Ecology 86:674–689.

Damman, H., and M.L. Cain. 1998. Population growth and viability analyses of the clonal woodland shrub, *Asarum canadense*. Journal of Ecology 86:13–26.

Daubenmire, R.F. 1966. Vegetation: identification of typal communities. Science 151: 291–298.

Davis, M.A., J.P. Grime, and K. Thompson. 2000. Fluctuating resources in plant communities: a general theory of invasibility. Journal of Ecology 88:528–534.

Davis, M.A., and K. Thompson. 2000. Eight ways to be a colonizer; two ways to be an invader: a proposed nomenclature scheme for invasion ecology. Bulletin of the Ecological Society of America 81:226–230.

Davis, M.A., K.J. Wrage, and P.B. Reich. 1998. Competition between tree seedlings and herbaceous vegetation: support for a theory of resource supply and demand. Journal of Ecology 86:652–661.

Davis, M.A., K.J. Wrage, P.B. Reich, M.G. Tjoelker, T. Schaeffer, and C. Muermann. 1999. Survival, growth, and photosynthesis of tree seedlings competing with herbaceous vegetation along a multiple resource gradient. Plant Ecology 145: 341–350.

Davison, S.E., and R.T.T. Forman. 1982. Herb and shrub dynamics in a mature oak forest: a thirty-year study. Bulletin of the Torrey Botanical Club 109:64–73.

Day, G.M. 1953. The Indian as an ecological factor in the northeast forest. Ecology 34:329–346.

Day, J.R., and H.P. Possingham. 1995. A stochastic metapopulation model with variability in patch size and position. Theoretical Population Biology 48:333–360.

DeAngelis, D.L., R.H. Gardner, and H.H. Shugart. 1981. Productivity of forest ecosystems studied during the IBP: the woodlands data set. Pages 567–672 in D.E. Reichle, ed., Dynamic properties of forest ecosystems, International Biological Programme 23. Cambridge: Cambridge University Press.

de Castro, F. 2000. Light spectral composition in a tropical forest: measurements and model. Tree Physiology 20:49–56.

DeFerrari, C.M., and R.J. Naiman. 1994. A multi-scale assessment of the occurrence of exotic plants on the Olympic Peninsula, Washington. Journal of Vegetation Science 5:247–258.

de Freitas, C.R., and N.J. Enright. 1995. Microclimatic differences between and within canopy gaps in a temperate rainforest. International Journal of Biometeorology 38:188–193.

De Grandpré, L., and Y. Bergeron. 1997. Diversity and stability of understory communities following disturbance in the southern boreal forest. Journal of Ecology 85:777–784.

De Grandpré, L., D. Gagnon, and Y. Bergeron. 1993. Changes in the understory of Canadian southern boreal forest after fire. Journal of Vegetation Science 4:803–810.

De Grandpré, L., J. Morissette, and S. Gauthier. 2000. Long-term post-fire changes in the northeastern boreal forest of Quebec. Journal of Vegetation Science 11:791–800.

Degreef, J., J.P. Baudoin, and O.J. Rocha. 1997. Case studies on breeding systems and its consequences for germplasm conservation. 2. Demography of wild Lima bean populations in the Central Valley of Costa Rica. Genetic Resources and Crop Evolution 44:429–438.

de Kroon, H., B. Fransen, J.W.A. van Rheenen, A. van Dijk, and R. Kreulen. 1996. High levels of inter-ramet water translocation in two rhizomatous *Carex* species, as quantified by deuterium labelling. Oecologia 106:73–84.

de Kroon, H., A. Plaisier, J. van Groenendael, and H. Caswell. 1986. Elasticity: the relative contribution of demographic parameters to population growth rate. Ecology 67:1427–1431.

de Kroon, H., J. van Groenendael, and J. Ehrlén. 2000. Elasticities: a review of methods and model limitations. Ecology 81:607–618.

Delcourt, H.R., and P.A. Delcourt. 1996. Eastern deciduous forests. Pages 357–395 in M.G. Barbour and D.W. Billings, eds., North American terrestrial vegetation. Cambridge: Cambridge University Press.

DeLuca, T.H., D.R. Keeney, and G.W. McCarty. 1992. Effect of freeze-thaw events on mineralization of soil nitrogen. Biology and Fertility of Soils 14:116–120.

DeLucia, E.H., K. Nelson, T.C. Vogelmann, and W.K. Smith. 1996. Contribution of internal reflectance to light absorption and photosynthesis of shade leaves. Plant, Cell & Environment 19:159–170.

DeLucia, E.H., H.D. Shenoi, S.L. Naidu, and T.A. Day. 1991. Photosynthetic symmetry of sun and shade leaves of different orientations. Oecologia 87:51–57.

DeMars, B., and R.E.J. Boerner. 1997. Foliar phosphorus and nitrogen resorption in three woodland herbs of contrasting phenology. Castanea 62:43–54.

Demmig-Adams, B., W.W. Adams, D.H. Barker, B.A. Logan, D.R. Bowling, and A.S. Verhoeven. 1996. Using chlorophyll fluorescence to assess the fraction of absorbed light allocated to thermal dissipation of excess excitation. Physiologia Plantarum 98:253–264.

Denevan, W.M. 1992. The pristine myth, the landscape of the Americas in 1492. Annals of the Association of American Geographers 34:329–346.

Denslow, J.S. 1980. Patterns of plant species diversity during succession under different disturbance regimes. Oecologia 46:18–21.

Denslow, J.S. 1985. Disturbance-mediated coexistence of species. Pages 309–323 in S.T.A. Pickett and P.S. White, eds., The ecology of natural disturbance and patch dynamics. New York: Academic Press.

Denslow, J.S., E. Newell, and A.M. Ellison. 1991. The effect of understory palms and cyclanths on the growth and survival of *Inga* seedlings. Biotropica 23:225–234.

DePamphilis, C.W., and H.S. Neufeld. 1989. Phenology and ecophysiology of *Aesculus sylvatica*, a vernal understory tree. Canadian Journal of Botany 67:2161–2167.

DePamphilis, C.W., N.D. Young, and A.D. Wolfe. 1997. Evolution of plastid gene rps2 in a lineage of hemiparasitic and holoparasitic plants: many losses of photosynthesis and complex patterns of rate variation. Proceedings of the National Academy of Sciences of the United States of America 94:7367–7372.

De Steven, D. 1991a. Experiments on mechanisms of tree establishment in old-field succession: seedling emergence. Ecology 72:1066–1075.

De Steven, D. 1991b. Expcrimcnts on mechanisms of tree establishment in old-field succession: seedling survival and growth. Ecology 72:1076–1088.

Diamond, J. M. 1984. Historic extinctions: A Rosetta stone for understanding prehistoric extinctions. Pages 824–862 in P. S. Martin and R. F. Klein, eds., Quaternary extinctions: a prehistoric revolution. Tucson: University of Arizona Press.

Díaz, S., and M. Cabido. 1997. Plant functional types and ecosystem function in relation to global change. Journal of Vegetation Science 8:463–474.

Dickens, R.S. 1976. Cherokee prehistory: the Pisgah phase in the Appalachian summit region. Knoxville: University of Tennessee Press.

Dix, R.L., and J.M.A. Swan. 1971. The roles of disturbance and succession in upland forest at Candle Lake, Saskatchewan. Canadian Journal of Botany 49:657–676.

Doak, D. F., and W. Morris. 1999. Detecting population-level consequences of ongoing environmental change without long-term monitoring. Ecology 80:1537–1551.

Doucet, R., M. Pineau, J.-C. Ruel, and G. Sheedy. 1996. Sylviculture appliquée. Pages 965–1004 in J.A. Bérard, and M. Côté, eds., Manuel de Foresterie. Sainte-Foy, Quebec: Presses de l'Université Laval.

Drew, A.P. 1988. Interference of black cherry by ground flora of the Allegheny uplands. Canadian Journal of Forest Research 18:652–656.

Drury, W.H., and I.C.T. Nisbet. 1973. Succession. Journal of the Arnold Arboretum 54:331–368.

Duffy, D.C. 1993a. Letter: Herbs and clearcutting: reply to Elliot and Loftis and Steinbeck. Conservation Biology 7:221–223.

Duffy, D.C. 1993b. Seeing the forest for the trees: response to Johnson et al. Conservation Biology 7:436–439.

Duffy, D.C., and A.J. Meier. 1992. Do Appalachian herbaceous understories ever recover from clearcutting? Conservation Biology 6:196–201.

Duncan, R.P., H.L. Buckley, S.C. Urlich, G.H. Stewart, and J. Geritzlehner. 1998. Small-scale species richness in forest canopy gaps: the role of niche limitation versus the size of the species pool. Journal of Vegetation Science 9:455–460.

Dunn, C.P., G.R. Gutenspergen, and J.R. Dorney. 1983. Catastrophic wind disturbance in an old-growth hemlock-hardwood forest. Canadian Journal of Botany 61:211–217.

Dwyer, L.M., and G. Merriam. 1984. Decomposition of natural litter mixtures in a deciduous forest. Canadian Journal of Botany 62:2340–2344.

Dzwonko, Z., and S. Loster. 1989. Distribution of vascular plant species in small woodlands on the Western Carpathian foothills. Oikos 56:77–86.

Easley, M.C., and W.S. Judd. 1990. Vascular flora of the southern upland property of Paynes Prairie State Preserve, Alachua County, Florida. Castanea 55:142–186.

Easterling, M.R. 1998. Integral projection model: theory, analysis, and application. Ph.D. dissertation. Raleigh: North Carolina State University.
Easterling, M.R., S.P. Ellner, and P.M. Dixon. 2000. Size-specific sensitivity: applying a new structured population model. Ecology 81:694–708.
Ebinger, J., D. O'Connel, S. Turner, F. Catchpole, and W. McClain. 1997. Vegetation survey of Elkhart Woods, Logan County, Illinois. Castanea 62:74–81.
Egler, F.E. 1954. Vegetational science concepts. I. Initial floristic composition, a factor in old-field vegetation development. Vegetatio 4:412–417.
Ehleringer, J.R., and H.A. Mooney. 1978. Leaf hairs: Effects on physiological activity and adaptive value to a desert shrub. Oecologia 37:183–200.
Ehleringer, J.R., and K.S. Werk. 1986. Modifications of solar radiation absorption patterns and implications for carbon gain at the leaf level. Pages 57–82 in T.J. Givnish, ed., On the economy of plant form and function. Cambridge: Cambridge University Press.
Ehrenfeld, J.G. 1980. Understory response to canopy gaps of varying size in a mature oak forest. Bulletin of the Torrey Botanical Club 107:29–41.
Ehrenfeld, J.G. 1997. Invasion of deciduous forest preserves in the New York metropolitan region by Japanese barberry (*Berberis thunbergii* DC.). Journal of the Torrey Botanical Society 124:210–215.
Ehrlén, J. 1995. Demography of the perennial herb *Lathyrus vernus*. II. Herbivory and population dynamics. Journal of Ecology 83:297–308.
Ehrlén, J. 1996. Spatiotemporal variation in predispersal seed predation intensity. Oecologia 108:708–713.
Ehrlén, J., and O. Eriksson. 2000. Dispersal limitation and patch occupancy in forest herbs. Ecology 81:1667–1674.
Eickmeier, W.G., and E.E. Schussler. 1993. Responses of the spring ephemeral *Claytonia virginica* L. to light and nutrient manipulations and implications for the "vernal-dam" hypothesis. Bulletin of the Torrey Botanical Club 120:157–165.
Eliáš, P. 1981. Some ecophysiological leaf characteristics of components of spring synuzium in temperate deciduous forests. Biológia 36:841–849.
Eliáš, P. 1983. Water relations pattern of understory species influenced by sunflecks. Biologia Plantarum 25:68–74.
Eliáš, P., and E. Masarovičová. 1986a. Chlorophyll content in leaves of spring geophytes in two temperate deciduous forest. Biologica (Bratislava) 41:477–485.
Eliáš, P., and E. Masarovičová. 1986b. Seasonal changes in leaf chlorophyll content of *Mercurialis perennis* growing in deciduous and coniferous forests. Photosynthetica 20:181–186.
Elliott, K.J., L.R. Boring, W.T. Swank, and B.R. Haines. 1997. Successional changes in plant species diversity and composition after clearcutting a Southern Appalachian watershed. Forest Ecology and Management 92:67–85.
Elliott, K.J., and D.L. Loftis. 1993. Letter: Vegetation diversity after logging in the southern Appalachians. Conservation Biology 7:220–221.
Elton, C.S. 1958. The ecology of invasions of animals and plants. London: Methuen.
Emery, K. M., P. Beuselinck, and J. T. English. 1999. Evaluation of the population dynamics of the forage legume *Lotus corniculatus* using matrix population models. New Phytologist 144:549–560.
Enright, N., and J. Ogden. 1979. Application of transition matrix models in forest dynamics: *Araucaria* in Papua New Guinea and Nothofagus in New Zealand. Australian Journal of Zoology 4:3–23.
Enright, N.J., R. Marsula, B.B. Lamont, and C. Wissel. 1998. The ecological significance

of canopy seed storage in fire-prone environments: a model for resprouting shrubs. Journal of Ecology 86:960–973.

Eriksson, O. 1988. Ramet behavior and population-growth in the clonal herb *Potentilla anserina*. Journal of Ecology 76:522–536.

Eriksson, O. 1992. Population structure and dynamics of the clonal dwarf-shrub *Linnaea borealis*. Journal of Vegetation Science 3:61–68.

Eriksson, O. 1995. Asynchronous flowering reduces seed predation in the perennial forest herb *Actaea spicata*. Acta Oecologica-International Journal of Ecology 16: 195–203.

Eriksson, O. 1997. Colonization dynamics and relative abundance of three plant species (*Antennaria dioica, Hieracium pilosella* and *Hypochoeris maculata*) in dry semi-natural grassland. Ecography 20:559–568.

Eriksson, O., and J. Ehrlén. 1992. Seed and microsite limitation of recruitment in plant populations. Oecologia 91:360–364.

Eriksson, O., and K. Kiviniemi. 1999. Site occupancy, recruitment and extinction thresholds in grassland plants: an experimental study. Biological Conservation 87:319–325.

Eshleman, K.N., R.P. Morgan II, J.R. Webb, F.A. Deviney, and J.N. Galloway. 1998. Temporal patterns of nitrogen leakage from mid-Appalachian forested watersheds: role of insect defoliation. Water Resources Research 34:2017–2030.

Evans, G.C. 1956. An area survey method of investigating the distribution of light intensity in woodlands, with particular reference to sunflecks. Journal of Ecology 44:391–428.

Evans, J.R. 1986. A quantitative analysis of light distribution between the two photosystems, considering variation in both the relative amounts of the chlorophyll-protein complexes and the spectral quality of light. Photobiochemistry and Photobiophysics 10:135–147.

Evans, J.R. 1996. Carbon dioxide profiles do reflect light absorption profiles in leaves. Australian Journal of Plant Physiology 22:79–84.

Evans, J.R. 1999. Leaf anatomy enables more equal access to light and CO_2 between chloroplasts. New Phytologist 143:93–104.

Evans, L. S., J. H. Adamski, and J. R. Renfro. 1996a. Relationships between cellular injury, visible injury of leaves, and ozone exposure levels for several dicotyledonous plant species at Great Smoky Mountains National Park. Environmental and Experimental Botany 36:229–237.

Evans, L.S., K. Albury, and N. Jennings. 1996b. Relationships between anatomical characteristics and ozone sensitivity of leaves of several herbaceous dicotyledonous plant species at Great Smoky Mountains National Park. Environmental and Experimental Botany 36:413–420.

Evans, M.E.K., R.W. Dolan, E.S. Menges, and D.R. Gordon. 2000. Genetic diversity and reproductive biology in *Warea carteri* (Brassicaceae), a narrowly endemic Florida scrub annual. American Journal of Botany 87:372–381.

Everett, R.L., S.H. Sharrow, and R.O. Meeuwig. 1983. Pinyon-juniper woodland understory distribution patterns and species associations. Bulletin of the Torrey Botanical Club 110:454–463.

Everham, E.M. III, and N.V.L. Brokaw. 1996. Forest damage and recovery from catastrophic wind. The Botanical Review 62:114–185.

Everson, D.A., and D.H. Boucher. 1998. Tree species-richness and topographic complexity along the riparian edge of the Potomac River. Forest Ecology and Management 109:305–314.

Fajvan, M.A., and J.M. Wood. 1996. Stand structure and development after gypsy moth defoliation in the Appalachian Plateau. Forest Ecology and Management 89:79–88.

Falinski, J.B. 1978. Uprooted trees, their distribution and influence in the primeval forest biotope. Vegetatio 38:175–183.

Falk, D.A., and K.H. Holsinger, eds. 1991. Genetics and conservation of rare plants. Oxford: Oxford University Press.

Fangmeier, A. 1989. Effects of open-top fumigations with SO_2, NO_2 and ozone on the native herb layer of a beech forest. Environmental and Experimental Botany 29: 199–213.

Farnsworth, E.J., J. Nunez-Farfan, S.A. Careaga, and F.A. Bazzaz. 1995. Phenology and growth of three temperate forest life forms in response to artificial soil warming. Journal of Ecology 83:967–977.

Ferguson, W.S., and E.R. Armitage. 1944. The chemical composition of Bracken (*Pteridium aquilinum*). Journal of Agricultural Science 34:165–171.

Fike, J., and W.A. Niering. 1999. Four decades of old field vegetation development and the role of *Celastrus orbiculatus* in the northeastern United States. Journal of Vegetation Science 10:483–492.

Fleming, P., and R. Kanal. 1995. Annotated checklist of vascular plants of Rock Creek Park, National Park Service, Washington D.C. Castanea 60:283–316.

Flint, A.L., and S.W. Childs. 1987. Calculation of solar radiation in mountainous terrain. Agricultural and Forest Meteorology 40:233–249.

Ford, W.M., R.H. Odom, P.E. Hale, and B.R. Chapman. 2000. Stand-age, stand characteristics, and landform effects on understory herbaceous communities in southern Appalachian cove-hardwoods. Biological Conservation 93:237–246.

Formann, R.T.T., and M. Godron. 1986. Landscape ecology. New York: John Wiley & Sons.

Foster, D.R. 1985. Vegetation development following fire in *Picea mariana* (black spruce)—*Pleurozium* forests of south-eastern Labrador, Canada. Journal of Ecology 73:517–534.

Foster, D.R. 1988a. Species and stand response to catastrophic wind in central New England, U.S.A. Journal of Ecology 76:135–151.

Foster, D.R. 1988b. Disturbance history, community organization and vegetation dynamics of the old-growth Pisgah Forest, southwestern New Hampshire, U.S.A. Journal of Ecology 76:105–134.

Foster, D.R., and E.R. Boose. 1992. Patterns of forest damage resulting from catastrophic wind in central New England, U.S.A. Journal of Ecology 80:79–98.

Foster, D.R., and G.A. King. 1986. Vegetation pattern and diversity in S.E. Labrador, Canada: *Betula papyrifera* (Birch) forest development in relation to fire history and physiography. Journal of Ecology 74:465–483.

Fowler, W.B. 1974. Microclimate. Pages N-1 to N-18 in O.P. Cramer, ed., Environmental effects of forest residues management in the Pacific Northwest: a state of knowledge compendium, General Technical Report PNW-24. Portland, OR: U.S. Forest Service, Pacific Northwest Forest Experiment Station.

Francis, C.M., M.J.W. Austen, J.M. Bowles, and W.B. Draper. 2000. Assessing floristic quality in Southern Ontario woodlands. Natural Areas Journal 20:66–77.

Franco, M., and J. Silvertown. 1990. Plant demography: what do we know? Evolutionary Trends in Plants 4:74–76.

Franco, M., and J. Silvertown. 1994. On trade-offs, elasticities and the comparative method—a reply to Shea, Rees and Wood. Journal of Ecology 82:958–958.

Franco, M., and J. Silvertown. 1996. Life history variation in plants: an exploration of the fast-slow continuum hypothesis. Philosophical Transactions of the Royal Society of London Series B-Biological Sciences 351:1341–1348.

Frank, E.C., and R. Lee. 1966. Potential solar beam irradiation on slopes: tables for 30° to 50° latitude, Research Paper RM-18. U.S.D.A. Forest Service.

Franklin, J.F., R.E. Jenkins, and R.M. Romancier. 1972. Research natural areas: contributors to environmental quality programs. Journal of Environmental Quality 1:133–139.

Fredericksen, T.S., B.D. Ross, W. Hoffman, M.L. Morrison, J. Beyea, B.N. Johnson, M.B. Lester and E. Ross. 1999. Short-term understory plant community responses to timber-harvesting intensity on non-industrial forestlands in Pennsylvania. Forest Ecology and Management 116:129–139.

Frelich, L.E., and P.B. Reich. 1995. Spatial patterns and succession in a Minnesota southern-boreal forest. Ecological Monographs 65:325–356.

Fritz, R., and G. Merriam. 1994. Fencerow and forest edge vegetation structure in eastern Ontario farmland. Ecoscience 1:160–172.

Fritz-Sheridan, J.K. 1988. Reproductive biology of *Erythronium grandiflorum* varieties *grandiflorum* and *candidum*. American Journal of Botany 75:1–14.

Gagnon, D., and G.E. Bradfield. 1986. Relationships among forest strata and environment in southern coastal British Columbia. Canadian Journal of Forest Research 16:1264–1271.

Gagnon, D., A. LaFond, and L.P. Amiot. 1958. Mineral nutrient content of some forest plant leaves and the humus layer as related to site quality. Canadian Journal of Botany 36:209–220.

Gaiser, R.N. 1951. Random sampling within circular plots by means of polar coordinates. Journal of Forestry 49:916–917.

Garrett, H.E., G.S. Cox, and J.E. Roberts. 1978. Spatial and temporal variations in carbon dioxide concentrations in an oak-hickory forest ravine. Forest Science 24: 180–190.

Garten, C.T. 1976. Correlations between concentrations of elements in plants. Nature 261:686–688.

Garten, C.T. 1978. Multivariate perspectives on the ecology of plant mineral element composition. American Naturalist 112:533–544.

Gates, F.C. 1930. Aspen association in northern lower Michigan. Botanical Gazette 90:233–259.

Gauthier, S., L. De Grandpré, and Y. Bergeron. 2000. Differences in forest composition in two boreal forest ecoregions of Quebec. Journal of Vegetation Science 11:781–790.

Geber, M. A., H. de Kroon, and M. A. Watson. 1997a. Organ preformation in mayapple as a mechanism for historical effects on demography. Journal of Ecology 85:211–223.

Geber, M. A., M. S. Watson, and H. de Kroon. 1997b. Organ preformation, development, and resource allocation in perennials. Pages 113–142 in F. A. Bazzaz and J. Grace, eds., Plant resource allocation. San Diego, CA: Academic Press.

Gehlhausen, S.M., M.W. Schwartz, and C.K. Augspurger. 2000. Vegetation and microclimatic edge effects in two mixed-mesophytic forest fragments. Plant Ecology 147:21–35.

George, L.O. 1996. The understory as an ecological filter. Ph.D. dissertation. Cambridge, MA: Harvard University.

George, L.O., and F.A. Bazzaz. 1999a. The fern understory as an ecological filter: emergence and establishment of canopy-tree seedlings. Ecology 80:833–845.

George, L.O., and F.A. Bazzaz. 1999b. The fern understory as an ecological filter: growth and survival of canopy tree seedlings. Ecology 80:846–856.

Gerloff, G.C., D.D. Morre, and T.T. Curtis. 1964. Mineral content of native plants of Wisconsin. Madison: College of Agriculture, University of Wisconsin. Research Report no. 14.

Germino, M.J., and W.K. Smith. 2000. High resistance to low-temperature photoinhibition in two alpine, snowbank species. Physiologia Plantarum 110:89–95.

Ghent, A.W., D.A. Fraser, and J.B. Thomas. 1957. Studies of regeneration in forest stands devastated by the spruce budworm. I. Evidence of trends in forest succession during the first decade following budworm devastation. Forest Science 3: 184–208.

Gibson, D.J., E.D. Adams, J.S. Ely, D.J. Gustafson, D. McEwan, and T.R. Evans. 2000. Eighteen years of herbaceous layer recovery of a recreation area in a mesic forest. Journal of the Torrey Botanical Society 127:230–239.

Gibson, D.J., J.S. Ely, and S.L. Collins. 1999. The core-satellite species hypothesis provides a theoretical basis for Grime's classification of dominant, subordinate, and transient species. Journal of Ecology 87:1064–1067.

Gildner, B.S., and D.W. Larson. 1992. Seasonal changes in photosynthesis in the desiccation-tolerant fern *Polypodium virginianum*. Oecologia 89:383–389.

Gill, D.S., and P.L. Marks. 1991. Tree and shrub seedling colonization of old fields in central New York. Ecological Monographs 61:183–205.

Gilliam, F.S. 1988. Interactions of fire with nutrients in the herbaceous layer of a nutrient-poor coastal plain forest. Bulletin of the Torrey Botanical Club 115:265–271.

Gilliam, F.S. 1991. The significance of fire in an oligotrophic forest ecosystem. In S.C. Nodvin and T.A. Waldrop, eds., Fire and the environment: ecological and cultural perspectives: Proceedings of an international symposium, 20–24 March 1990, Knoxville, TN. General Technical Report SE-69. Asheville, NC: U.S. Forest Service, Southeastern Forest Experiment Station.

Gilliam, F.S. 2002. Effects of harvesting on herbaceous layer diversity of a central Appalachian hardwood forest. Forest Ecology and Management 155:33–43.

Gilliam, F.S., and M.B. Adams. 1996a. Wetfall deposition and precipitation chemistry for a central Appalachian forest. Journal of the Air and Waste Management Association 46:978–984.

Gilliam, F.S., and M.B. Adams. 1996b. Plant and soil nutrients in young versus mature central Appalachian hardwood stands. Pages 109–118 in K.W. Gottschalk and S.L.C. Fosbroke, eds., 10th Central Hardwood Forest Conference. General Technical Report NE-197. Radnor, PA: U. S. Forest Service, Northeastern Forest Experiment Station.

Gilliam, F.S., M.B. Adams, and B.M. Yurish. 1996. Ecosystem nutrient responses to chronic nitrogen inputs at Fernow Experimental Forest, West Virginia. Canadian Journal of Forest Research 26:196–205.

Gilliam, F.S., and N.L. Christensen. 1986. Herb-layer response to burning in pine flatwoods of the lower Coastal Plain of South Carolina. Bulletin of the Torrey Botanical Club 113:42–45.

Gilliam, F.S., and M.R. Roberts. 1995. Impacts of forest management on plant diversity. Ecological Applications 5:911–912.

Gilliam, F.S., and N.L. Turrill. 1993. Herbaceous layer cover and biomass in a young versus a mature stand of a central Appalachian hardwood forest. Bulletin of the Torrey Botanical Club 120:445–450.

Gilliam, F.S., N.L. Turrill, and M.B. Adams. 1995. Herbaceous-layer and overstory species in clear-cut and mature central Appalachian hardwood forests. Ecological Applications 5:947–955.

Gilliam, F.S., N.L. Turrill, S.D. Aulick, D.K. Evans, and M.B. Adams. 1994. Herbaceous layer and soil response to experimental acidification in a central Appalachian hardwood forest. Journal of Environmental Quality 23:835–844.

Gilliam, F.S., B.M. Yurish, and M.B. Adams. 2001. Temporal and spatial variation of nitrogen transformations in nitrogen-saturated soils of a Central Appalachian hardwood forest. Canadian Journal of Forest Research 31:1768–1785.

Gilliam, F.S., B.M. Yurish, and L.M. Goodwin. 1993. Community composition of an old-growth longleaf pine forest: relationship to soil texture. Bulletin of the Torrey Botanical Club 120:287–294.

Gilpin, M.S. 1991. The genetic effective size of a metapopulation. Biological Journal of the Linnean Society 42:165–175.

Gilpin, M.S., and I. Hanski. 1991. Metapopulation dynamics: empirical and theoretical investigations. London: Academic Press.

Gitzendanner, M.A., and P. S. Soltis. 2000. Patterns of genetic variation in rare and widespread plant congeners. American Journal of Botany 87:783–792.

Givnish, T.J. 1982. On the adaptive significance of leaf height in forest herbs. American Naturalist 120:353–381.

Givnish, T.J. 1986. Biomechanical constraints on crown geometry in forest herbs. Pages 525–583 in T.J. Givnish, ed., On the economy of plant form and function. Cambridge: Cambridge University Press.

Givnish, T.J. 1987. Comparative studies of leaf form: assessing the relative roles of selective pressures and phylogenetic constraints. New Phytologist 106:131–160.

Givnish, T.J., and G.J. Vermeij. 1976. Sizes and shapes of liana leaves. American Naturalist 110:743–778.

Gleason, H. 1926. The individualistic concept of the plant association. Bulletin of the Torrey Botanical Club 53:1–20.

Gleason, H.A., and A. Cronquist. 1991. Manual of vascular plants of the northeastern United States and adjacent Canada, 2nd ed. New York: New York Botanical Garden.

Gliessman, S.R. 1976. Allelopathy in a broad spectrum of environments as illustrated by bracken. Botanical Journal of the Linnean Society 73:95–104.

Glitzenstein, J.S., P.A. Harcombe, and D.R. Streng. 1986. Disturbance, succession, and maintenance of species diversity in an east Texas forest. Ecological Monographs 56:243–258.

Glover, G.R., J.L. Creighton, and D.H. Gjerstad. 1989. Herbaceous weed control increases loblolly pine growth. Journal of Forestry 87:47–50.

Godt, M.J.W., and J.L. Hamrick. 1996. Genetic structure of two endangered pitcher plants, *Sarracenia jonesii* and *Sarracenia oreophila* (Sarraceniaceae). American Journal of Botany 83:1016–1023.

Godt, M.J.W., and J.L. Hamrick. 1998a. Allozyme diversity in the endangered pitcher plant *Sarracenia rubra* ssp. *alabamensis* (Sarraceniaceae) and its close relative *S. rubra* ssp. *rubra*. American Journal of Botany 85:802–810.

Godt, M.J.W., and J.L. Hamrick. 1998b. Low allozyme diversity in *Schwalbea americana* (Scrophulariaceae), an endangered plant species. Journal of Heredity 89:89–93.

Godt, M.J.W., J.L. Hamrick, and S. Bratton. 1995. Genetic diversity in a threatened wetland species, *Helonias bullata* (Liliaceae). Conservation Biology 9:596–604.

Godt, M.J., B.R. Johnson, and J.L. Hamrick. 1996. Genetic diversity and population size in four rare southern Appalachian plant species. Conservation Biology 10: 796–805.

Goebel, P.C., D.M. Hix, and A.M. Olivero. 1999. Seasonal ground-flora patterns and site factor relationships of second-growth and old-growth south-facing forest ecosystems, southeastern Ohio, USA. Natural Areas Journal 19:12–21.

Goldblum, D. 1997. The effects of treefall gaps on understory vegetation in New York, USA. Journal of Vegetation Science 8:125–132.

Goldblum, D. 1998. Regeneration in unmanaged conifer plantations, upstate New York. Northeastern Naturalist 5:343–358.

Goldblum, D., and S.W. Beatty. 1999. Influence of an old field/forest edge on a northeastern United States deciduous forest understory community. Journal of the Torrey Botanical Society 126:335–343.

Gonzalez, V.C. 1972. The ecology of *Hexastylis arifolia*, an evergreen herb in the North Carolina deciduous forest. Ph.D. dissertation. Durham, NC: Duke University (Dissertation Abstracts 33:5246-B).

Gordon, D.R. 1998. Effects of invasive, non-indigenous plants species on ecosystem processes: lessons from Florida. Ecological Applications 8:975–989.

Gosz, J.R., G.E. Likens, and F.H. Bormann. 1973. Nutrient release from decomposing leaf and branch litter in the Hubbard Brook forest, New Hampshire. Ecological Monographs 43:173–191.

Gould, A.M.A., and D.L. Gorchov. 2000. Effects of the exotic invasive shrub *Lonicera maackii* on the survival and fecundity of three species of native annuals. American Midland Naturalist 144:36–50.

Gould, K.S., K.R. Markham, R.H. Smith, and J.J. Goris. 2000. Functional role of anthocyanins in the leaves of *Quintinia serrata* A. Cunn. Journal of Experimental Botany 51:1107–1115.

Gould, K.S., and B.D. Quinn. 1999. Do anthocyanins protect leaves of New Zealand native species from UV-B? New Zealand Journal of Botany 37:175–178.

Grace, J., and D. Tilman. 1990. Perspectives in plant competition. New York: Academic Press.

Grace, J.B., and R.G. Wetzel. 1981. Habitat partitioning and competitive displacement in cattails (*Typha*): experimental field studies. American Naturalist 118:463–474.

Grant, A., and T.G. Benton. 2000. Elasticity analysis for density-dependent populations in stochastic environments. Ecology 81:680–693.

Grant, R.H. 1997. Partitioning of biologically active radiation in plant canopies. International Journal of Meteorology 40:26–40.

Gratani, L. 1997. Canopy structure, vertical radiation profile and photosynthetic function in a *Quercus ilex* evergreen forest. Photosynthetica 33:139–149.

Gravatt, D.A., and C.E. Martin. 1992. Comparative ecophysiology of five species of *Sedum* (Crassulaceae) under well-watered and drought-stressed conditions. Oecologia 92:532–541.

Graves, J.D. 1990. A model of the seasonal pattern of carbon acquisition in two woodland herbs, *Mercurialis perennis* L. and *Geum urbanum* L. Oecologia 83:479–484.

Green, R.E., P.E. Osborne, and E.J. Sears. 1994. The distribution of passerine birds in hedgerows during the breeding-season in relation to characteristics of the hedgerow and adjacent farmland. Journal of Applied Ecology 31:677–692.

Greig-Smith, P. 1952. The use of random and contiguous quadrats in the study of plant communities. Annals of Botany, New Series 16:293–316.

Greller, A.M., D.C. Locke, V. Kilanowski, and G.E. Lotowycz. 1990. Changes in vegetation composition and soil acidity between 1922 and 1985 at a site on the North Shore of Long Island, New York. Bulletin of the Torrey Botanical Club 117:450–458.

Grewal, J.S., and S.N. Singh. 1980. Effect of potassium nutrition on frost damage and yield of potato plants on alluvial soils of the Punjab (Inida). Plant and Soil 57: 105–110.

Griffin, S.R., K. Mavraganis, and C.G. Eckert. 2000. Experimental analysis of protogyny in *Aquilegia canadensis* (Ranunculaceae). American Journal of Botany 87:1246–1256.

Grigal, D.F., and L.F. Ohmann. 1980. Seasonal change in nutrient concentrations in forest herbs. Bulletin of the Torrey Botanical Club 107:47–50.

Grime, J.P. 1977. Evidence for the existence of three primary strategies in plants and its relevance to ecological and evolutionary theory. American Naturalist 11: 1169–1194.

Grime, J.P. 1979. Plant strategies and vegetation processes. Chichester, UK: John Wiley & Sons.

Grime, J.P. 1998. Benefits of plant diversity to ecosystems: immediate, filter and founder effects. Journal of Ecology 86:902–910.

Grimm, N.B., M. Grove, S.T.A. Pickett, and C.L. Redman. 2000. Integrated approaches to long-term studies of urban ecological systems. BioScience 50:571–584.

Grisez, T.J., and M.R. Peace. 1973. Requirements for advance reproduction in Allegheny hardwoods—an interim guide. Research Note NE-180. Upper Durby, PA: USDA Forest Service.

Groffman, P.M., D.R. Zak, S. Christensen, A. Mosier, and J.M. Tiedje. 1993. Early spring nitrogen dynamics in a temperate forest landscape. Ecology 74:1579–1585.

Grondin, P., L. De Grandpré, Y. Bergeron, L. Bélanger, G. Lessard and J.-F. Bergeron 1996. Écologie forestière. Pages 134–279 in J.A. Bérard and M. Côté, eds., Manuel de Foresterie. Sainte-Foy, Quebec: Presses de l'Université Laval.

Groninger, J.W., and L.H. McCormick. 1992. Effects of soil disturbance on hayscented fern establishment. Northern Journal of Applied Forestry 9:29–31.

Gross, K., J.R. Lockwood, C.C. Frost, and M.F. Morris. 1998. Modeling controlled burning and trampling reduction for conservation of *Hudsonia montana*. Conservation Biology 12:1291–1301.

Grubb, P.J. 1977. The maintenance of species-richness in plant communities: the importance of the regeneration niche. Biological Review 52:107–145.

Haberlandt, G. 1914. Physiological plant anatomy. London: Macmillan.

Halpern, C.B. 1988. Early successional pathways and the resistance and resilience of forest communities. Ecology 69:1703–1715.

Halpern, C.B. 1989. Early successional patterns of forest species: interactions of life history traits and disturbance. Ecology 70:704–720.

Halpern, C.B., and T.A. Spies. 1995. Plant species diversity in natural and managed forests of the Pacific Northwest. Ecological Applications 5:913–934.

Hamilton, M.B. 1994. Ex-situ conservation of wild plant species: time to reassess the genetic assumptions and implications of seed banks. Conservation Biology 8:39–49.

Hammond, D. N., D.W. Smith, S. M. Zedaker, D. K. Wright, and J.W. Thompson. 1998. Floral diversity following harvest on southern Appalachian mixed oak sites. Pages 461–465 in Proceedings of the Ninth Southern Biennial Silvicultural Research Conference, Clemson, SC. General Technical Report SRS-20 Asheville, NC: USDA Forest Service.

Handel, S.N. 1976. Dispersal ecology of *Carex pedunculata* (Cyperaceae), a new North American myrmecochore. American Journal of Botany 63:1071–1079.

Handel, S.N. 1978. The competitive relationship of three woodland sedges and its bearing on the evolution of ant-dispersal of *Carex pedunculata*. Evolution 32:151–163.

Handel, S.N., S.B. Fisch, and G.E. Schatz. 1981. Ants disperse the majority of herbs in a mesic forest community in New York State. Bulletin of the Torrey Botanical Club 108:430–437.

Hanlon, T.J., C.E. Williams, and W.J. Moriarity. 1998. Species composition of soil seed banks of Allegheny Plateau riparian forests. Journal of the Torrey Botanical Club 125:199–215.

Hanski, I. 1982. Dynamics of regional distribution: the core and satellite species hypothesis. Oikos 38:210–221.

Hanski, I. 1991. Single species metapopulation dynamics: concepts, models and observation. Biological Journal of the Linnean Society 42:17–38.

Harcombe, P.A., and P.L. Marks. 1977. Understory structure of a mesic forest in southeast Texas. Ecology 58:1144–1151.

Harper, J. L. 1977. Population biology of plants. London: Academic Press.

Harper, J.L., J.N. Clatsworthy, I.H.M. Naughton, and G.R. Sagar. 1961. The evolution and ecology of closely related species living in the same area. Evolution 15:209–227.

Harris, W.F., P. Sollins, N.T. Edwards, B.E. Dinger, and H.H. Shugart. 1975. Analysis of carbon flow and productivity in a temperate deciduous forest ecosystem. Pages 116–122 in D.E. Reichle, J.F. Franklin, and D.W. Goodall, eds., Productivity of World Ecosystems. Washington, DC: National Academy Press.

Harrison, S. 1991. Local extinction in a metapopulation context—an empirical evaluation. Biological Journal of the Linnean Society 42:73–88.

Harrison, S., and E. Bruna. 1999. Habitat fragmentation and large-scale conservation: what do we know for sure? Ecography 22:225–232.

Harrison, S., and J.F. Quinn. 1989. Correlated environments and the persistence of metapopulations. Oikos 56:1–6.

Hartnett, D.C., and D.R. Richardson. 1989. Population biology of *Bonamia grandiflora* (Convolvulaceae)—effects of fire on plant and seed bank dynamics. American Journal of Botany 76:361–369.

Hartshorn, G.S. 1975. A matrix model of tree population dynamics. Pages 41–51 in F. B. Golley and E. Medina, eds., Tropical ecological systems. New York: Springer-Verlag.

Harvey, B.D., and Y. Bergeron. 1989. Site patterns of natural regeneration following clear-cutting in northwestern Quebec. Canadian Journal of Forest Research 19: 1458–1469.

Harvey, B.D., A. Leduc, and Y. Bergeron. 1995. Early postharvest succession in relation to site type in the southern boreal forest of Quebec. Canadian Journal of Forest Research 25:1658–1672.

Harvey, G.W. 1980. Seasonal alteration of photosynthetic unit sizes in three herb layer components of a deciduous forest community. American Journal of Botany 67:293–299.

Hawkes, C.V., and E.S. Menges. 1995. Density and seed production of a Florida endemic, *Polygonella basiramia*, in relation to time since fire and open sand. American Midland Naturalist 113:138–148.

Hayek, L.C., and M.A. Buzas. 1997. Surveying natural populations. New York: Columbia University Press.

Hayek, L.C., and M.A. Buzas. 1998. SHE analysis: an integrated approach to the analysis of forest biodiversity. Pages 311–321 in F. Dallmeier and J.J. Comiskey, eds., Forest biodiversity research, monitoring, and modelling: conceptual background and old-world case studies. Washington, DC: Smithsonian Institution.

He, F., and H.J. Barclay. 2000. Long-term response of understory plant species to thinning and fertilization in a Douglas-fir plantation on southern Vancouver Island, British Columbia. Canadian Journal of Forest Research 30:566–572.

Heibert, R.D. 1997. Prioritizing invasive plants and planning for management. Pages 195–212 J.O. Luken and J.W. Thieret, eds., Assessment and management of plant invasions. New York: Springer-Verlag.

Heilman, P.E. 1966. Change in distribution and availability of nitrogen with forest succession on north slopes in interior Alaska. Ecology 47:825–831.

Heilman, P.E. 1968. Relationship of availability of phosphorus and cations to forest succession and bog formation in interior Alaska. Ecology 49:331–336.

Heinen, J.T., and R.C.D. Currey. 2000. A 22-year study on the effects of mammalian browsing on forest succession following a clearcut in northern lower Michigan. American Midland Naturalist 144:243–252.

Heinselman, M.L. 1981. Fire and succession in the conifer forests of northern North America. Pages 374–406 in D.C. West and D.B. Botkin, eds., Forest succession: concepts and application. New York: Springer-Verlag.

Heithaus, E.R. 1981. Seed predation by rodents on three ant-dispersed plants. Ecology 63:136–145.

Helvey, J.D., J.D. Hewlett, and J.E. Douglass. 1972. Predicting soil moisture in the southern Appalachians. Soil Science Society of America Proceedings 36:954–959.

Henry, D.G. 1973. Foliar nutrient concentrations of some Minnesota forest species. Minnesota Forestry Research Notes No. 241. St. Paul, MN: University of Minnesota.

Heppell, S.S., H. Caswell, and L.B. Crowder. 2000a. Life histories and elasticity patterns: perturbation analysis for species with minimal demographic data. Ecology 81: 654–665.

Heppell, S.S., C.A. Pfister, and H. de Kroon. 2000b. Elasticity analysis in population biology: methods and applications. Ecology 81:605–606.

Hermy, M. 1988. Correlation between forest layers in mixed deciduous forests in Flanders (Belgium). Pages 77–85 in H.J. During, M.J.A. Werger, and J.H. Willems, eds., Diversity and pattern in plant communities. The Hague: SPB Academic Publishing.

Herring, B.J., and W.S. Judd. 1995. A floristic study of Ichetucknee Springs State Park, Suwannee and Columbia counties, Florida. Castanea 60:318–369.

Hibbs, D.E. 1982. White pine in the transition hardwood forest. Canadian Journal of Botany 60:2046–2053.

Hibbs, D.E. 1983. Forty years of forest succession in central New England. Ecology 64:1394–1401.

Hicks, D.J. 1980. Intrastand distribution patterns of southern Appalachian cove forest herbaceous species. American Midland Naturalist 104:209–222.

Hicks, D.J., and B.F. Chabot. 1985. Deciduous forest. Pages 257–277 in B.F. Chabot and H.A. Mooney, eds., Physiological ecology of North American plant communities. New York: Chapman and Hall.

Hill, J.D. 1996. Population dynamics of hayscented fern (*Dennstaedtia punctilobula*) and its effects on the composition, structure, and dynamics of a northeastern forest. Ph.D. dissertation. Storrs: University of Connecticut.

Hill, J.D., C.D. Canham, and D.M. Wood. 1994. Patterns and causes of resistance to tree invasion in rights-of-way. Ecological Applications 5:459–470.

Hill, J.D., and J.A. Silander, Jr. 2001. Distribution and dynamics of two ferns: *Dennstaedtia punctilobula* (Dennstaedtiaceae) and *Thelypteris noveboracensis* (Thelypteridaceae) in a northeast mixed hardwoods—hemlock forest. American Journal of Botany 88:894–902.

Hill, M.O. 1979a. DECORANA: A FORTRAN program for detrended correspondence analysis and reciprocal averaging. Ithaca, NY: Section of Ecology and Systematics, Cornell University.

Hill, M.O. 1979b. The development of a flora in even-aged plantations. Pages 175–192 in E.D. Ford, D.C. Malcolm, and J. Atterson, eds., The ecology of even-aged forest plantations. Cambridge, UK: Institute of Terrestrial Ecology.

Hobbs, R.J., and L.F. Huenneke. 1992. Disturbance, diversity and invasion: implications for conservation. Conservation Biology 6:324–337.

Hobbs, R.J., and S.E. Humphries. 1995. An integrated approach to the ecology and management of plant invaders. Conservation Biology 9:761–770.

Holt, R.D., and R. Gomulkiewicz. 1997. How does immigration influence local adaptation? A reexamination of a familiar paradigm. American Naturalist 149:563–572.

Hommels, C.H., P.J.C. Kuiper, and O.G. Tanczos. 1989. Luxury uptake and specific utilization rates of three macroelements in two *Taraxacum* microspecies of contrasting mineral ecology. Physiologia Plantarum 77:569–578.

Hooper, D.U., and P.M. Vitousek. 1998. Effects of plant composition and diversity on nutrient cycling. Ecological Monographs 68:121–149.

Hooper, M.C., K. Arii, and M.J. Lechowicz. 2001. Impact of a major ice storm on an old-growth hardwood forest. Canadian Journal of Botany 79:70–75.

Hoppes, W.G. 1988. Seedfall pattern of several species of bird-dispersed plants in an Illinois woodland. Ecology 69:320–329.

Hori, Y., and T. Yokoi. 1999. Population structure and dynamics of an evergreen shade herb, *Ainsliaea apiculata* (Asteraceae), with special reference to herbivore effects. Ecological Research 14:39–48.

Horn, H.S. 1971. The adaptive geometry of trees. Princeton, NJ: Princeton University Press.

Horn, H.S. 1975. Forest succession. American Scientist 232:90–98.

Horsley, S.B. 1977a. Allelopathic inhibition of black cherry by fern, grass, goldenrod, and aster. Canadian Journal of Forest Research 7:205–216.

Horsley, S.B. 1977b. Allelopathic inhibition of black cherry. II. Inhibition by woodland grass, ferns, and club moss. Canadian Journal of Forest Research 7:515–519.

Horsley, S.B. 1988. How vegetation can influence regeneration? Pages 38–55 in H.C. Smith, A.W. Perkey, and W.E. Kidd, eds., Guidelines for regenerating Appalachian hardwood stands. Morgantown, WV: USDA Forest Service.

Horsley, S.B. 1993a. Mechanisms of interference between hayscented fern and black cherry. Canadian Journal of Forest Research 23:2059–2069.

Horsley, S.B. 1993b. Role of allelopathy in hayscented fern interference with black cherry regeneration. Journal of Chemical Ecology 19:2737–2755.

Horsley, S.B. 1994. Regeneration success and plant species diversity of Allegheny hardwood stands after Roundup application and shelterwood cutting. Northern Journal of Applied Forestry 11:109–116.

Horsley, S.B., and D.A. Marquis. 1983. Interference by weeds and deer with Allegheny hardwood reproduction. Canadian Journal of Forest Research 13:61–69.

Horton, J.L., and H.S. Neufeld. 1998. Photosynthetic responses of *Microstegium vimineum* (Trin.) A. Camus, a shade-tolerant, C4 grass, to variable light environments. Oecologia 114:11–19.

Horvitz, C.C., J.B. Pascarella, S. McMann, A. Freedman, and R.H. Hofstetter. 1998. Functional roles of invasive non-indigenous plants in hurricane-affected subtropical hardwood forests. Ecological Applications 8:947–974.

Horvitz, C.C., D.W. Schemske, and H. Caswell. 1997. The relative "importance" of life-history stages to population growth: prospective and retrospective analyses. Pages 247–271 in S. Tuljapurkar and H. Caswell, eds., Structured population models in marine, terrestrial and freshwater systems. New York: Chapman and Hall.

Host, G.E., and K.S. Pregitzer. 1992. Geomorphic influences on ground-flora and overstory composition in upland forests of northwestern lower Michigan. Canadian Journal of Forest Research 22:1547–1555.

Hubbell, S.P. 1997. A unified theory of biogeography and relative species abundance and its application to tropical rain forests and coral reefs. Coral Reefs (suppl. 16): S9–S21.

Hubbell, S.P., R.B. Foster, S.T. O'Brien, K.E. Harms, R. Condit, B. Wechsler, S. J. Wright, and S. Loo de Lao. 1999. Light-gap disturbances, recruitment limitation, and tree diversity in a neotropical forest. Science 283:554–557.

Huenneke, L.F. 1983. Understory response to gaps caused by the death of *Ulmus americana* in central New York. Bulletin of the Torrey Botanical Club 110:170–175.

Hughes, J.W., and T.J. Fahey. 1991. Colonization dynamics of herbs and shrubs in a disturbed northern hardwood forest. Journal of Ecology 79:605–616.

Hughes, J.W., T.J. Fahey, and F.H. Bormann. 1988. Population persistence and reproductive ecology of a forest herb: *Aster acuminatus*. American Journal of Botany 75:1057–1064.

Hull, J.C. 2002. Photosynthetic responses to sunflecks of deciduous forest understory herbs with different phenologies. International Journal of Plant Science 163:913–924.

Hungerford, R.D., and R.E. Babbitt. 1987. Overstory removal treatments affect soil surface, air, and soil temperature: implications for seedling survival. Ogden, UT: USDA Forest Service Research Paper INT-377.

Hunter, M.L. 1989. What constitutes an old-growth stand? Journal of Forestry 87: 33–35.

Hurlbert, S.H. 1984. Pseudoreplication and the design of ecological field experiments. Ecological Monographs 54:187–211.

Husband, B.C., and S.C.H. Barrett. 1996. A metapopulation perspective in plant population biology. Journal of Ecology 84:461–469.

Huston, M. 1979. A general hypothesis of species diversity. American Naturalist 113: 81–101.
Huston, M.A. 1994. Biological diversity: the coexistence of species on changing landscapes. Cambridge: Cambridge University Press.
Huston, M.A. 1997. Hidden treatments in ecological experiments: re-evaluating the ecosystem function of biodiversity. Oecologia 110:449–460.
Huston, M., and T. Smith. 1987. Plant succession: life history and competition. American Naturalist 130:168–198.
Hutchings, M.J., and J.P. Barkham. 1976. An investigation of shoot interactions in *Mercurialis perennis* L., a rhizomatous perennial herb. Journal of Ecology 64:723–743.
Hutchinson, G.E. 1959. Homage to Santa Rosalia, or why are there so many kinds of animals? American Naturalist 93:145–159.
Hutchinson, T.F., and J.L. Vankat. 1997. Invasibility and effects of Amur honeysuckle in southwestern Ohio forests. Conservation Biology 11:1117–1124.
Hutchison, B.A., and D.R. Matt. 1976. Beam enrichment of diffuse radiation in a deciduous forest. Agricultural Meteorology 17:93–110.
Hutchison, B.A., and D.R. Matt. 1977. The distribution of solar radiation within a deciduous forest. Ecological Monographs 47:185–207.
Hutchison, B.A., D.R. Matt, R.T. McMillen, L.J. Gross, S.J. Tajchman, and J.M. Norman. 1986. The architecture of a deciduous forest canopy in eastern Tennessee. Journal of Ecology 74:635–646.
Ingvarsson, P. K., and S. Lundberg. 1995. Pollinator functional response and plant population dynamics—pollinators as a limiting resource. Evolutionary Ecology 9: 421–428.
Irland, L.C. 1998. Ice storm 1998 and the forests of the Northeast: a preliminary assessment. Journal of Forestry 96:32–40.
Irland, L.C. 1999. The northeast's changing forest. Boston: Harvard University Press.
Ives, A.R., M.G. Turner, and S. M. Pearson. 1998. Local explanations of landscape patterns: can analytical approaches approximate simulation models of spatial processes? Ecosystems 1:35–51.
Jackson, L.W.R. 1959. Relation of pine forest overstory opening diameter to growth of pine reproduction. Ecology 40:478–480.
Jahnke, L.S., and D.B. Lawrence. 1965. Influence of photosynthetic crown structure on potential productivity of vegetation, based primarily on mathematical models. Ecology 46:319–326.
Jandl, R., H. Kopeszki, and G. Glatzel. 1997. Effect of a dense *Allium ursinum* (L.) ground cover on nutrient dynamics and mesofauna of a *Fagus sylvatica* (L.) woodland. Plant and Soil 189:245–255.
Jenkins, M.A., and G.R. Parker. 1999. Composition and diversity of ground-layer vegetation in silvicultural openings of southern Indiana forests. American Midland Naturalist 142:1–16.
John, E., and R. Turkington. 1995. Herbaceous vegetation in the understorey of the boreal forest: does nutrient supply or snowshoe hare herbivory regulation species composition and abundance? Journal of Ecology 83:581–590.
Johnson, A.S., W.M. Ford, and P.E. Hale. 1993. The effects of clearcutting on herbaceous understories are still not fully known. Conservation Biology 7:433–435.
Johnson, D.W., and B. Schultz. 1999. Responses of carbon and nitrogen cycles to disturbance in forests and rangelands. Pages 545–569 in L.R. Walker, ed., Ecosystems of disturbed ground. Amsterdam: Elsevier.

Johnson, E.A. 1979. Fire recurrence in the subarctic and its implications for vegetation composition. Canadian Journal of Botany 57:1374–1379.

Johnston, M.H., and J.A. Elliott. 1996. Impacts of logging and wildfire on an upland black spruce community in northwestern Ontario. Environmental Monitoring and Assessment 39:283–297.

Johnstone, I.M. 1986. Plant invasion windows: a time-based classification of invasion potential. Biological Review 61:369–394.

Jones, J., J. Pither, R.D. DeBruyn, and R.J. Robertson. 2001. Modeling ice storm damage to a mature, mixed-species hardwood forest in eastern Ontario. Ecoscience 8:513–521.

Joyce, L.A., and R.L. Baker. 1987. Forest overstory-understory relationships in Alabama forests. Forest Ecology and Management 18:49–59.

Joyner, J.M., and E.W. Chester. 1994. The vascular flora of Cross Creeks National Wildlife Refuge, Stewart County, Tennessee. Castanea 59:117–145.

Jules, E. S., and B. J. Rathcke. 1999. Mechanisms of reduced *Trillium* recruitment along edges of old-growth forest fragments. Conservation Biology 13:784–793.

Jurdant, M., J.-L. Bélair, V. Gérardin, and J.-P. Ducruc. 1977. L'inventaire du capital nature: méthode de classification et de cartographie du territoire (3e approximation). Service des Études écologiques régionales, Dir. Gén. des Terres, Environnement Canada, Quebec.

Jurik, T.W. and B.F. Chabot. 1986. Leaf dynamics and profitability in wild strawberries. Oecologia 69:296–304.

Kalisz, S., F.M. Hanzawa, S.J. Tonsor, D.A. Thiede, and S. Voigt. 1999. Ant-mediated seed dispersal alters pattern of relatedness in a population of *Trillium grandiflorum*. Ecology 80:2620–2634.

Kalisz, S., and M. McPeek. 1992. Demography of an age-structured annual: resampled projection matrices, elasticity analyses, and seed bank effects. Ecology 73:1082–1093.

Kalisz, S., and M.A. McPeek. 1993. Extinction dynamics, population growth and seedbanks. Oecologia 95:314–320.

Kawano, S. 1970. Species problems viewed from productive and reproductive biology I. Ecological life histories of some representative members associated with temperate deciduous forests in Japan. Journal of the College of Liberal Arts, Toyama University, Natural Sciences 3:181–213.

Kawano, S., H. Takasu, and Y. Nagai. 1978. The productive and reproductive biology of flowering plants. IV. Assimilation behavior of some temperate woodland herbs. Journal of the College of Liberal Arts, Toyama University, Natural Sciences 11: 33–60.

Kearns, C.A., D.W. Inouye, and N.M. Waser. 1998. Endangered mutualisms: the conservation of plant-pollinator interactions. Annual Review of Ecology and Systematics 29:83–112.

Kearsley, J.B. 1999. Inventory and vegetation classification of floodplain forest communities in Massachusetts. Rhodora 101:105–135.

Keddy, P.A. 1992. Assembly and response rules: two goals for predictive community ecology. Journal of Vegetation Science 3:157–164.

Keddy, P.A., and C.G. Drummond. 1996. Ecological properties for the evaluation, management, and restoration of temperate deciduous forest ecosystems. Ecological Applications 6:748–762.

Keenan, R.J., and J.P. Kimmins. 1993. The ecological effects of clear-cutting. Environmental Reviews 1:121–144.

Keever, C. 1950. Causes of succession on old fields of the piedmont, North Carolina. Ecological Monographs 20:229–250.

Keever, C. 1983. Retrospective view of old-field succession after 35 years. American Midland Naturalist 110:397–404.

Kikuzawa, K. 1984. Leaf survival of woody plants in deciduous broad-leaved forests. 2. Small trees and shrubs. Canadian Journal of Botany 62:2551–2556.

Kikuzawa, K. 1991. A cost-benefit analysis of leaf habit and leaf longevity of trees and their geographical pattern. American Naturalist 138:1250–1263.

Kikuzawa, K., and D. Ackerly. 1999. Significance of leaf longevity in plants. Plant Species Biology 14:39–45.

Kilburn, P.D. 1957. Historical development and structure of the aspen, jack pine, and oak vegetation types on sandy soils in northern lower Michigan. PhD dissertation. Ann Arbor: University of Michigan.

Killingbeck, K.T. 1996. Nutrients in senesced leaves: keys to the search for potential resorption and resorption proficiency. Ecology 77:1716–1727.

Kimmins, J.P. 1996. Forest ecology: a foundation for sustainable management. 2nd ed. Upper Saddle River, NJ: Prentice-Hall.

Kinerson, R.S. 1979. Studies of photosynthesis and diffusion resistance in paper birch (*Betula papyrifera* Marsh) with synthesis through computer simulation. Oecologia 39:37–49.

Kirkman, L.K., M.G. Drew, and D. Edwards. 1998. Effects of experimental fire regimes on the population dynamics of *Schwalbea americana* L. Plant Ecology 137:115–137.

Kjellsson, G. 1985. Seed fate in a population of *Carex pilulifera* L. II. Seed predation and its consequences for dispersal and seed bank. Oecologia 67:424–429.

Knight, G.J. 1964. Distribution of *Osmunda cinnamomea* L. and *Osmunda claytoniana* L. in relation to natural soil drainage. MS thesis. Cambridge, MA: Harvard University.

Kochenderfer, J.N., and G.W. Wendel. 1983. Plant succession and hydrologic recovery on a deforested and herbicided watershed. Forest Science 29:545–558.

Koizumi, H. 1985. Studies on the life history of an evergreen herb, *Pyrola japonica*, population on a forest floor in a warm temperate region. 1. Growth, net production and matter economy. Botanical Magazine of Tokyo 98:383–392.

Koizumi, H. 1989. Studies on the life history of an evergreen herb, *Pyrola japonica*, population on a forest floor in a warm temperate region. 2. Photosynthesis, respiration and gross production. Botanical Magazine of Tokyo 102:521–532.

Koizumi, H., and Y. Oshima. 1985. Seasonal changes in photosynthesis of four understory herbs in deciduous forest. Botanical Magazine of Tokyo 98:1–13.

Koizumi, H., and Y. Oshima. 1993. Light environment and carbon gain of understory herbs associated with sunflecks in a warm temperate deciduous forest in Japan. Ecological Research 8:135–142.

Kolb, T.E., T.W. Bowersox, and L.H. McCormick. 1990. Influences of light intensity on weed-induced stresses of tree seedlings. Canadian Journal of Forest Research 20:503–507.

Kolb, T.E., T.W. Bowersox, L.H. McCormick, and K.C. Steiner. 1989. Effects of shade and herbaceous vegetation on first-year germination and growth of direct-seeded northern red oak, white ash, white pine, and yellow-poplar. Pages 156–161 in G. Rink and C.A. Budelsky, eds., Proceedings of the Seventh Central Hardwood Conference, Southern Illinois University, Carbondale. St. Paul, MN: North Central Forest Experimental Station. General Technical Report NC-132.

Königer, M., G.C. Harris, A. Virgo, and K. Winter. 1995. Xanthophyll-cycle pigments and photosynthetic capacity in tropical forest species: a comparative field study on canopy, gap and understory plants. Oecologia 104:280–290.

Körner, C. 1994. Scaling from species to vegetation: the usefulness of functional groups. Pages 117–140 in E.-D. Schulze and H.A. Mooney, eds., Biodiversity and ecosystem function. Berlin: Springer-Verlag.

Körner, Ch., J.A. Scheel, and H. Bauer. 1979. Maximum leaf diffusive conductance in vascular plants. Photosynthetica 13:45–82.

Koroleff, A. 1954. Leaf litter as a killer. Journal of Forestry 52:178–187.

Koyama, H., and S. Kawano. 1973. Biosystematic studies on *Maianthemum* (Liliaceae-Polygonatae). VII. Photosynthetic behavior of *M. dilatatum* under changing temperate woodland environments and its biological implications. Botanical Magazine of Tokyo 86:89–101.

Kriebitzsch, W.U. 1984. Seasonal course of chlorophyll content in leaves of *Mercurialis perennis* L. Flora 175:111–115.

Kriebitzsch, W.U. 1988. Seasonal changes in leaf water potential in understorey plants in a submontane beech forest on limestone. Flora 181:363–370.

Kriebitzsch, W.U. 1992a. CO_2 and H_2O gas exchange of understory plants in a submontane beech forest on limestone. I. Seasonal changes of the photosynthetic response to light. Flora 186:67–85.

Kriebitzsch, W.U. 1992b. CO_2 and H_2O gas exchange of understory plants in a submontane beech forest on limestone. II. Influence of temperature and air humidity. Flora 186:87–103.

Kriebitzsch, W.U. 1992c. CO_2 and H_2O gas exchange of understory plants in a submontane beech forest on limestone. III. CO_2 balances and net primary production. Flora 187:135–158.

Kriebitzsch, W.U. 1993. Der Wasserumsatz von Pflanzen in der Krautschicht eines Kalkbuchenwaldes. Phytocoenologia 23:35–50.

Kriebitzsch, W.U., and J. Regel. 1982. Bestimmung der Netto-Primärproduktion von Mercurialis perennis über die CO_2-Bilanz bzw. über die Ernte-Methode—Ein Methodenvergleich. Kurzmtlg. SFB 135 1:9–16.

Kriedeman, P.E., T.F. Neales, and D.H. Ashton. 1964. Photosynthesis in relation to leaf orientation and light interception. Australian Journal of Biological Science 17:591–600.

Kupfer, J.A., and G.P. Malanson. 1993. Structure and composition of a riparian forest edge. Physical Geography 14:154–170.

Küppers, M. 1994. Canopy gaps: Competitive light interception and economic space filling—A matter of whole-plant allocation. Pages 111–144 in M.M. Caldwell and R.W. Pearcy, eds., Exploitation of environmental heterogeneity by plants: ecophysiological processes above- and belowground. San Diego, CA: Academic Press.

Küppers, M., H. Timm, F. Orth, J. Stegemann, R. Stöber, H. Schneider, K. Paliwal, K.S.T.K. Karunaichamy, and R. Ortiz. 1996. Effects of light environment and successional status on lightfleck use by understory trees of temperate and tropical forests. Tree Physiology 16:69–80.

Kurmis, V., and E. Sucoff. 1989. Population density and height distribution of *Corylus cornuta* in undisturbed forest of Minnesota: 1965–1984. Canadian Journal of Botany 67:2409–2413.

Kursar, T.A., and P.D. Coley. 1993. Photosynthetic induction times in shade-tolerant species with long and short-lived leaves. Oecologia 93:165–170.

Laatsch, J.R., and R.C. Anderson. 2000. An evaluation of oak woodland management in northeastern Illinois, USA. Natural Areas Journal 20:211–220.

Lack, D. 1947. Darwin's finches: an essay on the general biological theory of evolution. Cambridge: Cambridge University Press.

Lambers, H. 1985. Respiration in intact plants and tissues: Its regulation and dependence on environmental factors, metabolism and invaded organisms. Pages 418–473 in R. Douce and D.A. Day, eds., Encyclopedia of plant physiology, ns, vol. 18. Berlin: Springer-Verlag.

Lambers, H., F.S. Chapin III, and T.L. Pons. 1998. Plant physiological ecology. New York: Springer.

Lambert, B.B. and E.S. Menges. 1996. The effects of light, soil disturbance and presence of organic litter on the field germination and survival of the Florida golden aster, *Chrysopsis floridana* Small. Florida Scientist 59:121–137.

Lande, R. 1988. Genetics and demography of biological conservation. Science 241: 1455–1460.

Lande, R. 1993. Risks of population extinction from demographic and environmental stochasticity and random catastrophes. American Naturalist 142:911–927.

Landhäusser, S.M., K.J. Stadt, and V.J. Lieffers. 1997. Photosynthetic strategies of summergreen and evergreen herbs of the boreal mixedwood forest. Oecologia 112: 173–178.

Lanza, J., M. A. Schmitt, and A. B. Awad. 1992. Comparative chemistry of elaiosomes of three species of *Trillium*. Journal of Chemical Ecology 18:209–221.

Lapointe, L., and J. Molard. 1997. Cost and benefits of mycorrhizal infection in a spring ephemeral, *Erythronium americanum*. New Phytologist 135:491–500

Lau, R.R., and D.R. Young. 1988. Influence of physiological integration on survivorship and water relations in a clonal herb. Ecology 69:215–219.

Lautenschlager, R.A., and C. Nielsen. 1999. Ontario's forest science efforts following the 1998 ice storm. Forestry Chronicle 75:633–641.

Lawton, R.O. 1990. Canopy gaps and light penetration into a wind-exposed tropical lower montane rain forest. Canadian Journal of Forest Research 20:659–667.

Leach, M.K., and T.J. Givnish. 1999. Gradients in the composition, structure, and diversity of remnant oak savannas in southern Wisconsin. Ecological Monographs 69:353–374.

Leckie, S., M. Vellend, G. Bell, M.J. Waterway, and M.J. Lechowicz. 2000. The seed bank in an old-growth, temperate deciduous forest. Canadian Journal of Botany 78:181–192.

Lee, D.W., and R. Graham. 1986. Leaf optical properties of rainforest sun and extreme-shade plants. American Journal of Botany 73:1100–1108.

Lefkovitch, L. P. 1965. The study of population growth in organisms grouped by stages. Biometrics 21:1–18.

Lei, T.T., and T. Koike. 1998. Some observations of phenology and ecophysiology of *Daphne kamtschatica* Maxim. var. jezoensis (Maxim.) Ohwi, a shade deciduous shrub, in the forest of northern Japan. Journal of Plant Research 111:207–212.

Lemon, P.C. 1961. Forest ecology of ice storms. Bulletin of the Torrey Botanical Club 88:21–29.

Lertzman, K.P. 1992. Patterns of gap-phase replacement in a subalpine, old-growth forest. Ecology 73:657–669.

Lertzman, K., and J. Fall. 1998. From forest stands to landscapes: spatial scales and the roles of disturbances. Pages 339–368 in D.L. Peterson and V.T. Parker, eds., Ecological scale: theory and applications. New York: Columbia University Press.

Leslie, P.H. 1945. On the use of matrices in certain population mathematics. Biometrika 33:183–212.

Levey, D. J., and M. M. Byrne. 1993. Complex ant-plant interactions: rain-forest ants as secondary dispersers and post-dispersal seed predators. Ecology 74:1802–1812.

Levin, D.A. and A.C. Wilson. 1976. Rates of evolution in seed plants: net increase in diversity of chromosome numbers and species numbers through time. Proceedings of the National Academy of Sciences of the United States of America 73:2086–2090.

Levin, P.S., and D.A. Levin. 2001. The real biodiversity crisis. American Scientist 90: 6–8.

Levine, J.M. 2000. Species diversity and biological invasions: relating process to community pattern. Science 288:852–854.

Levine, J.M., and C.M. D'Antonio. 1999. Elton revisited: a review of the evidence linking diversity and invasibility. Oikos 87:15–26.

Levins, R. 1968. Evolution in changing environments. Princeton, NJ: Princeton University Press.

Levins, R. 1970. Extinctions. Pages 75–107 in M. Gerstenhaber, ed., Lectures on mathematics in the life sciences, vol. 2. Providence, RI: American Mathematical Society.

Likens, G.E., and F.H. Bormann. 1970. Chemical analyses of plant tissues from the Hubbard Brook Ecosystem in New Hampshire. School of Forestry Bulletin no. 79. New Haven, CT: Yale University.

Likens, G.E., F.H. Bormann, R.S. Pierce, J.S. Eaton, and N.M. Johnson. 1977. Biogeochemistry of a forested ecosystem. New York: Springer-Verlag

Lindenmayer, D.B., M.A. Murgman, H.R. Akcakaya, R.C. Lacy, and H. P. Possingham. 1995. A review of the generic computer-programs ALEX, RAMAS/SPACE and VORTEX for modeling the viability of wildlife metapopulations. Ecological Modelling 82:161–174.

Lippincott, C.L. 2000. Effects of *Imperata cylindrica* (L.) Beauv. (cogongrass) invasion on fire regime in Florida sandhill (USA). Natural Areas Journal 20:140–149.

Lippmaa, T. 1939. The unistratal concept of plant communities (the unions). American Midland Naturalist 21:111–143.

Lipscomb, M.V. 1986. The influence of water and light on the physiology and spatial distributions of three shrubs in the southern Appalachian Mountains. M.S. thesis. Blacksburg, VA: Virginia Polytechnic Institute and State University.

Lipscomb, M.V., and E.T. Nilsen. 1990a. Environmental and physiological factors influencing the natural distribution of evergreen and deciduous ericaceous shrubs on northeast and southwest slopes of the southern Appalachian Mountains. I. Irradiance tolerance. American Journal of Botany 77:108–115.

Lipscomb, M.V., and E.T. Nilsen. 1990b. Environmental and physiological factors influencing the natural distribution of evergreen and deciduous ericaceous shrubs on northeast and southwest slopes of the southern Appalachian Mountains. II. Water relations. American Journal of Botany 77:517–526.

Lipson, D.A., W.D. Bowman, and R.K. Monson. 1996. Luxury uptake and storage of nitrogen in the rhizomatous alpine herb, *Bistorta bistortoides*. Ecology 77:1277–1285.

Lobstein, M. B., and L. L. Rockwood. 1993. Influence of elaiosome removal on germination in five ant-dispersed plant species. Virginia Journal of Science 44:59–72.

Lockhart, C.S., D.F. Austin, W.E. Jones, and L.A. Downey. 1999. Invasion of carrotwood (*Cupaniopsis anacardioides*) in Florida natural areas (USA). Natural Areas Journal 19:254–262.

Lodge, D.M. 1993. Biological invasions: lessons for ecology. Trends in Ecology and Evolutionary Biology 8:133–137.
Loehle, C. 2000. Strategy space and the disturbance spectrum: a life-history model for tree species coexistence. American Naturalist 156:14–33.
Logan, B.A., D.H. Barker, W.W. Adams III, and B.D. Demmig-Adams. 1997. The response of xanthophyll cycle-dependent energy dissipation in *Alocasia brisbanensis* to sunflecks in a subtropical rainforest. Australian Journal of Plant Physiology 24:27–33.
Lonn, M., H.C. Prentice, and K. Bengtsson. 1996. Genetic structure, allozyme habitat associations and reproductive fitness in *Gypsophila fastigiata* (Caryophyllaceae). Oecologia 106:308–316.
Lonsdale, W.M. 1999. Global patterns of plant invasions and the concept of invasibility. Ecology 80:1522–1536.
Lorimer, C. G. 1977. The presettlement forest and natural disturbance cycle of northeastern Maine. Ecology 58:139–148.
Lubbers, A.E., and M.J. Lechowicz. 1989. Effects of leaf removal on reproduction vs. belowground storage in *Trillium grandiflorum*. Ecology 70:85–96.
Luettge, U. 2000. Light stress and Crassulacean acid metabolism. Phyton 40:65–82.
Luken, J.O. 1988. Population structure and biomass allocation of the naturalized shrub *Lonicera maackii* (Rupr.) Maxim. in forest and open habitats. American Midland Naturalist 119:258–267.
Luken, J.O. 1994. Valuing plants in natural areas. Natural Areas Journal 14:295–299.
Luken, J.O. 1997. Management of plant invasions: implicating ecological succession. Pages 133–144 in J.O. Luken and J.W. Thieret, ed., Assessment and management of plant invasions. New York: Springer-Verlag.
Luken, J.O., L.M. Kuddes, and T.C. Tholemeier. 1997a. Response of understory species to gap formation and soil disturbance in *Lonicera maackii* thickets. Restoration Ecology 5:229–235.
Luken, J.O., L.M. Kuddes, T.C. Tholemeier, and D.M. Haller. 1997b. Comparative responses of *Lonicera maackii* (Amur honeysuckle) and *Lindera benzoin* (spicebush) to increased light. American Midland Naturalist 138:331–343.
Luken, J.O., and D.T. Mattimiro. 1991. Habitat-specific resilience of the invasive shrub Amur honeysuckle (*Lonicera maackii*) during repeated clipping. Ecological Applications 1:104–109.
Luken, J.O., and M. Shea. 2000. Repeated prescribed burning at Dinsmore Woods State Nature Preserve (Kentucky, USA): responses of the understory community. Natural Areas Journal 20:150–158.
Luken, J.O., and J.W. Thieret. 1996. Amur honeysuckle, its fall from grace. Bioscience 46:18–24.
Luken, J.O., and J.W. Thieret, eds. 1997. Assessment and management of plant invasions. New York: Springer-Verlag.
Lundberg, S., and P. K. Ingvarsson. 1998. Population dynamics of resource limited plants and their pollinators. Theoretical Population Biology 54:44–49.
Lundegardh, H. 1921. Ecological studies in the assimilation of certain forest-plants and shore plants. Svensk Botanisk Tidskrift 15:45–95.
Lyford, W.H., and D.W. MacLean. 1966. Mound and pit microrelief in relation to soil disturbance and tree distribution in New Brunswick, Canada. Harvard Forest Paper 15:1–18.
Lyon, J., and C.L. Sagers. 1998. Structure of herbaceous plant assemblages in a forested riparian landscape. Plant Ecology 138:1–16.

Lyon, L.J., and P.F. Stickney. 1976. Early vegetal succession following northern Rocky Mountain wildfires. Pages 355–375 in Proceedings of the Montana Tall Timbers Fire Ecology Conference and Fire and Land Management Symposium, no. 14. Tallahassee, FL: Tall Timbers Research Station.

Machado, J.L., and P.B. Reich. 1999. Evaluation of several measures of canopy openness as predictors of photosynthetic photon flux density in deeply shaded conifer-dominated forest understory. Canadian Journal of Forest Research 29:1438–1444.

Mack, R.N., D. Simberloff, W.M. Lonsdale, H. Evans, M. Clout, and F.A. Bazzaz. 2000. Biotic invasions: causes, epidemiology, global consequences, and control. Ecological Applications 10:689–710.

MacLean, D., and R.W. Wein. 1977. Changes in understory vegetation with increasing stand age in New Brunswick forests: species composition, cover, biomass and nutrients. Canadian Journal of Botany 55:2818–2831.

MacLean, D.A., and R.W. Wein. 1978. Weight loss and nutrient changes in decomposing litter and forest floor material in New Brunswick forest stands. Canadian Journal of Botany 56:2730–2749.

MacLean, D.A., S.J. Woodley, M.G. Weber, and R.W. Wein. 1983. Fire and nutrient cycling. Pages 111–132 in R.W. Wein and D.A. MacLean, eds., The role of fire in northern circumpolar ecosystems. New York: John Wiley & Sons.

Maguire, D.A., and R.T.T. Forman. 1983. Herb cover effects on tree seedling patterns in a mature hemlock-hardwood forest. Ecology 64:1367–1380.

Mahall, B.E., and F.H. Bormann. 1978. A quantitative description of the vegetative phenology of herbs in a northern hardwood forest. Botanical Gazette 139:467–481.

Malanson, G P. 1984. Intensity as a third factor of disturbance regime and its effect on species diversity. Oikos 43:411–413.

Malanson, G.P. 1997a. Effects of feedbacks and seed rain on ecotone patterns. Landscape Ecology 12:27–38.

Malanson, G.P. 1997b. Simulated responses to hypothetical fundamental niches. Journal of Vegetation Science 8:307–316.

Malik, C.P., Gill, R.K., and M.B. Singh. 1983. Nonphotosynthetic carbon dioxide fixation in *Monotropa uniflora*, a colorless angiosperm. Indian Journal of Experimental Biology 21:405–407.

Manion, P.D. 1991. Tree disease concepts, 2nd ed. Englewood Cliffs, NJ: Prentice-Hall.

Mann, L. K., A. W. King, V. H. Dale, W. W. Hargrove, R. Washington-Allen, L. R. Pounds, and T. L. Ashwood. 1999. The role of soil classification in geographic information system modeling of habitat pattern: threatened calcareous ecosystem. Ecosystems 2:524–538.

Marks, P.L., and F.H. Bormann. 1972. Revegetation following forest cutting: mechanisms for return to steady-state nutrient cycling. Science 176:914–915.

Marquis, D.A., and R. Brenneman 1981. The impact of deer on forest vegetation in Pennsylvania. General Technical Report NE-65. Broomall, PA: U.S. Forest Service.

Marschner, H. 1995. Mineral nutrition of higher plants, 2nd ed. London: Academic Press.

Martin, J.L. 1955. Observations on the origin and early development of a plant community following a forest fire. Forestry Chronicle 31:154–161.

Martin, W.H. 1992. Characteristics of old-growth mixed mesophytic forests. Natural Areas Journal 12:127–135.

Masarovičová, E. and P. Eliáš 1986. Photosynthetic rate and water relations in some forest herbs in spring and summer. Photosynthetica 20:187–195.

Matlack, G. 1994a. Plant demography, land-use history, and the commercial use of forests. Conservation Biology 8:298–299.
Matlack, G. 1994b. Plant species migration in a mixed-history forest landscape in Eastern North America. Ecology 75:1491–1502.
Mattson, W.J., and N.D. Addy. 1975. Phytophagous insects as regulators of forest primary production. Science 190:515–522.
McCarron, J.K., III. 1995. Ecophysiological adaptations of *Galax aphylla* to the understory of southern Appalachian forests. M.S. thesis. Boone, NC: Appalachian State University,
McCarthy, B.C. 1995. Eastern old-growth forests. The Ohio Woodland Journal 2:8–10.
McCarthy, B.C. 1997. Response of a forest understory community to experimental removal of an invasive nonindigenous plant (*Alliaria petiolata*, Brassicaceae). Pages 117–130 in J.O. Luken and J.W. Thieret, eds., Assessment and management of plant invasions. New York: Springer-Verlag.
McCarthy, B.C., and D.R. Bailey. 1996. Composition, structure and disturbance history of Crabtree Woods: an old-growth forest in western Maryland. Bulletin of the Torrey Botanical Club 123:350–365.
McCarthy, B.C., and J.M. Facelli. 1990. Microdisturbances in oldfields and forests: implications for woody seedling establishment. Oikos 58:55–60.
McCarthy, B.C., C.A. Hammer, G.L. Kauffman, and P.D. Cantino. 1987. Vegetation patterns and structure of an old-growth forest in southeastern Ohio. Bulletin of the Torrey Botanical Club 114:33–45.
McCarthy, B.C., C.J. Small, and D.L. Rubino. 2001. Composition, structure, and dynamics of Dysart Woods, an old-growth mixed mesophytic forest of southeastern Ohio. Forest Ecology and Management 140:193–213.
McCauley, D.E., J. Raveill, and J. Antonovics. 1995. Local founding events as determinants of genetic structure in a plant metapopulation. Heredity 75:630–636.
McCook, L.J. 1994. Understanding ecological community succession: causal models and theories, a review. Vegetatio 110:115–147.
McCune, B., and J.A. Antos. 1981. Correlations between forest layers in the Swan Valley, Montana. Ecology 62:1196–1204.
McDonald, D., and D.A. Norton. 1992. Light environments in temperate New Zealand Podocarp rainforests. New Zealand Journal of Ecology 16:15–22.
McDonnell, M.J. 1986. Old field vegetation height and the dispersal pattern of bird-disseminated woody plants. Bulletin of the Torrey Botanical Club 113:6–11.
McEvoy, P.B., and E.M. Coombs. 1999. Biological control of plant invaders: regional patterns, field experiments, and structured population models. Ecological Applications 9:387–401.
McGee, C.E., and R.C. Smith. 1967. Undisturbed rhododendron thickets are not spreading. Journal of Forestry 65:334–336.
McGee, G.G. 2001. Stand-level effects on the role of decaying logs as vascular plant habitat in Adirondack northern hardwood forests. Journal of the Torrey Botanical Club 128:370–380.
McGrady-Steed, J., P.M. Harris, and P.J. Morin. 1997. Biodiversity regulates ecosystem predictability. Nature 390:162–165.
McInnes, P.F., R.J. Naiman, J. Pastor, and Y. Cohen. 1992. Effects of moose browsing on vegetation and litter of the boreal forest, Isle Royale, Michigan, USA. Ecology 73:2059–2075.

McInnis, B.G., and Roberts, M.R. 1995. Seedling microenvironment in full-tree and tree-length logging slash. Canadian Journal of Forest Research 25:128–136.

McIntosh, R.P. 1985. The background of ecology: concept and theory. New York: Cambridge University Press.

McIntyre, S., S. Lavorel, and R.M. Tremont. 1995. Plant life-history attributes: their relationship to disturbance response in herbaceous vegetation. Journal of Ecology 83:31–44.

McKnight, B.N., ed. 1993. Biological pollution: the control and impact of invasive exotic species. Indianapolis: Indiana Academy of Science.

McLachlan, S.M., and B.R. Bazely. 2001. Recovery patterns of understory herbs and their use as indicators of deciduous forest regeneration. Conservation Biology 15: 98–110.

McLean, A. 1969. Fire resistance of forest species as influenced by root systems. Journal of Range Management 22:120–122.

McLendon, J.H. 1992. Photographic survey of the occurrence of bundle sheath extensions in deciduous dicots. Plant Physiology 99:677–1699.

McMillen, C.G., and J.H. McClendon. 1979. Leaf angle: an adaptive feature of sun and shade leaves. Botanical Gazette 140:437–442.

McNaughton, S.J. 1988. Mineral nutrition and spatial concentrations of African ungulates. Nature 334:343–345.

McPherson, J.K., and G.L. Thompson. 1972. Competitive and allelopathic suppression of understory by Oklahoma oak forests. Bulletin of the Torrey Botanical Club 99: 293–300.

McWilliams, W.H., S.L. Stout, T.W. Bowersox, and L.H. McCormick. 1995. Adequacy of advance tree-seedling regeneration in Pennsylvania's forests. Northern Journal of Applied Forestry 12:187–191.

Meagher, T.R. 1982. The population biology of *Chamaelirium luteum*, a dioecious member of the lily family: two-sex population projection and stable population structure. Ecology 63:1701–1711.

Meagher, T.R. 1978. The evolutionary consequences of dioecy in *Chamaelirium luteum*, a perennial plant species. Ph.D. dissertation. Durham, NC: Duke University.

Meagher, T.R., J. Antonovics, and R. Primack. 1978. Experimental ecological genetics in Plantago. III. Genetic variation and demography in relation to survival of *Plantago cordata*, a rare species. Biological Conservation 14:242–257.

Meekins, J., and B.C. McCarthy. 1999. Competitive ability of *Alliaria petiolata* (garlic mustard, Brassicaceae), an invasive, nonindigenous forest herb. International Journal of Plant Sciences 160:743–752.

Meekins, J.F., and B.C. McCarthy. 2000. Responses of the biennial forest herb, *Alliaria petiolata*, to variation in population density, nutrient addition and light availability. Journal of Ecology 88:447–463.

Meentemeyer, V. 1989. Geographical perspectives of space, time, and scale. Landscape Ecology 3:163–173.

Meentemeyer, V., and E.O. Box. 1987. Scale effects in landscape studies. Pages 15–36 in M.G. Turner, ed., Landscape heterogeneity and disturbance. New York: Springer-Verlag.

Mehroff, III, L.A. 1983. Pollination in the genus, *Isotria* (Orchidaceae). American Journal of Botany 70:1444–1453.

Meier, A.J., S.P. Bratton, and D.C. Duffy. 1995. Possible ecological mechanisms for loss of vernal-herb diversity in logged eastern deciduous forests. Ecological Applications 5:935–946.

Meier, A.J., S.P. Bratton, and D.C. Duffy. 1996. Biodiversity in the herbaceous layer and salamanders in Appalachian primary forests. Pages 49–64 in M.B. Davis, ed., Eastern old-growth forests. Washington, DC: Island Press.

Melillo, J.M., J.D. Aber, A.E. Linkins, A. Ricca, B. Fry, and K.J. Nadelhoffer. 1989. Carbon and nitrogen dynamics along the decay continuum: Plant litter to soil organic matter. Plant and Soil 115:189–198.

Melin, E. 1930. Biological decomposition of some types of litter from North American forests. Ecology 11:72–101.

Menge, J.A., D. Steirle, D.J. Bagyaraj, E.L.V. Johnson, and R.T. Leonard. 1978. Phosphorus concentrations in plants responsible for inhibition of mycorrhizal infection. New Phytologist 80:575–578.

Menges, E.S. 1987. Biomass allocation and geometry of the clonal forest herb *Laportea canadensis*: adaptive responses to the environment or allometric constraints? American Journal of Botany 74:551–563.

Menges, E.S. 1990. Population viability analysis for an endangered plant. Conservation Biology 4:41–62.

Menges, E.S. 2000a. Applications of population viability analyses in plant conservation. Ecological Bulletins 48:73–84.

Menges, E.S. 2000b. Population viability analyses in plants: challenges and opportunities. Trends in Ecology and Evolution 15:51–56.

Menges, E.S., and R.W. Dolan. 1998. Demographic viability of populations of *Silene regia* in midwestern prairies: relationships with fire management, genetic variation, geographic location, population size, and isolation. Journal of Ecology 86: 63–78.

Menges, E.S., and D.R. Gordon. 1996. Three levels of monitoring intensity for rare plant species. Natural Areas Journal 16:227–237.

Menges, E.S., and C.V. Hawkes. 1998. Interactive effects of fire and microhabitat on plants of Florida scrub. Ecological Applications 8:935–946.

Merrens, E.J., and D.R. Peart. 1992. Effects of hurricane damage on individual growth and stand structure in a hardwood forest in New Hampshire, USA. Journal of Ecology 80:787–795.

Milligan, B.G., J. Leebensmack, and A.E. Strand. 1994. Conservation genetics: beyond the maintenance of marker diversity. Molecular Ecology 3:423–435.

Mills, L.S., D.F. Doak, and M.J. Wisdom. 1999. Reliability of conservation actions based on elasticity analysis of matrix models. Conservation Biology 13:815–829.

Minckler, L.S., J.D. Woerheide, and R.C. Schlesinger. 1973. Light, soil moisture, and tree reproduction in hardwood forest openings. Research Paper NC-89. St. Paul, MN: U.S. Forest Service, North Central Forest Experiment Station.

Ministère des Ressources naturelles du Québec. 1994. Le point d'observation écologique: normes techniques. Québec: Service de l'inventaire forestier, ministère des Ressources naturelles du Québec.

Mladenoff, D.J. 1990. The relationship of the soil seed bank and understory vegetation in old-growth northern hardwood-hemlock treefall gaps. Canadian Journal of Botany 68:2714–2721.

Molnar, E., and S. Bokros. 1996. Studies on the demography and life history of *Taraxacum serotinum* (Waldst et Kit) Poir. Folia Geobotanica & Phytotaxonomica 31:453–464.

Moloney, K.A. 1988. Fine-scale spatial and temporal variation in the demography of a perennial bunchgrass. Ecology 69:1588–1598.

Monk, C.D. 1966. An ecological significance of evergreenness. Ecology 47:504–505.

Monk, C.D., D.T. McGinty, and F.P. Day Jr. 1985. The ecological importance of *Kalmia latifolia* and *Rhododendron maximum* in the deciduous forest of the southern Appalachians. Bulletin of the Torrey Botanical Club 112:187–193.

Montalvo, A. M., S. L. Williams, K. J. Rice, S. L. Buchmann, C. Cory, S. N. Handel, G. P. Nabhan, R. Primack, and R. H. Robichaux. 1997. Restoration biology: a population biology perspective. Restoration Ecology 5:277–290.

Mooney, H.A., and J.A. Drake, eds. 1986. Ecology of biological invasions. New York: Springer-Verlag.

Moore, J.M., and R.W. Wein. 1977. Viable seed populations by soil depth and potential site recolonization after disturbance. Canadian Journal of Botany 55:2408–2412.

Moore, M.R., and J.L. Vankat. 1986. Responses of the herb layer to the gap dynamics of a mature beech-maple forest. American Midland Naturalist 115:336–347.

Morgan, M.D. 1968. Life history and energy relationships of *Hydrophyllum appendiculatum*. Ph.D. dissertation. Urbana: University of Illinois.

Morley, T. 1982. Flowering frequency and vegetative reproduction in Erythronium albidum and E. propullans, and related observations. Bulletin of the Torrey Botanical Club 109:169–176.

Morneau, C., and S. Payette. 1989. Postfire lichen-spruce woodland recovery at the limit of the boreal forest in northern Quebec. Canadian Journal of Botany 67: 2770–2782.

Morris, L.A., S.A. Moss, and W.S. Garbett. 1993. Competitive interference between selected herbaceous and woody plants and *Pinus taeda* L. during two growing seasons following planting. Forest Science 39:166–187.

Morris, W.F., and D.F. Doak. 1998. Life history of the long-lived gynodioecious cushion plant *Silene acaulis* (Caryophyllaceae), inferred from size-based population projection matrices. American Journal of Botany 85:784–793.

Mosse, B., and J.M. Phillips. 1971. The influence of phosphate and other nutrients on the development of vesicular-arbuscular mycorrhiza in culture. Journal of General Microbiology 69:157–166.

Motten, A.F. 1986. Pollination ecology of the spring wildflower community of a temperate deciduous forest. Ecological Monographs 56:21–42.

Muir, A.M. 1995. The cost of reproduction to the clonal herb *Asarum canadense* (wild ginger). Canadian Journal of Botany 73:1683–1686.

Mulder, C.P.H. 1999. Vertebrate herbivores and plants in the Arctic and subarctic: effects on individuals, populations, communities and ecosystems. Perspectives in Plant Ecology, Evolution and Systematics 2:29–55.

Muller, R.N. 1978. The phenology, growth and ecosystem dynamics of *Erythronium americanum* in the northern hardwood forest. Ecological Monographs 48:1–20.

Muller, R.N. 1979. Biomass accumulation and reproduction in *Erythronium albidum*. Bulletin of the Torrey Botanical Club 106:276–283.

Muller, R.N. 1990. Spatial interrelationships of deciduous forest herbs. Bulletin of the Torrey Botanical Club 117:101–105.

Muller, R.N., and F.H. Bormann. 1976. Role of *Erythronium amercanum* Ker. in energy flow and nutrient dynamics of a northern hardwood forest ecosystem. Science 193:1126–1128.

Muraoka, H., A. Takenaka, Y. Tang, H. Koizumi, and I. Washitani. 1998. Flexible leaf orientations of *Arisaema heterophyllum* maximize light capture in a forest understorey and avoid excess irradiance at a deforested site. Annals of Botany 82:297–307.

Murphy, S.D., and L. Vasseur. 1995. Pollen limitation in a northern population of *Hepatica acutiloba*. Canadian Journal of Botany 73:1234–1241.

Muzika, R.M., and A.M. Liebhold. 1999. Changes in radial increment of host and nonhost tree species with gypsy moth defoliation. Canadian Journal of Forest Research 29:1365–1373.

Nakashizuka, T. 1987. Regeneration dynamics of beech forests in Japan. Vegetatio 69:169–175.

Nakashizuka, T. 1989. Role of uprooting in composition and dynamics of an old growth forest in Japan. Ecology 70:1273–1278.

Nantel, P., and D. Gagnon. 1999. Variability in the dynamics of northern peripheral versus southern populations of two clonal plant species, *Helianthus divaricatus* and *Rhus aromatica*. Journal of Ecology 87:748–760.

Nantel, P., D. Gagnon, and A. Nault. 1996. Population viability analysis of American ginseng and wild leek harvested in stochastic environments. Conservation Biology 10:608–621.

Nault, A., and D. Gagnon. 1988. Seasonal biomass and nutrient allocation patterns in wild leek (*Allium tricoccum* Ait.), a spring geophyte. Bulletin of the Torrey Botanical Club. 115:45–54.

Nault, A., and D. Gagnon. 1993. Ramet demography of *Allium tricoccum*, a spring ephemeral, perennial forest herb. Journal of Ecology 81:101–119.

Naumburg, E., and D.S. Ellsworth. 2000. Photosynthetic sunfleck utilization potential of understory saplings growing under elevated CO_2 in FACE. Oecologia 122:163–174.

Nekola, J.C. 1999. Paleorefugia and neorefugia: the influence of colonization history on community pattern and process. Ecology 80:2459–2473.

Nemati, N., and H. Goetz. 1995. Relationships of overstory to understory cover variables in a Ponderosa pine/Gambel oak ecosystem. Vegetatio 119:15–21.

Newell, C.L., and R.K. Peet. 1998. Vegetation of Linville Gorge Wilderness, North Carolina. Castanea 63:275–332.

Newell, S.J., and E.J. Tramer. 1978. Reproductive strategies in herbaceous plant communities during succession. Ecology 59:228–234.

Nguyen-Xuan, T., Y. Bergeron, D. Simard, J.W. Fyles, and D. Paré. 2000. The importance of forest floor disturbance in the early regeneration patterns of the boreal forest of western and central Quebec: a wildfire versus logging comparison. Canadian Journal of Forest Research 30:1353–1364.

Niering, W.A., and F.E. Egler. 1955. A shrub community of *Viburnum lentago*, stable for twenty-five years. Ecology 36:356–360.

Nilsen, E.T., and D.M. Orcutt. 1996. The physiology of plants under stress. New York: John Wiley & Sons.

Nilsen, E.T., D.A. Stetler, and D.A. Gassman. 1988. Influence of age and microclimate on the photochemistry of *Rhododendron maximum* leaves: II. Chloroplast structure and photosynthetic light response. American Journal of Botany 75:1526–1534.

Nilsen, E.T., J.F. Walker, O.K. Miller, S.W. Semones, T.T. Lei, and B.D. Clinton. 1999. Inhibition of seedling survival under *Rhododendron maximum* (Ericaceae): could allelopathy be a cause? American Journal of Botany 86:1597–1605.

Noble, I.R., and R. Dirzo. 1997. Forests as human-dominated ecosystems. Science 277: 522–525.

Noble, I.R., and H. Gitay. 1996. A functional classification for predicting the dynamics of landscapes. Journal of Vegetation Science 7:329–336.

Noble, I.R., and R.O. Slatyer. 1980. The use of vital attributes to predict successional changes in plant communities subject to recurrent disturbances. Vegetatio 43:5–21.

Nowacki, G.J., and P.A. Trianosky. 1993. Literature on old-growth forests of eastern North America. Natural Areas Journal 13:87–107.

Nuñez, M. 1980. The calculation of solar radiation in mountainous terrain. Journal of Biogeography 7:173–186.

Nuzzo, V.A., W. McClain, and T. Strole. 1996. Fire impact on groundlayer flora in a Sand Forest 1990–1994. American Midland Naturalist 136:207–221.

Oberhuber, W., and H. Bauer. 1991. Photoinhibition of photosynthesis under natural conditions in ivy (*Hedera helix* L.) growing in an understory of deciduous trees. Planta 185:545–553.

Odum, E.P. 1943. Vegetation of the Edmund Niles Huyck Preserve, New York. American Midland Naturalist 29:72–88.

Odum, E.P. 1969. The strategy of ecosystem development. Science 164:262–270.

Ögren, E., and U. Sundin. 1996. Photosynthetic responses to variable light: a comparison of species from contrasting habitats. Oecologia 106:18–27.

Ohlson, M., L. Söderström, G. Hörnberg, O. Zackrisson, and J. Hermansson. 1997. Habitat qualities versus long-term continuity as determinants of biodiversity in boreal old-growth swamp forests. Biological Conservation 81:221–231.

Ohmann, L., and D.F. Grigal. 1979. Early revegetation and nutrient dynamics following the 1971 Little Sioux forest fire in northeastern Minnesota. Forest Science Monograph 21, Washington, DC.

Oinonen, E. 1967a. Sporal regeneration of bracken (*Pteridium aquilinum* (L.) Kuhn.) in Finland in light of the dimensions and the age of its clones. Acta Forestalia Fennica 83:1–96.

Oinonen, E. 1967b. Sporal regeneration of ground pine (*Lycopodium complanatum* L.) in southern Finland in the light of the size and the age of its clones. Acta Forestalia Fennica 83:76–85.

Oláh, R. and E. Masarovičová. 1997. Response of CO2 uptake, chlorophyll content, and some productional features of forest herb *Smyrnium perfoliatum* L. (Apiaceae) to different light conditions. Acta Physiologiae Plantarum 19:285–293.

Olesen, T. 1992. Daylight spectra (400–700 nm) beneath sunny, blue skies in Tasmania, and the effect of a forest canopy. Australian Journal of Ecology 17:451–461.

Olesen, J. M. 1996. From naiveté to experience: Bumblebee queens (*Bombus terrestris*) foraging on *Corydalis cava* (Fumariaceae). Journal of the Kansas Entomological Society 69:274–286.

Oliver, C.D. 1982. Stand development—its uses and methods of study. Pages 100–112 in J.E. Means, ed., Forest development and stand research in the Northwest. Corvallis: Forest Research Laboratory, Oregon State University.

Oliver, C.D., and B.C. Larson. 1996. Forest stand dynamics. Update edition. New York: John Wiley & Sons.

Olivero, A.M., and D.M. Hix. 1998. Influence of aspect and stand age on ground flora of southeastern Ohio forest ecosystems. Plant Ecology 139:177–187.

Olivieri, I., D. Couvert, and P.H. Gouyon. 1990. The genetics of transient populations: research at the metapopulation level. Trends in Ecology and Evolution 5:207–210.

Olivieri, I., Y. Michalakis, and P.H. Gouyon. 1995. Metapopulation genetics and the evolution of dispersal. American Naturalist 146:202–228.

Olmstead, N.W., and J.D. Curtis. 1947. Seeds of the forest floor. Ecology 28:49–52.

Olson, J.S. 1958. Rates of succession and soil changes on southern Lake Michigan sand dunes. Botanical Gazette 119:125–169.

O'Neill, R.V., and A.W. King. 1998. Homage to St. Michael; or, Why are there so many books on scale? Pages 3–16 in D.L. Peterson and V.T. Parker, ed., Ecological scale: theory and applications. New York: Columbia University Press.

Oostermeijer, J.G.B., M.L. Brugman, E.R. DeBoer, and H.C.M. DenNijs. 1996. Temporal and spatial variation in the demography of *Gentiana pneumonanthe*, a rare perennial herb. Journal of Ecology 84:153–166.

Oosting, H. F., and M. E. Humphreys. 1940. Buried viable seeds in successional series of old fields and forest soils. Bulletin of the Torrey Botanical Club 67:253–273.

Oosting, H.J. 1942. An ecological analysis of the plant communities of Piedmont, North Carolina. American Midland Naturalist 28:1–126.

Oosting, H.J. 1956. The study of plant communities, 2nd ed. San Francisco: W.H. Freeman.

Orwig, D.A., and D.R. Foster. 1998. Forest response to the introduced hemlock woolly adelgid in southern New England, USA. Journal of the Torrey Botanical Society. 125:60–73.

Osawa, A. 1994. Seedling responses to forest canopy disturbance following a spruce budworm outbreak in Maine. Canadian Journal of Forest Research 24:850–859.

Oshima, K., Y. Tang, and I. Washitani. 1997. Spatial and seasonal patterns of microsite light availability in a remnant fragment of deciduous riparian forest and their implication in the conservation of *Arisaema heterophyllum*, a threatened plant species. Journal of Plant Research 110:321–327.

Ovington, J.D. 1962. Quantitative ecology and the woodland ecosystem concept. Advances in Ecological Research 1:103–192.

Pacala, S.W. 1986. Neighborhood models of plant population dynamics. II. Multispecies models of annuals. Theoretical Population Biology 29:262–292.

Pacala, S.W., C.D. Canham, J. Saponara, J.A. Silander Jr., R.K. Kobe, and E. Ribbens. 1996. Forest models defined by field measurements: estimation, error analysis and dynamics. Ecological Monographs 66:1–43.

Pacala, S.W., C.D. Canham, and J.A. Silander Jr. 1993. Forest models defined by field measurements: I. The design of a northeastern forest simulator. Canadian Journal of Forest Research 23:1980–1988.

Pacala, S.W., C.D. Canham, J.A. Silander Jr., and R.K. Kobe. 1994. Sapling growth as a function of resources in a north temperate forest. Canadian Journal of Forest Research 24:2172–2183.

Pakeman, R.J. 2001. Plant migration rates and seed dispersal mechanisms. Journal of Biogeography 28:795–800.

Paliwal, K., M. Küppers, and H. Schneider. 1994. Leaf gas exchange in lightflecks of plants of different successional range in the under-storey of a central European beech forest. Current Science 67:29–34.

Palmer, M.W. 1990. The estimation of species richness by extrapolation. Ecology 71: 1195–1198.

Palmer, M.W. 1993. Putting things in even better order: the advantages of canonical correspondence analysis. Ecology 74:2215–2230.

Palmer, M.W., and T.A. Maurer. 1997. Does diversity beget diversity? A case study of crops and weeds. Journal of Vegetation Science 8:235–240.

Panetta, F.D., and N.D. Mitchell. 1991. Homocline analysis and the prediction of weediness. Weed Research 31:273–284.

Pardy, A.B. 1997. Forest succession following a severe spruce budworm outbreak at Cape Breton Highlands National Park. M.Sc. thesis. Fredericton, New Brunswick, Canada: University of New Brunswick.

Parent, S., and C. Messier. 1996. A simple and efficient method to estimate microsite light availability under a forest canopy. Canadian Journal of Forest Research 26: 151–154.

Parish, T., K.H. Lakhani, and T.H. Sparks. 1994. Modeling the relationship between bird population variables and hedgerow and other field margin attributes. 1. Species richness of winter, summer, and breeding birds. Journal of Applied Ecology 31:764–775.

Parker, G.G. 1995. Structure and microclimate of forest canopies. Pages 73–106 in M.D. Lowman and N.M. Nadkarni, eds., Forest canopies. San Diego, CA: Academic Press.

Parker, G.G., and M.J. Brown. 2000. Forest canopy stratification—is it useful? The American Naturalist 155:473–484.

Parker, G.R. 1989. Old-growth forests of the central hardwood region. Natural Areas Journal 9:5–11.

Parker, K.C., and J. Bendix. 1996. Landscape-scale geomorphic influences on vegetation patterns in four environments. Physical Geography 17:113–141.

Parker, P. G., A. A. Snow, M. D. Schug,.G. C. Booton, and P. A. Fuerst. 1998. What molecules can tell us about populations: choosing and using a molecular marker. Ecology 79:361–382

Parkhurst, D.G., and O.L. Loucks. 1972. Optimal leaf size in relation to environment. Journal of Ecology 60:505–537.

Parrish, J.A.D., and F.A. Bazzaz. 1982. Competitive interactions in plant communities of different successional ages. Ecology 63:314–320.

Pascarella, J.B., and C.C. Horvitz. 1998. Hurricane disturbance of the population dynamics of a tropical understory shrub: megametric elasticity analysis. Ecology 79: 547–563.

Pastor, J., and S.D. Bridgham. 1999. Nutrient efficiency along nutrient availability gradients. Oecologia 118:50–58.

Pastor, J., and M. Broschart. 1990. The spatial pattern of a northern conifer-hardwood landscape. Landscape Ecology 4:55–68.

Patterson, W.A., III, and K.E. Sassaman. 1988. Indian fires in the prehistory of New England. Pages 107–135 in G. P. Nicholas, ed., Holocene human ecology in northeastern North America. New York: Plenum Press.

Pearcy, R.W. 1983. The light environment and growth of C_3 and C_4 tree species in the understory of a Hawaiian forest. Oecologia 58:19–25.

Pearcy, R.W. 1987. Photosynthetic gas exchange responses of Australian tropical forest trees in canopy, gap and understory microenvironments. Functional Ecology 1:169–178.

Pearcy, R.W. 1988. Photosynthetic utilization of lightflecks by understory plants. Australian Journal of Plant Physiology 15:223–238.

Pearcy, R.W. 1998. Acclimation to sun and shade. Pages 250–263 in A. S. Raghavendra, ed., Photosynthesis: a comprehensive treatise. Cambridge: Cambridge University Press.

Pearcy, R.W. 1999. Responses of plants to heterogeneous light environments. Pages 269–314 in F.I. Pugnaire, and F. Valladares, eds., Handbook of functional plant ecology. New York: Marcel Dekker.

Pearcy, R.W., O. Bjorkman, M.M. Caldwell, J.E. Keeley, R.K. Monson, and B.R. Strain. 1987. Carbon gain by plants in natural environments. BioScience 37:21–29.

Pearcy, R.W., R.L. Chazdon, L.J. Gross, and K.A. Mott. 1994. Photosynthetic utilization of sunflecks: A temporally patchy resource on a time scale of seconds to minutes. Pages 175–208 in M.M. Caldwell and R.W. Pearcy, eds., Exploitation of environmental heterogeneity by plants: ecophysiological processes above-and belowground. San Diego: CA: Academic Press.

Pearcy, R.W., and W.A. Pfitsch. 1991. Influence of sunflecks on the $\delta^{13}C$ of *Adenocaulon bicolor* plants occurring in contrasting forest understory microsites. Oecologia 86: 457–462.

Pearcy, R.W., J.S. Roden, and J.A. Gamon. 1990. Sunfleck dynamics in relation to canopy structure in a soybean (*Glycine max* (L.) Merr.) canopy. Agricultural and Forest Meteorology 52:359–372.

Pearcy, R.W., and D.A. Sims 1994. Photosynthetic acclimation to changing light environments: scaling from the leaf to the whole plant. Pages 223–234 in M.M. Caldwell and R.W. Pearcy, eds., Exploitation of environmental heterogeneity by plants: ecophysiological processes above-and belowground. San Diego, CA: Academic Press.

Pearcy, R.W., and W. Yang. 1996. A three-dimensional shoot architecture model for assessment of light capture and carbon gain by understory plants. Oecologia 108: 1–12.

Peet, R.K., and N.L. Christensen. 1980. Succession: a population process. Vegetatio 43:131–140.

Peet, R.K., and N.L. Christensen. 1987. Competition and tree death. BioScience 37: 586–595.

Peet, R.K., and N.L. Christensen. 1988. Changes in species diversity during secondary forest succession on the North Carolina Piedmont. Pages 233–245 in H.J. During, M.J.A. Werger, and J.H. Willems, eds., Diversity and pattern in plant communities. The Hague: SPB Academic Publishing.

Peine, H. 1989. Spruce budworm defoliation and growth loss in young balsam fir: recovery of growth in spaced stands. Canadian Journal of Forest Research 19: 1616–1624.

Peloquin, R.L., and R.D. Hiebert. 1999. The effects of black locust (*Robinia pseudoacacia* L.) on species diversity and composition of black oak savanna/woodland communities. Natural Areas Journal 19:121–131.

Peroni, P.A., and W.G. Abrahamson. 1985. A rapid method for determining losses of native vegetation. Natural Areas Journal 5:20–24.

Perry, J.N. 1998. Measures of spatial pattern for counts. Ecology 79:1008–1017.

Perry, J.N., and J.L. Gonzalez-Andujar. 1993. Dispersal in a metapopulation neighborhood model of an annual plant with a seedbank. Journal of Ecology 81:453–463.

Peterken, G.F., and M. Game. 1984. Historical factors affecting the number and distribution of vascular plant species in the woodlands of central Lincolnshire. Journal of Ecology 72:155–182.

Peterson, C.J. 2000. Damage and recovery of tree species after two different tornadoes at the same old growth forest: a comparison of infrequent wind disturbances. Forest Ecology and Management 135:237–252.

Peterson, C.J., and W.P. Carson. 1996. Generalizing forest regeneration models: the dependence of propagule availability on disturbance history and stand size. Canadian Journal of Forest Research 26:45–52.

Peterson, C.J., W.P. Carson, B.C. McCarthy, and S.T.A. Pickett. 1990. Microsite variation and soil dynamics within newly created treefall pits and mounds. Oikos 58:39–46.

Peterson, C.J., and S.T.A. Pickett. 1991. Treefall and resprouting following catastrophic windthrow in an old-growth hemlock-hardwoods forest. Forest Ecology and Management 42:205–217.

Peterson, C.J., and S.T.A. Pickett. 1995. Forest reorganization: a case study in an old-growth forest catastrophic blowdown. Ecology 76:763–774.

Peterson, C.J., and A.J. Rebertus. 1997. Tornado damage and initial recovery in three adjacent, lowland temperature forests in Missouri. Journal of Vegetation Science 8:559–564.

Peterson, D.L., and G.L. Rolfe. 1982. Nutrient dynamics of herbaceous vegetation in upland and floodplain forest communities. American Midland Naturalist 107:325–339.

Pfister, C.A. 1998. Patterns of variance in stage-structured populations: evolutionary predictions and ecological implications. Proceedings of the National Academy of Sciences of the United States of America 95:213–218.

Pfitsch, W.A., and R.W. Pearcy. 1989a. Daily carbon gain by *Adenocaulon bicolor*, a redwood forest understory herb, in relation to its light environment. Oecologia 80:465–470.

Pfitsch, W.A., and R.W. Pearcy. 1989b. Steady-state and dynamic photosynthetic responses of *Adenocaulon bicolor* in its redwood forest habitat. Oecologia 80:471–476.

Pfitsch, W.A., and R.W. Pearcy. 1992. Growth and reproductive allocation of *Adenocaulon bicolor* following experimental removal of sunflecks. Ecology 73:2109–2117.

Pfitsch, W.A., and R.W. Pearcy. 1995. The consequences of sunflecks for photosynthesis and growth of forest understory plants. Pages 343–359 in E.-D. Schulze and M.M. Caldwell, ed., Ecophysiology of photosynthesis. Berlin: Springer-Verlag.

Phillips, D.L., and W.H. Murdy. 1985. Effects of rhododendron (*Rhododendron maximum* L.) on regeneration of southern Appalachian hardwoods. Forest Science 31:226–233.

Pickett, S.T.A. 1980. Non-equilibrium coexistence of plants. Bulletin of the Torrey Botanical Club 107:238–248.

Pickett, S.T.A., and J.S. Kempf. 1980. Branching patterns in forest shrubs and understory trees in relation to habitat. New Phytologist 86:219–228.

Pickett, S.T.A., J. Kolasa, J.J. Armesto, and S.L. Collins. 1989. The ecological concept of disturbance and its expression at various hierarchical levels. Oikos 54:129–136.

Pickett, S.T.A., and M.J. McDonnell. 1989. Seed bank dynamics in temperate deciduous forest. Pages 123–147 in M.A. Leck, V.T. Parker, and R.L. Simpson, eds., Ecology of soil seed banks. San Diego; CA: Academic Press.

Pickett, S.T.A., V.T. Parker, and P.L. Fiedler. 1992. The new paradigm in ecology: implications for conservation biology above the species level. Pages 65–88 in P.L. Fiedler and S.K. Jain, eds., The theory and practice of nature conservation, preservation and management. New York: Chapman and Hall.

Pickett, S.T.A., and P.S. White. 1985. The ecology of natural disturbance and patch dynamics. New York: Academic Press.

Pielou, E.C. 1969. An introduction to mathematical ecology. New York: Wiley-Interscience.

Pimental, D., L. Lach, R. Zuniga, and A.D. Morrison. 2000. Environmental and economic costs of non-indigenous species in the United States. BioScience 50:53–65.

Pimm, S.L., and M.E. Gilpin. 1989. Theoretical issues in conservation biology. Pages 287–305 in J. Roughgarden, R. M. May, and S. A. Levin, eds., Perspectives in ecological theory. Princeton, NJ: Princeton University Press.

Piqueras, J., and L. Klimes. 1998. Demography and modelling of clonal fragments in the pseudoannual plant *Trientalis europaea* L. Plant Ecology 136:213–227.

Pitelka, L.F., and W.F. Curtis. 1986. Photosynthetic responses to light in an understory herb, *Aster acuminatus*. American Journal of Botany 73:535–540.

Pitelka, L.F., J.W. Ashmun, and R.L. Brown. 1985. The relationships between seasonal variation in light intensity, ramet size, and sexual reproduction in natural and experimental populations of *Aster acuminatus* (Compositae). American Journal of Botany 72:311–319.

Pitelka, L.F., D.S. Stanton, and M.O. Peckenham. 1980. Effects of light and density on resource allocation patterns in a forest herb, *Aster acuminatus* (Compositae). American Journal of Botany 67:942–948.

Pitman, N.C.A., J. Terborgh, M.R. Silman, and P.V. Nunez. 1999. Tree species distributions in an upper Amazonian forest. Ecology 80:2651–2661.

Planty-Tabacchi, A.M., E. Tabacchi, R.J. Naiman, C. Deferrari, and H. Decamps. 1995. Invasibility of species-rich communities in riparian zones. Conservation Biology 10:598–607.

Plašilová, J. 1970. A study of the root systems and root ecology of perennial herbs in the undergrowth of deciduous forests. Preslia 42:136–152.

Platt, W.J., and I.M. Weis. 1977. Resource partitioning and competition within a guild of fugitive prairie plants. American Naturalist 111:479–513.

Pleasants, J. M., and J. F. Wendel. 1989. Genetic diversity in a clonal narrow endemic, Erythronium propullans, and in its widespread progenitor, *Erythronium albidum*. American Journal of Botany 76:1136–1151.

Pockman, W.T., and J.S. Sperry. 2000. Vulnerability to xylem cavitation and the distribution of Sonoran desert vegetation. American Journal of Botany 87:1287–1299.

Pons, T.L. 1977. An ecophysiological study in the field layer of ash coppice. II. Experiments with *Geum urbanum* and *Cirsium palustre* in different light intensities. Acta Botanica Neerlandica 26:29–42.

Poorter, H. 1994. Construction costs and payback time of biomass: a whole plant perspective. Pages 111–127 in J. Roy and E. Garnier, eds., A whole-plant perspective on carbon-nitrogen interactions. The Hague: SPB Academic Publishing.

Poorter, H., and J.R. Evans. 1998. Photosynthetic nitrogen-use efficiency of species that differ inherently in specific leaf area. Oecologia 116:26–37.

Poorter, L., and S.F. Oberbauer. 1993. Photosynthesis induction responses of two rainforest tree species in relation to light environment. Oecologia 96:193–199.

Powles, S.B. 1984. Photoinhibition of photosynthesis induced by visible light. Annual Review of Plant Physiology 35:15–44.

Pregitzer, K.S., and B.V. Barnes. 1992. The use of ground flora to indicate edaphic factors in upland ecosystems of the McCormick Experimental Forest, Upper Michigan. Canadian Journal of Forest Research 12:661–672.

Primack, R. 1980. Phenotypic variation of rare and widespread species of *Plantago*. Rhodora 82:87–96.

Provencher, L., B.J. Herring, D.R. Gordon, H.L. Rodgers, G.W. Tanner, L.A. Brennan, and J.L. Hardesty. 2000. Restoration of northwest Florida sandhills through harvest of invasive *Pinus clausa*. Restoration Ecology 8:175–185.

Putz, F.E., and C.D. Canham. 1992. Mechanisms of arrested succession in shrublands: root and shoot competition between shrubs and tree seedlings. Forest Ecology and Management 49:267–275.

Pyle, L.L. 1995. Effects of disturbance on herbaceous exotic plant species on the floodplain of the Potomac River. American Midland Naturalist 134:244–253.

Pysck, P. 1995. On the terminology used in plant invasion studies. Pages 71–81 in P. Pysek, K. Prach, M. Rejmanek, and M. Wade, eds., Plant invasions—general aspects and special problems. Amsterdam: SPB Academic Publishing.

Pysek, P., and K. Prach. 1993. Plant invasions and the role of riparian habitats: a comparison of four species alien to central Europe. Journal of Biogeography 20: 413–420.

Qian, H., K. Klinka, and B. Sivak. 1997. Diversity of the understory vascular vegetation in 40 year-old and old-growth forest stands on Vancouver Island, British Columbia. Journal of Vegetation Science 8:773–780.

Quintana-Ascencio, R. F., and E. S. Menges. 1996. Inferring metapopulation dynamics from patch-level incidence of Florida scrub plants. Conservation Biology 10:1210–1219.

Quintana-Ascencio, R. F., and M. Morales-Hernandez. 1997. Fire-mediated effects of shrubs, lichens and herbs on the demography of *Hypericum cumulicola* in patchy Florida scrub. Oecologia 112:263–271.

Radford, A.E., H.E. Ahles, and C. R. Bell. 1968. Manual of the flora of the Carolinas. Chapel Hill: University of North Carolina Press.

Raghavendra, A.A., J.M. Rao, and V.S.R. Das. 1976. Replaceability of potassium by sodium for stomatal opening in epidermal strips of *Commelina benghalensis*. Zurich Pflanzenphysiologie 80:36–42.

Ramovs, B.V. 2001. Understory plant composition, microenvironment and stand structure of maturing plantations and naturally regenerated forests. Master's thesis. Fredericton: University of New Brunswick.

Randall, J.M. 1997. Defining weeds of natural areas. Pages 18–25 in J.O. Luken and J.W. Thieret, eds., Assessment and management of plant invasions. New York: Springer-Verlag.

Randall, W.J. 1952. Interrelations of autoecological characteristics of forest herbs. Ph.D. dissertation. Madison: University of Wisconsin.

Raunkiaer, C. 1934. The life forms of plants and statistical plant geography. Oxford: Clarendon Press.

Raven, J.A. 1989. Fight or flight: the economics of repair and avoidance of photoinhibition of photosynthesis. Functional Ecology 3:5–19.

Reader, R.J., and B.D. Bricker. 1992a. Response of five deciduous forest herbs to partial canopy removal and patch size. American Midland Naturalist 127:149–157.

Reader, R.J., and B.D. Bricker. 1992b. Value of selectively cut deciduous forest for understory herb conservation: an experimental assessment. Forest Ecology and Management 51:317–327.

Rebertus, A.J., S.R. Shifley, R.H. Richards, and L.M. Roovers. 1997. Ice storm damage to an old-growth oak-hickory forest in Missouri. American Midland Naturalist 137:48–61.

Reich, P.B., P. Bakken, D. Carlson, L.E. Frelich, S.K. Friedman, and D.F. Grigal. 2001. Influence of logging, fire, and forest type on biodiversity and productivity in southern boreal forests. Ecology 82:2731–2748.

Reichard, S.H. 1997. Prevention of invasive plant introductions on national and local levels. Pages 215–227 in J.O. Luken and J.W. Thieret, eds., Assessment and management of plant invasions. New York: Springer-Verlag.

Reichard, S.H., and C.W. Hamilton. 1997. Predicting invasions of woody plants introduced into North America. Conservation Biology 11:193–203.

Reid, A. 1964. Light intensity and herb growth in white oak forests. Ecology 45:396–398.

Reifsnyder, W.E., G.M. Furnival, and J.L. Horowitz. 1971. Spatial and temporal distribution of solar radiation beneath forest canopies. Agricultural Meteorology 9: 21–37.

Reiners, W.A. 1992. Twenty years of ecosystem reorganization following experimental deforestation and regrowth suppression. Ecological Monographs 62:503–523.

Reznicek, A.A., and P.M. Catling. 1989. Flora of Long Point, Regiona Municipality of Haldimand-Norfolk, Ontario. Michigan Botanist 28:99–175.

Rich, P.M., D.B. Clark, D.A. Clark, and S.F. Oberbauer. 1993. Long-term study of solar radiation regimes in a tropical wet forest using quantum sensors and hemispherical photography. Agricultural and Forest Meteorology 65:107–127.

Ricketts, T.H., E. Dinerstein, D.M. Olson, and C. Loucks. 1999. Who's where in North America? BioScience 49:369–381.

Roberts, M.R. 2002. Effects of forest plantation management on herbaceous-layer composition and diversity. Canadian Journal of Botany 80:378–389.

Roberts, M.R., and N.L. Christensen. 1988. Vegetation variation among mesic successional forest stands in northern lower Michigan. Canadian Journal of Botany 66:1080–1090.

Roberts, M.R., and H. Dong. 1993. Effects of soil organic layer removal on regeneration after clearcutting a northern hardwood stand in New Brunswick. Canadian Journal of Forest Research 23:2093–2100.

Roberts, M.R., and F.S. Gilliam. 1995a. Patterns and mechanisms of plant diversity in forested ecosystems: implications for forest management. Ecological Applications 5:969–977.

Roberts, M.R., and F.S. Gilliam. 1995b. Disturbance effects on herbaceous layer vegetation and soil nutrients in *Populus* forests of northern lower Michigan. Journal of Vegetation Science 6:903–912.

Roberts, M.R., and C.J. Richardson. 1985. Forty-one years of population change and community succession in aspen forests on four soil types, northern lower Michigan, U.S.A. Canadian Journal of Botany 63:1641–1651.

Roberts, M.R., and L. Zhu. 2002. Early response of the herbaceous layer to harvesting in a mixed coniferous-deciduous forest in New Brunswick, Canada. Forest Ecology and Management 155:17–31.

Roberts, T.L., and J.L. Vankat. 1991. Floristics of a chronosequence corresponding to old field-deciduous forest succession in southwestern Ohio. II. Seed banks. Bulletin of the Torrey Botanical Club 118:377–384.

Robertson, C. 1895. The philosophy of flower seasons, and the phaenological relations of the entomophilous flora and the anthrophilous insect fauna. American Naturalist 29:97–117.

Robichaux, R.H., E.A. Friar, and D.W. Mount. 1997. Molecular genetic consequences of a population bottleneck associated with reintroduction of the Mauna Kea silversword (*Argyroxiphium sandwichense* ssp. *sandwichense* (Asteraceae)). Conservation Biology 11:114–1146.

Robinson, G.R., M.E. Yurlina, and S.N. Handel. 1994. A century of change in the Staten Island flora: ecological correlates of species losses and invasions. Bulletin of the Torrey Botanical Club 121:119–129.

Rockwood, L.L., and M.B. Lobstein. 1994. The effects of experimental defoliation on reproduction in four species of herbaceous perennials from northern Virginia. Castanea 59:41–50.

Rogers, R.S. 1981. Mature mesophytic hardwood forest: community transitions, by laycr, from cast-central Minnesota to southeastern Michigan. Ecology 62:1634–1647.

Rogers, R.S. 1982. Early spring herb communities in mesophytic forests of the Great Lakes region. Ecology 63:1050–1063.

Rogers, R.S. 1983. Small-area coexistence of vernal forest herbs: does functional similarity of plants matter? American Naturalist 121:834–850.

Rogers, R.S. 1985. Local coexistence of deciduous-forest groundlayer species growing in different seasons. Ecology 66:701–707.

Rooney, T.P., and W.J. Dress. 1997a. Species loss over sixty-six years in the groundlayer vegetation of Heart's Content, an old-growth forest in Pennsylvania, USA. Natural Areas Journal 17:297–305.

Rooney, T.P., and W.J. Dress. 1997b. Patterns of plant diversity in overbrowsed primary and mature secondary hemlock northern hardwood forest stands. Journal of the Torrey Botanical Club 124:43–51.

Rose, R.J., R.T. Clarke, and S.B. Chapman. 1998. Individual variation and the effects of weather, age and flowering history on survival and flowering of the long-lived perennial *Gentiana pneumonanthe*. Ecography 21:317–326.

Rothstein, D.E. 2000. Spring ephemeral herbs and nitrogen cycling in a northern hardwood forest: an experimental test of the vernal dam hypothesis. Oecologia 124:446–453.

Rothstein, D.E., and D.R. Zak. 2001. Photosynthetic adaptation and acclimation to exploit seasonal periods of direct irradiance in three temperate, deciduous-forest herbs. Functional Ecology 15:722–731.

Rowe, J.S. 1983. Concepts of fire effects on plant individual and species. Pages 134–154 in R.W. Wein, and D.A. MacLean, eds., The role of fire in northern circumpolar ecosystems. New York: John Wiley & Sons.

Rowe, J.S. 1956. Uses of undergrowth plant species in forestry. Ecology 37:461–473.

Ruben, J.A., Bolger, D.T., Peart, D.R., and M.P. Ayres. 1999. Understory herb assemblages 25 and 60 years after clearcutting of a northern hardwood forest, USA. Biological Conservation 90:203–215.

Rubino, D.L., and B.C. McCarthy. 2000. Dendroclimatological analysis of white oak (*Quercus alba* L., Fagaceae) from an old-growth forest of southeastern Ohio, USA. Journal of the Torrey Botanical Society 127:240–250.

Ruel, J.-C. 2000. Factors influencing windthrow in balsam fir forests: from landscape studies to individual tree studies. Forest Ecology and Management 135:169–178.

Runkle, J.R. 1981. Gap regeneration in some old-growth forest of the eastern United States. Ecology 62:1041–1051.

Runkle, J.R. 1982. Patterns of disturbance in some old-growth mesic forests of eastern North America. Ecology 63:1533–1546.

Runkle, J.R. 1985. Disturbance regimes in temperate forests. Pages 17–33 in S.T.A.

Pickett and P.S. White, eds., The ecology of natural disturbance and patch dynamics. Orlando, FL: Academic Press.

Russell, E.W.B. 1983. Indian-set fires in the forests of the northeastern United States. Ecology 64:78–88.

Russell, N.H. 1958. Vascular flora of the Edmund Niles Huyck Preserve, New York. American Midland Naturalist 59:138–145.

Rust, R.W., and R.R. Roth. 1981. Seed production and seedling establishment in the mayapple *Podophyllum peltatum* L. American Midland Naturalist 105:51–60.

Ryan, M.G. 1991. Effects of climate change on plant respiration. Ecological Applications 1:157–167.

Saetra, P., L.S. Saetra, P.-O. Brandtberg, H. Lundkvist, and J. Bengtsson. 1997. Ground vegetation composition and heterogeneity in pure Norway spruce and mixed Norway spruce—birch stands. Canadian Journal of Forest Research 27:1110–1116.

Sage, R.F. 1993. Light-dependent modulation of ribulose-1,5-bisphosphate carboxylase/oxygenase activity in the genus *Phaseolus*. Photosynthesis Research 35:219–226.

Sage, R.F., and J.R. Seemann. 1993. Regulation of ribulose-1,5-bisphosphate carboxylase/oxygenase activity in response to reduced light intensity in C-4 plants. Plant Physiology 102:21–28.

Sage, R.F., D.A. Wedin, and M Li. 1999. The biogeography of C4 photosynthesis: Patterns and controlling factors. Pages 313–373 in R.F. Sage and R.K. Monson, eds., C4 plant biology. New York: Academic Press.

Sagers, C.L., and J. Lyon. 1997. Gradient analysis in a riparian landscape: contrasts among forest layers. Forest Ecology and Management 96:13–26.

Salisbury, E.J. 1928. On the causes and ecological significance of stomatal frequency with special reference to the woodland flora. Philosophical Transactions of the Royal Society Series B 216:1–65.

Salomonson, A., M. Ohlson, and L. Ericson. 1994. Meristem activity and biomass production as response mechanisms in two forest herbs. Oecologia 100: 29–37.

Sarukhán, J., and M. Gadgil. 1974. Studies on plant demography: *Ranunculus repens* L., *R. bulbosus* L., and *R. acris* L. III. A mathematical model incorporating multiple modes of reproduction. Journal of Ecology 62:921–936.

SAS Institute Inc. 1985. SAS user's guide: statistics. Cary, NC: SAS Institute, Inc.

Sassenrath-Cole, G.F., and R.W. Pearcy. 1992. The role of ribulose-1,5-bisphosphate regeneration in the induction requirement of photosynthetic carbon dioxide exchange under transient light conditions. Plant Physiology 99:227–234.

Sato, T., and A. Sakai. 1980. Phenological study of the leaf of pteridophyte in Hokkaido. Japanese Journal of Ecology 30:369–375.

Saucier, J.-P., J.-P. Berger, H. D'Avignon, and P. Racine. 1994. Le point d'observation écologique, normes techniques. Québec: Ministère des Ressources naturelles du Québec, Direction de la gestion des stocks, Service des inventaires forestiers.

Saucier, J.-P., J.-F. Bergeron, P. Grondin, and A. Robitaille. 1998. The land regions of southern Québec (3rd version): One element in the hierarchical land classification system developed by the Ministère des Ressources naturelles du Québec. Internal report. Québec: Ministère des Ressources naturelles du Québec.

Scanlan, M.J. 1981. Biogeography of forest plants in the prairie-forest ecotone in western Minnesota. Pages 97–124 in R.L. Burgess and D.M. Sharpe, eds., Forest island dynamics in man-dominated landscapes. New York: Springer-Verlag.

Schaetzl, R.J., S.F. Burns, D.L. Johnson, and T.W. Small. 1989. Tree uprooting: review of impacts on forest ecology. Vegetatio 79:165–176.

Schaetzl, R.J., and L.R. Follmer. 1990. Longevity of treethrow microtopography: implications for mass wasting. Geomorphology 3:113–123.

Schafale, M., and N.L. Christensen. 1986. Vegetational variation among old fields in Piedmont North Carolina. Bulletin of the Torrey Botanical Club 113:413–420.

Schemske, D.W., B.C. Husband, M.H. Ruckelshaus, C. Goodwillie, I.M Parker, and J.G. Bishop. 1994. Evaluating approaches to the conservation of rare and endangered plants. Ecology 75:584–606.

Scheu, S. 1997. Effects of litter (beech and stinging nettle) and earthworms (*Octolasion lacteum*) on carbon and nutrient cycling in beech forests on a basalt-limestone gradient: a laboratory experiment. Biology and Fertility of Soils 24:384–393.

Schiefthaler, U., A.W. Russell, H.R. Bolhár-Nordenkampf, and C. Critchley. 1999. Photoregulation and photodamage in *Schefflera arboricola* leaves adapted to different light environments. Australian Journal of Plant Physiology 26:485–494.

Schierenbeck, K.A., R.N. Mack, and R.R. Sharitz. 1994. Effects of herbivory on growth and biomass allocation in native and introduced species of *Lonicera*. Ecology 75: 1661–1672.

Schoener, T.W. 1983. Field experiments on interspecific competition. American Naturalist 122:240–285.

Schulze, E.D. 1972. Die Wirkung von Licht und Temperatur auf den CO_2-Gaswechsel verschiedener Lebensformen aus der Krautschicht eines montanen Buchenwaldes. Oecologia 9:235–258.

Schumm, S.A., and R.W. Lichty. 1965. Time, space, and causality in geomorphology. American Journal of Science 263:110–119.

Schwartz, M.W. 1997. Defining indigenous species: an introduction. Pages 7–17 in J.O. Luken and J.W. Thieret, eds., Assessment and management of plant invasions. New York: Springer-Verlag.

Schwartz, M.W., and J.R. Heim. 1996. Effects of a prescribed fire on degraded forest vegetation. Natural Areas Journal 16:184–191.

Scott, D.R.M. 1955. Amount and chemical composition of the organic matter contributed by overstory and understory vegetation to forest soil. Yale University School of Forestry. Bulletin no. 62. New Haven, CT: Yale University.

Seybold, A., and K. Eagle. 1937. Lichtfeld und Blattfarbstoffe. I. Planta 26:491–515.

Shafi, M.I., and G.A. Yarranton. 1973. Vegetational heterogeneity during a secondary (postfire) succession. Canadian Journal of Botany 51:73–90.

Shaver, G.R., and J.M. Melillo. 1984. Nutrient budgets of marsh plants: efficiency concepts and relation to availability. Ecology 65:1491–1510.

Shea, K., M. Rees, and S.N. Wood. 1994. Trade-offs, elasticities and the comparative method. Journal of Ecology 82:951–957.

Shugart, H.H. 1997. Plant and ecosystem functional types. Pages 20–43 in Smith, T.M., H.H. Shugart, and F.I. Woodward, eds., Plant functional types: their relevance to ecosystem properties and global change. Cambridge: Cambridge University Press.

Siccama, T.G., F.H. Bormann, and G.E. Likens. 1970. The Hubbard Brook ecosystem study: productivity, nutrients and phytosociology of the herbaceous layer. Ecological Monographs 40:389–402.

Siccama, T.G., G. Weir, and K. Wallace. 1976. Ice damage in a mixed hardwood forest in Connecticut in relation to *Vitis* infection. Bulletin of the Torrey Botanical Club 103:180–183.

Silva, J. 1978. Studies on the population biology of *Maianthemum canadense* Desf. Ph.D. dissertation. Cambridge, MA: Harvard University.

Silvertown, J.W. 1987. Introduction to plant population ecology, 2nd ed. New York: Longman Scientific and Technical.

Silvertown, J.W. 1991. Dorothy's dilemma and the unification of plant population biology. Trends in Ecology and Evolution 6:346–348.

Silvertown, J., M. Franco, and E. Menges. 1996. Interpretation of elasticity matrices as an aid to the management of plant populations for conservation. Conservation Biology 10:591–597.

Silvertown, J., M. Franco, I. Pisanty, and A. Mendoza. 1993. Comparative plant demography and relative importance of life-cycle components to the finite rate of increase in woody and herbaceous perennials. Journal of Ecology 81:465–476.

Silvertown, J., and R. Law. 1987. Do plants need niches? Some recent developments in plant community ecology. Trends in Ecology and Evolution 2:24–26.

Silvertown, J.W., and J. Lovett Doust. 1993. Introduction to plant population biology. Oxford: Blackwell Scientific.

Simonovich, V. 1973. Study of the root biomass in the herb layer of an oak-hornbeam forest. Biologia 28:11–22.

Sims, D.A., and R.W. Pearcy. 1993. Sunfleck frequency and duration affects growth rate of the understory plant, *Alocasia macrorrhiza*. Functional Ecology 7:683–689.

Singleton, R., S. Gardescu, P.L. Marks, and M.A. Geber. 2001. Forest herb colonization of postagricultural forests in central New York State, USA. Journal of Ecology 89: 325–338.

Sirois, L., and S. Payette. 1989. Postfire black spruce establishment in subarctic and boreal Quebec. Canadian Journal of Forest Research 19:1571–1580.

Skillman, J.B., B.R. Strain, and C.B. Osmond. 1996. Contrasting patterns of photosynthetic acclimation and photoinhibition in two evergreen herbs from a winter deciduous forest. Oecologia 107:446–455.

Skinner, W.R., and E.S. Telfer. 1974. Spring, summer, and fall foods of deer in New Brunswick. Journal of Wildlife Management 38:210–214.

Skre, O., F.E. Wielgolaski, and B. Moe. 1998. Biomass and chemical response of common forest plants in response to fire in western Norway. Journal of Vegetation Science 9:501–510.

Skutch, A.F. 1929. Early stages of plant succession following forest fires. Ecology 10: 177–190.

Small, C.J., and B.C. McCarthy. 2002. The influence of spatial and temporal variability on understory diversity in an eastern deciduous forest. Plant Ecology 164:37–48.

Smith, D. 2000. The population dynamics and community ecology of root hemiparasitic plants. American Naturalist 155:13–23.

Smith, E.P., and G. van Belle. 1984. Nonparametric estimation of species richness. Biometrics 43:793–803.

Smith, F.E. 1972. Spatial heterogeneity, stability, and diversity in ecosystems. Transactions of the Connecticut Academy of Arts and Sciences 44:309–355.

Smith, M., and S. Stocker. 1992. Effect of temperature on rhizome regrowth and biomass in *Muhlenbergia sobolifera*, a shade-tolerant C-4 grass. Transactions of the Illinois State Academy of Science 85:19–28.

Smith, M., and Y. Wu. 1994. Photosynthetic characteristics of the shade-adapted C4 grass *Muhlenbergia sobolifera* (Muhl.) Trin.: control of development of photorespiration by growth temperature. Plant, Cell & Environment 17:763–769.

Smith, T.L. 1989. An overview of old-growth forests in Pennsylvania. Natural Areas Journal 9:40–44.
Smith, T.M., and M.A. Huston. 1989. A theory of the spatial and temporal dynamics of plant communities. Vegetatio 83:49–69.
Smith, T.M., H.H. Shugart, and F.I. Woodward, eds. 1997. Plant functional types: their relevance to ecosystem properties and global change. Cambridge: Cambridge University Press.
Smith, W.G. 1975. Dynamics of pure and mixed populations of *Desmodium glutinosum* and *D. nudiflorum* in natural oak-forest communities. American Midland Naturalist 94:99–107.
Smith, W.H. 1981. Air pollution and forests: interactions between air contaminants and forest ecosystems. New York: Springer-Verlag.
Smith, W.K., A.K. Knapp, and W.A. Reiners. 1989. Penumbral effects on sunlight penetration in plant communities. Ecology 70:1603–1609.
Smith, W.K., T.C. Vogelmann, E.H. DeLucia, D.T. Bell, and K.A. Shepard. 1997. Leaf form and photosynthesis. BioScience 47:785–793.
Snaydon, R.W. 1962. Micro-distribution of *Trifolium repens* L. and its relation to soil factors. Journal of Ecology 50:133–143.
Snellgrove, R.C., W.E. Splittstoesser, D.B. Stribley, and P.B. Tinker. 1982. The distribution of carbon and the demand of the fungal symbiont in leek plants with vesicular-arbuscular mycorrhizas. New Phytologist 92:75–87.
Snyder, J.D., and R.A. Janke. 1976. Impact of moose browsing on boreal-type forests of Isle Royale National Park. American Midland Naturalist 95:79–92.
Snyder, K.M., J.M. Baskin, and C.C. Baskin. 1994. Comparative ecology of the narrow endemic *Echinacea tennesseensis* and two geographically widespread congeners: relative competitive ability and growth characteristics. International Journal of Plant Science 155:57–65.
Sobey, D.G., and P. Barkhouse. 1977. The structure and rate of growth of the rhizomes of some forest herbs and dwarf shrubs of the New Brunswick—Nova Scotia border region. Canadian Field Naturalist 91:177–383.
Sobieraj, J.H. 2002. The effects of ageing on the hydraulic conductivity of *Galax urceolata* (Poir.). Masters Thesis, Appalachian State University, Boone, NC.
Sohn, J.J., and D. Policansky. 1977. The costs of reproduction in the mayapple *Podophyllum peltatum* (Berberidaceae). Ecology 58:1366–1374.
Solbrig, O. T., R. Sarandon, and W. Bossuyt. 1988. A density-dependent growth model of a perennial herb, *Viola frimbriatula*. American Naturalist 17:385–400.
Sousa, W.P. 1980. The response of a community to disturbance: the importance of successional age and species' life histories. Oecologia 45:72–81.
Sparling, J.H. 1964. Ontario's woodland flora. Ontario Naturalist 2:18–25.
Sparling, J.H. 1967. Assimilation rates of some woodland herbs in Ontario. Botanical Gazette 128:160–168.
Sparling, J.H., and M. Alt. 1965. The establishment of carbon dioxide gradients in Ontario woodlands. Canadian Journal of Botany 44:321–329.
Sperduto, M. B., and R. G. Congalton. 1996. Predicting rare orchid (small whorled pogonia) habitat using GIS. Photogrammetric Engineering and Remote Sensing 62:1269–1279.
Sperry, J.S., J.R. Donnelly, and M.T. Tyree. 1988. A method for measuring hydraulic conductivity and embolisms in xylem. Plant, Cell & Environment 11:35–40.
Spies, T.A. 1991. Plant species diversity and occurrence in young, mature, and old-growth Douglas-fir stands in western Oregon and Washington. Pages 111–121

in L.F. Ruggiero, K.B. Aubry, A.B. Carey, and M.H. Huff, coordinators, Wildlife and vegetation of unmanaged Douglas-fir forests. General Technical Report PNW-GTR-285. Portland, OR: U.S. Forest Service.

Spurr, S.H., and J.H. Zumberge. 1956. Late Pleistocene features of Cheboygan and Emmet Counties, Michigan. American Journal of Science 254:96–109.

Stapanian, M.A., S.D. Sundberg, G.A. Baumgardner, and A. Liston. 1998. Alien plant species composition and associations with anthropogenic disturbance in North American forests. Plant Ecology 139:49–62.

Stebbins, G.L. 1974. Flowering plants: evolution above the species level. Cambridge: Harvard University Press.

Stein, B.A., L.S. Kutner, and J.S. Adams. 2000. Precious heritage. Oxford: Oxford University Press.

Stephens, E.P. 1956. The uprooting of trees: a forest process. Soil Science Society of America Proceedings 20:113–116.

Steubing, L., A. Fangmeier, R. Both, and M. Frankenfeld. 1989. Effects of SO_2, NO_2 and O_3 on population development and morphological and physiological parameters of native herb layer species in a beech forest. Environmental Pollution 58: 281–302.

Stohlgren, T.J., D. Binkley, G.W. Chong, M.A. Kalkhan, L.D. Schell, K.A. Bull, Y. Otsuki, G. Newman, M. Bashkin, and Y. Son. 1999. Exotic species invade hot spots of native plant diversity. Ecological Monographs 69:25–46.

Stommel, H. 1963. Varieties of oceanographic experience. Science 139:572–576.

Stone, E.L., and P.J. Kalisz. 1991. On the maximum extent of tree roots. Forest Ecology and Management 46:59–102.

Stone, W.E., and M.L. Wolfe. 1996. Response of understory vegetation to variable tree mortality following a mountain pine beetle epidemic in lodgepole pine stands in northern Utah. Vegetatio 122:1–12.

Stoutjesdijk, P. 1974. The open shade as an interesting microclimate. Acta Botanica Neerlandica 23:125–130.

St-Pierre, H., R. Gagnon, and P. Bellefleur. 1992. Régénération après feu de l'épinette noire (Picea mariana) et du pin gris (Pinus banksiana) dans la forêt boréale, Québec. Canadian Journal of Forest Research 22:474–481.

Struik, G.J. 1965. Growth patterns of some native annual and perennial herbs in southern Wisconsin. Ecology 46:401–420.

Struik, G.J., and J.T. Curtis. 1962. Herb distribution in an *Acer saccharum* forest. American Midland Naturalist 68:285–296.

Stuefer, J.F., H.J. During and H. de Kroon. 1994. High benefits of clonal integration in two stoloniferous species, in response to heterogeneous light environments. Journal of Ecology 82:511–518.

Swanson, A.M., and J.L. Vankat. 2000. Woody vegetation and vascular flora of an old-growth mixed-mesophytic forest in southwestern Ohio. Castanea 65:36–55.

Sydes, C., and J.P. Grime. 1981a. Effects of tree leaf litter on herbaceous vegetation in deciduous woodland. I. Field investigations. Journal of Ecology 69:237–248.

Sydes, C., and J.P. Grime. 1981b. Effects of tree leaf litter on herbaceous vegetation in deciduous woodland. II. An experimental investigation. Journal of Ecology 69: 249–262.

Syvertsen, J.P., and G.L. Cunningham 1979. The effects of irradiating adaxial and abaxial leaf surfaces on the rate of net photosynthesis of *Peresia nana* and *Helianthus annuus*. Photosynthetica 13:287–293.

Tamm, C.O. 1948. Observations on reproduction and survival of some perennial herbs. Botaniska Notiser 1948, Hafte 3:305–321.

Tamm, C.O. 1956. Further observations on the survival and flowering of some perennial herbs. Oikos 7:274–292.

Tamm, C.O. 1972a. Survival and flowering of some perennial herbs: II. The behaviour of some orchids on permanent plots. Oikos 23:23–28.

Tamm, C.O. 1972b. Survival and flowering of some perennial herbs: III. The behaviour of *Primula veris* on permanent plots. Oikos 23:159–166.

Tang, Y., N. Kachi, A. Furukawa, and M.B. Awang. 1999. Heterogeneity of light availability and its effects on simulated carbon gain of tree leaves in a small gap and the understory in a tropical rain forest. Biotropica 31:268–278.

Tang, Y., I. Washitani, and H. Iwaki. 1992. Effects of microsite light availability on the survival and growth of oak seedlings within a grassland. Botanical Magazine of Tokyo 105:281–288.

Tansley, A.G. 1935. The use and abuse of vegetational concepts and terms. Ecology 16:284–307.

Taylor, A.H., and Z. Qin. 1992. Tree regeneration after bamboo die-back in Chinese *Abies-Betula* forests. Journal of Vegetation Science 3:253–260.

Taylor, A.H., Z. Qin, and J. Liu. 1995. Tree regeneration in an *Abies faxoniana* forest after bamboo dieback, Wang Lang Natural Reserve, China. Canadian Journal of Forest Research 25:2034–2039.

Taylor, B.R., D. Parkinson, and W.F.J. Parsons. 1989. Nitrogen and lignin content as predictors of litter decay rates: a microcosm test. Ecology 70:97–104.

Taylor, R.J., and R.W. Pearcy. 1976. Seasonal patterns of the CO_2 exchange characteristics of understory plants from a deciduous forest. Canadian Journal of Botany 54:1094–1103.

Taylor, S.J., T.J. Carleton, and P. Adams. 1987. Understorey vegetation change in a *Picea mariana* chronosequence. Vegetatio 73:63–72.

Tear, T.H., J.M. Scott, P.H. Hayward, and B. Griffith. 1995. Recovery plans and the endangered-species-act—are criticisms supported by data? Conservation Biology 9:182–195.

Templeton, A.R., and D.A. Levin. 1979. Evolutionary consequences of seed pools. American Naturalist 114:232–249.

ter Braak, C.J.F. 1986. Canonical correspondence analysis: a new eigenvector technique for multivariate direct gradient analysis. Ecology 67: 1167–1179.

ter Braak, C.J.F. 1987. CANOCO—a FORTRAN program for community ordination by [partial] [detrended] [canonical] correspondence analysis, principal components analysis and redundancy analysis, version 2.1. Wageningen, The Netherlands: Agricultural Mathematics Group.

ter Braak, C.J.F. 1990. Update notes: CANOCO, version 3.10. Wageningen, The Netherlands: Agricultural Mathematics Group.

Tester, J.R. 1996. Effects of fire frequency on plant species in oak savanna in east-central Minnesota. Bulletin of the Torrey Botanical Club 123:304–305.

Thomas, S.C., C.B. Halpern, D.A. Falk, D.A. Liguori, and K.A. Austin. 1999. Plant diversity in managed forests: understory responses to thinning and fertilization. Ecological Applications 9:864–879.

Thompson, I.D., W.J. Curran, J.A. Hancock, and C.E. Butler. 1992. Influence of moose browsing on successional forest growth on black spruce sites in Newfoundland. Forest Ecology and Management 47:29–37.

Thompson, J.D. 1986. Pollen transport and deposition by bees in *Erythronium*: influence of floral nectar and bee grooming. Journal of Ecology 74:1–13.

Tilman, D. 1982. Resource competition and community structure. Princeton, NJ: Princeton University Press.

Tilman, D. 1985. The resource-ratio hypothesis of plant succession. American Naturalist 125:827–852.

Tilman, D. 1994.Competition and biodiversity in spatially structured habitats. Ecology 75:2–16.

Tilman, D. 1996. Biodiversity: population versus ecosystem stability. Ecology 77:350–363.

Tilman, D. 1999. The ecological consequences of changes in biodiversity: A search for general principles. Ecology 80:1455–1474.

Tinoco-Ojanguren, C., and R.W. Pearcy. 1993a. Stomatal dynamics and its importance to carbon gain in two rainforest *Piper* species I. VPD effects on the transient stomatal response to lightflecks. Oecologia 94:388–394.

Tinoco-Ojanguren, C., and R.W. Pearcy. 1993b. Stomatal dynamics and its importance to carbon gain in two rainforest *Piper* species II. Stomatal versus biochemical limitations during photosynthetic induction. Oecologia 94:395–402.

Tissue, D.T., J.B. Skillman, E.P. McDonald, and B.R. Strain. 1995. Photosynthesis and carbon allocation in *Tipularia discolor* (Orchidaceae), a wintergreen understory herb. American Journal of Botany 82:1249–1256.

Tuljapurkar, S. 1984. Demography in stochastic environments. 1. Exact distributions of age structure. Journal of Mathematical Biology 9:335–350.

Turnbull, M.H., and D.J. Yates. 1993. Seasonal variation in the red/far-red ratio and photon flux density in an Australian sub-tropical rainforest. Agricultural and Forest Meteorology 64:111–127.

Turner, D.P., and E.H. Franz. 1986. The influence of canopy dominants on understory vegetation patterns in an old-growth cedar-hemlock forest. American Midland Naturalist 116:387–393.

Tyler, G. 1975. Soil factors controlling ion absorption in the wood anemone (*Anemone nemorosa*). Oikos 27:71–80.

Tyler, G. 1989. Interacting effects of soil acidity and canopy cover on the species composition of field-layer vegetation in oak/hornbeam forests. Forest Ecology and Management 28:101–114.

USDA Forest Service. 1993. Forest health assessment for the Northeastern Area. NA-TP-01-95. USDA Forest Service.

Uemura, S. 1994. Patterns of leaf phenology in forest understory. Canadian Journal of Botany 72:409–414.

U.S. Fish and Wildlife Service. 1993a. Recovery plan for *Geocarpon minimum* Mackenzie. Atlanta: USFWS.

U.S. Fish and Wildlife Service. 1993b. Small whorled pogonia (*Isotria medeoloides*) recovery plan. First revision. Newton Corner, MA.: USFWS.

Valladares, F. 1999. Architecture, ecology, and evolution of plant crowns. Pages 121–194 in F.I. Pugnaire and F. Valladares, eds., Handbook of functional plant ecology. New York: Marcel Dekker.

Valverde, T., and J. Silvertown. 1995. Spatial variation in the seed ecology of a woodland herb (*Primula vulgaris*) in relation to light environment. Functional Ecology 9:942–950.

Valverde, T., and J. Silvertown. 1997a. An integrated model of demography, patch dynamics and seed dispersal in a woodland herb, *Primula vulgaris*. Oikos 80:67–77.

Valverde, T., and J. Silvertown. 1997b. A metapopulation model for *Primula vulgaris*, a temperate forest understory herb. Journal of Ecology 85:193–210.

Valverde, T., and J. Silvertown. 1998. Variation in the demography of a woodland understorey herb (*Primula vulgaris*) along the forest regeneration cycle: projection matrix analysis. Journal of Ecology 86:545–562.

van Breemen, N., A.C. Finzi, and C.D. Canham. 1997. Canopy tree—soil interactions within temperate forests: effects of soil elemental composition and texture on species distributions. Canadian Journal of Forest Research 27:1110–1116.

Van Cleve, K., R. Barney, and R. Schlentner. 1981. Evidence of temperature control of production and nutrient cycling in two interior Alaska black spruce ecosystems. Canadian Journal of Forest Research 11:258–273.

van Deelan, T.R., K.S. Pregitzer, and J.B. Haufler. 1996. A comparison of presettlement and present-day forests in two northern Michigan deer yards. American Midland Naturalist 135:181–194.

van Dorp, D., P. Schippers, and J. M. van Groenendael. 1997. Migration rates of grassland plants along corridors in fragmented landscapes assessed with a cellular automation model. Landscape Ecology 12:39–50.

Van Dyke, O. 1999. A literature review of ice storm impacts on forests in eastern North America. Technical Report no. 112. North Bay, Ontario: Ontario Ministry of Natural Resources, Southcentral Sciences Section.

van Groenendael, J., H. de Kroon, S. Kalisz, and S. Tuljapurkar. 1994. Loop analysis: evaluating life history pathways in population projection matrices. Ecology 75: 2410–2415.

Vankat, J.L., and G.W. Snyder. 1991. Floristics of a chronosequence corresponding to old field-deciduous forest succession in southwestern Ohio II. Undisturbed vegetation. Bulletin of the Torrey Botanical Club 118:365–376.

van Tienderen, P. H. 2000. Elasticities and the link between demographic and evolutionary dynamics. Ecology 81:666–679.

Veblen, T.T. 1982. Growth patterns of *Chusquea* bamboos in the understory of Chilean *Nothofagus* forests and their influences in forest dynamics. Bulletin of the Torrey Botanical Club 109:474–487.

Veblen, T.T. 1989. Tree responses to gaps along a transandean gradient. Ecology 70: 541–543.

Veblen, T.T., D.H. Ashton, and F.M. Schlegel. 1979. Tree regeneration strategies in a lowland *Nothofagus*-dominated forest in south-central Chile. Journal of Biogeography 6:329–340.

Verheyen, K., and M. Hermy. 2001. The relative importance of dispersal limitation of vascular plants in secondary forest succession in Muizen Forest, Belgium. Journal of Ecology 89:829–840.

Viereck, L.A., and L.A. Schandelmeier. 1980. Effects of fire in Alaska and adjacent Canada: a literature review. Technical report 6. Anchorage: Bureau of Land Management, Alaska State Office.

Vierling, L.A., and C.A. Wessman. 2000. Photosynthetically active radiation heterogeneity within a monodominant Congolese rain forest canopy. Agricultural and Forest Meteorology 103:265–278.

Vincent, J.S., and L. Hardy. 1977. L'évolution et l'extinction des lacs glaciaires Barlow et Ojibway et territoires québécois. Géographie Physique Quaternaire 31:357–372.

Vitousek, P. 1982. Nutrient cycling and nutrient use efficiency. American Naturalist 119:553–572.

Vitousek, P.M. 1985. Community turnover and ecosystem nutrient dynamics. Pages 325–333 in S.T.A. Pickett and P.S. White, eds., The ecology of natural disturbance and patch dynamics. Orlando, FL: Academic Press.

Vitousek, P. M., D. M. D'Antonio, L. L. Loope, and R. Westbrooks. 1996. Biological invasions as global environmental change. American Scientist 84:468–478.

Vitousek, P.M., H.A. Mooney, J. Lubchenco, and J. Mellilo. 1997. Human domination of earth's ecosystem. Science 277:494–499.

Vogelmann, T.C. 1993. Plant tissue optics. Annual Review of Plant Physiology and Plant Molecular Biology 44:231–251.

Vogelmann, T.C., J.F. Bornman, and D.J. Yates, 1996. Focusing of light by leaf epidermal cells. Physiologia Plantarum 98:48–56.

Vogt, K.A., C.C. Grier, C.E. Meier, and M.R. Keyes. 1983. Organic matter and nutrient dynamics in forest floors of young and mature *Abies amabilis* stands in western Washington, as affected by fine-root input. Ecological Monographs 53:139–157.

von Caemmerer, S., and R.T. Furbank. 1999. Modeling C4 photosynthesis. Pages 173–211 in R.F. Sage and R.K. Monson, eds., C4 plant biology. New York: Academic Press.

Wada, N. 1993. Dwarf bamboos affect the regeneration of zoochorous trees by providing habitats to acorn-feeding rodents. Oecologia 94:403–407.

Walker, J.E. 1983. The travel notes of Joseph Gibbons, 1804. Ohio History 92:96–146.

Walker, L.R., and S.D. Smith. 1997. Impacts of invasive plants on community and ecosystem properties. Pages 69–86 in J.O. Luken and J.W. Thieret, eds., Assessment and management of plant invasions. New York: Springer-Verlag.

Walters, M.B., and P.B. Reich. 1997. Growth of *Acer saccharum* seedlings in deeply shaded understories of northern Wisconsin: effects of nitrogen and water availability. Canadian Journal of Forest Research 27:237–247.

Waring, R.H., and W.H. Schlesinger. 1985. Forest ecosystems: concepts and management. New York: Academic Press.

Washitani, I., and Y. Tang. 1991. Microsite variation in light availability and seedling growth of *Quercus serrata* in a temperate pine forest. Ecological Research 6:305–316.

Waterman, J.R., A.R. Gillespie, J.M. Vose, and W.T. Swank. 1995. The influence of mountain laurel on regeneration in pitch pine canopy gaps of the Coweeta Basin, North Carolina, USA. Canadian Journal of Forest Research 25:1756–1762.

Watkinson, A.R. 1986. Plant population dynamics. Pages 137–185 in M. J. Crawley, ed., Plant ecology. Oxford: Blackwell Scientific.

Watkinson, A.R. 1998. The role of the soil community in plant population dynamics. Trends in Ecology and Evolution 13:171–172.

Watling, J.R., S.A. Robinson, I.E. Woodrow, and C.B. Osmond. 1997. Responses of rainforest understorey plants to excess light during sunflecks. Australian Journal of Plant Physiology 24:17–25.

Watson, L.E., W.J. Elisens, and J.R. Estes. 1991. Electrophoretic and genetic evidence for allopolyploid origin of *Marshallia mohrii* (Asteraceae). American Journal of Botany 78:408–416.

Watt, A.S. 1947a. Pattern and process in the plant community. Journal of Ecology 35:1–22.

Watt, A.S. 1947b. Contributions to the ecology of bracken. (*Pteridium aquilinium*) IV. The structure of the community. New Phytologist 46:97–121.

Weaver, J.E., and F.E. Clements. 1938. Plant ecology, 2nd ed. New York: McGraw-Hill.

Weber, J.A., T.W. Jurik, J.D. Tenhunen, and D.M. Gates. 1985. Analysis of gas exchange in seedlings of *Acer saccharum*: integration of field and laboratory studies. Oecologia 65:338–347.

Weber, M.G., and K. Van Cleve. 1981. Nitrogen dynamics in the forest floor of interior Alaska black spruce ecosystems. Canadian Journal of Forest Research 11:743–751.

Weber, M.G., and K. Van Cleve. 1984. Nitrogen transformations in feather moss and forest floor layers of interior Alaska black spruce ecosystems. Canadian Journal of Forest Research 14:278–290.

Wein, R.W., and J.M. Moore. 1977. Fire history and rotations in the New Brunswick Acadian Forest. Canadian Journal of Forest Research 7:285–294.

Wells, B.W. 1932. The natural gardens of North Carolina. Chapel Hill: University of North Carolina Press.

Werger, M.J., and E.M. van Laar. 1985. Seasonal changes in the structure of the herb layer of a deciduous woodland. Flora 176:351–364.

Werner, P.A. 1978. Determination of age in *Liatris aspera* using cross-sections of corms—implications for past demographic studies. American Naturalist 112: 1113–1120.

Werner, P.A., and H. Caswell. 1977. Population growth rates and age versus stage-distribution models for teasel (*Dipsacus sylvestris* Huds.). Ecology 58:1103–1111.

Werner, P.A., and W.J. Platt. 1976. Ecological relationships of co-occurring goldenrods. American Naturalist 110:959–971.

Whigham, D.F. 1984. Biomass and nutrient allocation of *Tipularia discolor* (Orchidaceae). Oikos 2:303–313.

Whigham, D.F. 1990. The effect of experimental defoliation on the growth and reproduction of a woodland orchid, *Tipularia discolor*. Canadian Journal of Botany 68:1812–1816.

White, P.S. 1979. Pattern, process, and natural disturbance in vegetation. The Botanical Review 45:229–299.

White, P.S., and S.T.A. Pickett. 1985. Natural disturbance and patch dynamics: an introduction. Pages 3–13 in S.T.A. Pickett and P.S. White, eds., The ecology of natural disturbance and patch dynamics. Orlando, FL: Academic Press.

Whitney, G.G. 1990. The history and status of the hemlock-hardwood forests of the Allegheny plateau. Journal of Ecology 78:443–458.

Whitney, G.G. 1994. From coastal wilderness to fruited plain: a history of environmental change in temperate North America from 1500 to the present. Cambridge: Cambridge University Press.

Whitney, G.G., and D.R. Foster. 1988. Overstorey composition and age as determinants of the understorey flora of woods of central New England. Journal of Ecology 76: 867–876.

Whitney, G.W. 1987. Some reflections on the value of old-growth forests, scientific and otherwise. Natural Areas Journal 7:92–99.

Whitney, H.E., and W.C. Johnson. 1984. Ice storms and forest succession in southwestern Virginia. Bulletin of the Torrey Botanical Club 111:429–437.

Whittaker, R.H. 1956. Vegetation of the Great Smoky Mountains. Ecological Monographs 26:1–80.

Whittaker, R.H. 1969. Evolution of diversity in plant communities. Brookhaven Symposium in Biology 22:178–195.

Whittaker, R.H., and S.A. Levin, eds. 1975. Niche: theory and application. Stroudsburg, PA: Dowden, Hutchinson & Ross.

Whittaker, R.H., and S.A. Levin. 1977. The role of mosaic phenomena in natural communities. Theoretical Population Biology 12:117–139.

Widden, P.M. 1996. The morphology of vesicular-arbuscular mycorrhizae in *Clintonia borealis* and *Medeola virginiana.* Canadian Journal of Botany 74:679–685.

Wijesinghe, D.K., and D.F. Whigham. 1997. Costs of producing clonal offspring and the effects of plant size on population dynamics of the woodland herb *Uvularia perfoliata* (Liliaceae). Journal of Ecology 85:907–919.

Wilbur, R.B., and N.L. Christensen. 1983. Effects of fire on nutrient availability in a North Carolina coastal plain pocosin. American Midland Naturalist 110:54–61.

Wilcock, C.C., and S.B. Jennings. 1999. Partner limitation and restoration of sexual reproduction in the clonal dwarf shrub *Linnaea borealis* L. (Caprifoliaceae). Protoplasma 208:76–86.

Williams, C.E. 1996. Alien plant invasions and forest ecosystem integrity: a review. Pages 169–185 in S.K. Majumdar, E.W. Miller, and R.J. Brenner, eds., Forests—a global perspective. Easton, PA: The Pennsylvania Academy of Science.

Williams, C.E. 1997. Potential valuable ecological functions of nonindigenous plants. Pages 26–34 in J.O. Luken and J.W. Thieret, eds., Assessment and management of plant invasions. New York: Springer-Verlag.

Williams, C.E., W.J. Moriarity, and G.L. Walters. 1999a. Overstory and herbaceous layer of a riparian savanna in northwestern Pennsylvania. Castanea 64:90–97.

Williams, C.E., W.J. Moriarity, G.L. Walters, and L. Hill. 1999b. Influence of inundation potential and forest overstory on the ground-layer vegetation of Allegheny Plateau riparian forests. American Midland Naturalist 141:323–338.

Williams, C.E., E.V. Mosbacher, and W.J. Moriarity. 2000. Use of turtlehead (*Chelone glabra* L.) and other herbaceous plants to assess intensity of white-tailed deer browsing on Allegheny Plateau riparian forests, USA. Biological Conservation 92: 207–215.

Williams, K., C.B. Field, and H.A. Mooney. 1989. Relationships among leaf construction cost, leaf longevity, and light environment in rain-forest plants of the genus Piper. American Naturalist 133:198–211.

Williams, L.D. 1998. Factors affecting growth and reproduction in the invasive grass *Microstegium vimineum.* Master's thesis. Boone, NC: Appalachian State University.

Williams, M. 1989. Americans and their forests: a historical geography. New York: Cambridge University Press.

Williamson, M. 1996. Biological invasions. New York: Chapman & Hall.

Williamson, M., and A. Fitter. 1996. The varying success of invaders. Ecology 77: 1661–1666.

Willmot, A. 1989. The phenology of leaf life spans in woodland populations of the ferns *Dryopteris filix-mas* (L.) Schott and *D. dilatata* (Hoffm.) A. Gray in Derbyshire. Botanical Journal of the Linnean Society 99:387–395.

Wilson, A.D., and D.J. Shure. 1993. Plant competition and nutrient limitation during early succession in the Southern Appalachian Mountains. American Midland Naturalist 129:1–9.

Wilson, J.B. 1999a. Assembly rules in plant communities. Pages 130–164 in E. Weiher and P. Keddy, ed., Ecological assembly rules: perspectives, advances, retreats. Cambridge: Cambridge University Press.

Wilson, J.B. 1999b. Guilds, functional types, and ecological groups. Oikos 86:507–522.

Wilson, J.B., H. Gitay, J.B. Steel, and W.M. King. 1998. Relative abundance distributions in plant communities: effects of species richness and of spatial scale. Journal of Vegetation Science 9:213–220.

Wilson, J.B., J.B. Steel, W.M. King, and H. Gitay. 1999. The effect of spatial scale on evenness. Journal of Vegetation Science 10:463–468.

Winkler, E., M. Fischer, and B. Schmid. 1999. Modelling the competitiveness of clonal plants by complementary analytical and simulation approaches. Oikos 85:217–233.

Winn, A.A., and L.F. Pitelka. 1981. Some effects of density on the reproductive patterns and patch dynamics of *Aster acuminatus*. Bulletin of the Torrey Botanical Club 108:438–445.

Winter, K., M.R. Schmitt, and G.E. Edwards. 1982. *Microstegium vimineum*, a shade-adapted C4 grass. Plant Science Letters 24:311–318.

Wise, D.H., and M. Schaefer. 1994. Decomposition of leaf litter in a mull beech forest: comparison between canopy and herbaceous species. Pedobiologia 38:269–288.

Wiser, S.K., R.B. Allen, P.W. Clinton, and K.H. Platt. 1998. Community structure and forest invasion by an exotic herb over 23 years. Ecology 79:2071–2081.

Wistendahl, W.A. 1958 The flood plain of the Raritan River, New Jersey. Ecological Monographs 28:129–153.

Wolfe, J.N., R.T. Wareham, and H.T. Schofield. 1949. Microclimates and macroclimate of Neatoma, a small valley in central Ohio. Columbus: The Ohio State University.

Wolters, V. 1999. *Allium ursinum* litter triggering decomposition on a beech forest floor—the effect of earthworms. Pedobiologia 43:528–536.

Woods, K.D. 1993. Effects of invasion by *Lonicera tatarica* L. on herbs and tree seedlings in four New England forests. American Midland Naturalist 130:62–74.

Woods, K.D. 1997. Community response to plant invasion. Pages 56–68 in J.O. Luken and J.W. Thieret, ed., Assessment and management of plant invasions. New York: Springer-Verlag.

Wu, B., and F.E. Smeins. 2000. Multiple-scale habitat modeling approach for rare plant conservation. Landscape and Urban Planning 51:11–28.

Yanhong, T., K Hiroshi, S. Mitsumasa and I. Washitani. 1994. Characteristics of transient photosynthesis in *Quercus serrata* seedlings grown under lightfleck and constant light regimes. Oecologia 100:463–469.

Yarie, J. 1980. The role of understory vegetation in the nutrient cycle of forested ecosystems in the mountain hemlock biogeoclimatic zone. Ecology 61:1498–1514.

Ying, T. 1983. The floristic relationships of the temperate forest regions of China and the United States. Annals of the Missouri Botanical Garden 70:597–604.

Yoda, K., T. Kira, H. Ogawa, and K. Hozumi. 1963. Self-thinning in overcrowded pure stands under cultivated and natural conditions. Journal of Biology, Osaka City University 14:107–129.

Yorks, T.E. and S. Dabydeen. 1999. Seasonal and successional understory vascular plant diversity in second-growth hardwood clearcuts of western Maryland, USA. Forest Ecology and Management 119:217–230.

Yoshie, F., and S. Kawano. 1986. Seasonal changes in photosynthetic characteristics of *Pachysandra terminalis* (Buxaceae), an evergreen woodland chamaephyte, in the cool temperate regions of Japan. Oecologia 71:6–11.

Yoshie, F., and S. Yoshida. 1987. Seasonal changes in photosynthetic characteristics of *Anemone raddeana*, a spring-active geophyte, in the temperate region of Japan. Oecologia 72:202–206.

Young, A., T. Boyle, and T. Brown. 1996. The population genetic consequences of habitat fragmentation for plants. Trends in Ecology and Evolution 11:413–418.

Young, D.R., and W.K. Smith. 1979. Influence of sunflecks on the temperature and water relations of two subalpine understory congeners. Oecologia 43:195–205.

Zak, D.R., P.M. Groffman, K.S. Pregitzer, S. Christensen, and J.M. Tiedje. 1990. The vernal dam: plant-microbe competition for nitrogen in northern hardwood forests. Ecology 71:651–656.

Zak, D.R., and K.S. Pregitzer. 1988. Nitrate assimilation by herbaceous ground flora in late successional forests. Journal of Ecology 76:537–546.

Zak, D.R., K.S. Pregitzer, and G.E. Host. 1986. Landscape variation of nitrogen mineralization and nitrification. Canadian Journal of Forest Research 16:1258–1263.

Zampella, R.A., and K.J. Laidig. 1997. Effects of watershed disturbance on Pinelands stream vegetation. Journal of the Torrey Botanical Society 124:52–66.

Zar, J.H. 1996. Biostatistical analysis, 3rd ed. Englewood Cliffs, NJ:Prentice-Hall.

Zavitkovsky, J. 1976. Ground vegetation biomass, production, and efficiency of energy utilization in some northern Wisconsin forest ecosystems. Ecology 57:694–706.

Zimmerman, J.K., and D.F. Whigham. 1992. Ecological functions of carbohydrates stored in corms of *Tipularia discolor* (Orchidaceae). Functional Ecology 6:575–581.

Zogg, G.P., D.R. Zak, D.B. Ringelber, N.W. MacDonald, K.S. Pregitzer, and D.C. White. 1997. Compositional and functional shifts in microbial communities due to soil warming. Soil Science Society of America Journal 61:475–481.

Index

Printed in the United States
87314LV00002B/70-72/A

9 780195 140880